AF324846

Applied Mathematical Sciences
Volume 96

Editors
F. John J.E. Marsden L. Sirovich

Advisors
M. Ghil J.K. Hale J. Keller
K. Kirchgässner B.J. Matkowsky
J.T. Stuart A. Weinstein

Applied Mathematical Sciences

(continued following index)

Carlo Marchioro   Mario Pulvirenti

Mathematical Theory of Incompressible Nonviscous Fluids

With 85 Illustrations

Springer-Verlag
New York Berlin Heidelberg London Paris
Tokyo Hong Kong Barcelona Budapest

Carlo Marchioro
Department of Mathematics
University of Rome
 "La Sapienza"
Rome 00185
Italy

Mario Pulvirenti
Department of Mathematics
University of Rome
 "La Sapienza"
Rome 00185
Italy

Editors

F. John
Courant Institute of
  Mathematical Sciences
New York University
New York, NY 10012
USA

J.E. Marsden
Department of
  Mathematics
University of California
Berkeley, CA 94720
USA

L. Sirovich
Division of
  Applied Mathematics
Brown University
Providence, RI 02912
USA

Mathematics Subject Classification (1991): 76Cxx, 35Qxx

Library of Congress Cataloging-in-Publication Data
Marchioro, Carlo.
    Mathematical theory of incompressible nonviscous fluids/Carlo
Marchioro and Mario Pulvirenti.
        p. cm.—(Applied mathematical sciences; v. 96)
    Includes bibliographical references and index.
    ISBN 0-387-94044-8 (acid-free)
    1. Fluid dynamics. 2. Lagrange equations. I. Pulvirenti, M.
(Mario), 1946–   . II. Title. III. Series: Applied mathematical
sciences (Springer-Verlag New York, Inc.); v. 96.
QA1.A647 vol. 96
[QA911]
510 s—dc20   93-4683

Printed on acid-free paper.

© 1994 Springer-Verlag New York, Inc.
All rights reserved. This work may not be translated or copied in whole or in part without the written permission of the publisher (Springer-Verlag New York, Inc., 175 Fifth Avenue, New York, NY 10010, USA), except for brief excerpts in connection with reviews or scholarly analysis. Use in connection with any form of information storage and retrieval, electronic adaptation, computer software, or by similar or dissimilar methodology now known or hereafter developed is forbidden.
The use of general descriptive names, trade names, trademarks, etc., in this publication, even if the former are not especially identified, is not to be taken as a sign that such names, as understood by the Trade Marks and Merchandise Marks Act, may accordingly be used freely by anyone.

Production coordinated by Brian Howe and managed by Francine McNeill; manufacturing supervised by Vincent Scelta.
Typeset by Asco Trade Typesetting Ltd., Hong Kong.
Printed and bound by R.R. Donnelley & Sons, Harrisonburg, VA.
Printed in the United States of America.

9 8 7 6 5 4 3 2 1

ISBN 0-387-94044-8 Springer-Verlag New York Berlin Heidelberg
ISBN 3-540-94044-8 Springer-Verlag Berlin Heidelberg New York

Preface

Fluid dynamics is an ancient science incredibly alive today. Modern technology and new needs require a deeper knowledge of the behavior of real fluids, and new discoveries or steps forward pose, quite often, challenging and difficult new mathematical problems. In this framework, a special role is played by incompressible nonviscous (sometimes called *perfect*) flows. This is a mathematical model consisting essentially of an evolution equation (the Euler equation) for the velocity field of fluids. Such an equation, which is nothing other than the Newton laws plus some additional structural hypotheses, was discovered by Euler in 1755, and although it is more than two centuries old, many fundamental questions concerning its solutions are still open. In particular, it is not known whether the solutions, for reasonably general initial conditions, develop singularities in a finite time, and very little is known about the long-term behavior of smooth solutions. These and other basic problems are still open, and this is one of the reasons why the mathematical theory of perfect flows is far from being completed.

Incompressible flows have been attached, by many distinguished mathematicians, with a large variety of mathematical techniques so that, today, this field constitutes a very rich and stimulating part of applied mathematics. The idea of writing the present book was motivated by the fact that, although there are many interesting books on the subject, no recent one, to our knowledge, is oriented toward mathematical physics. By this we mean a book that is mathematically rigorous and as complete as possible without hiding the underlying physical ideas, presenting the arguments in a natural order, from basic questions to more sophisticated ones, proving everything and trying, at the same time, to avoid boring technicalities. This is our purpose.

The book does not require a deep mathematical knowledge. The required background is a good understanding of the classical arguments of mathematical analysis, including the basic elements of ordinary and partial differential equations, measure theory and analytic functions, and a few notions of potential theory and functional analysis.

The exposition is as self-contained as possible. Several appendices, devoted to technical or elementary classical arguments, are included. This does not mean, however, that the book is easy to read. In fact, even if we tried to present the topics in an elementary fashion and in the simplest cases, the style is, in general, purely mathematical and rather concise, so that the reader quite often is requested to spend some time in independent thinking during the most delicate steps of the exposition. Some exercises, with a varying degree of difficulty (the most difficult are marked by *), are presented at the end of many chapters. We believe solving them is the best test to see whether the basic notions have been understood.

The choice of arguments is classical and in a sense obligatory. The presentation of the material, the relative weight of the various arguments, and the general style reflect the tastes of the authors and their knowledge. It cannot be otherwise.

The material is organized as follows: In Chapter 1 we present the basic equations of motion of incompressible nonviscous fluids (the Euler equation) and their elementary properties. In Chapter 2 we discuss the construction of the solutions of the Cauchy problem for the Euler equation. In Chapter 3 we study the stability properties of stationary solutions. In Chapter 4 we introduce and discuss the vortex model. In Chapter 5 we briefly analyze the approximation schemes for the solutions of fluid dynamical equations. Chapter 6 is devoted to the time evolution of discontinuities such as the vortex sheets or the water waves. Finally, in Chapter 7 we discuss turbulent motions. This last chapter mostly contains arguments of current research and is essentially discursive.

The final section of each chapter is generally devoted to a discussion of the existing literature and further developments. We hope that this will stimulate the reader to study and research further.

The book can be read following the natural order of the chapters, but also along the following paths:

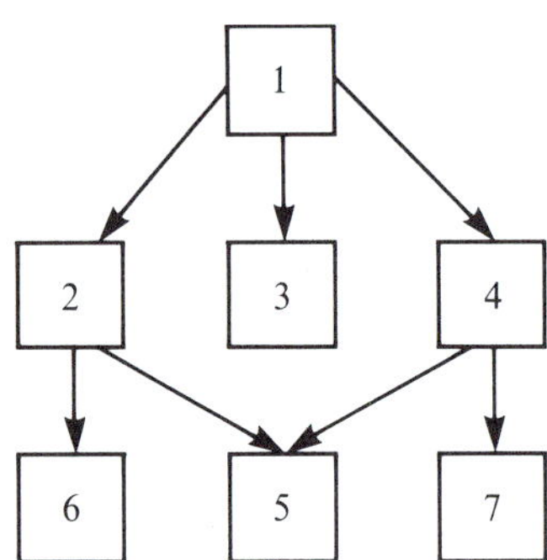

A possible criticism of the book is that two-dimensional flows are treated in much more detail than three-dimensional ones, which are, physically speaking, much more interesting. Unfortunately, for a mathematical treatise, it cannot be otherwise: The mathematical theory of a genuine three-dimensional flow is, at present, still poor compared with the rather rich analysis of the two-dimensional case to which we address many efforts.

It is a pleasure to thank D. Benedetto, E. Caglioti, A.J. Chorin, P. Drazin, R. Esposito, T. Kato, D. Levi, R. Robert, and R. Temam for useful suggestions and, particularly, P. Laurence and C. Maffei for their constructive criticism in reading some parts of the present book. We are also grateful to H. Aref for having sent us the MacVortex program. We finally thank C. Vaughn for her advice in improving our English.

Rome, Italy   CARLO MARCHIORO
                                                     MARIO PULVIRENTI

Contents

Contents

CHAPTER 1

General Considerations on the Euler Equation

This chapter has an introductory nature, wherein we discuss the fundamental equations describing the motion of an incompressible nonviscous fluid and establish some elementary properties.

1.1. The Equation of Motion of an Ideal Incompressible Fluid

In this section we establish the mathematical model of an ideal incompressible fluid, deriving heuristically the equation governing its motion.

Fluid mechanics studies the behavior of gases and liquids. The phenomena we want to study are macroscopic: we do not want to investigate the dynamics of the individual molecules constituting the fluid, but the gross behavior of many of them. For this purpose we assume the fluid as a continuum, a point of which is a very small portion of the real fluid, negligible with respect to the macroscopic size (for instance, the size of the vessel containing the fluid), but very large with respect to the molecular length. This small volume, a point in our mathematical description, will be called *fluid particle* or *element of fluid* later in this book. As a consequence, the physical state of a fluid will be described by properties of the fluid particles and not by the physical state of all the microscopic molecules. The macroscopic fields describing the state, as, for instance, the velocity field, $u = u(x)$, the density field $\rho = \rho(x)$, the temperature field, $T = T(x)$, etc., can be physically interpreted (and, in principle, calculated) by means of averages of suitable microscopic quantities. For example, the macroscopic velocity field in a point $u(x)$ means

$$u(x) = \frac{1}{N(x)} \sum_{i=1}^{N(x)} \mu_i, \tag{1.1}$$

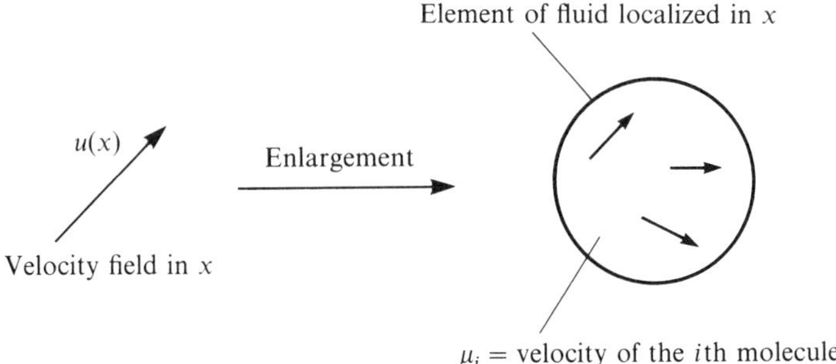

Figure 1.1

where $N(x)$ is the number of molecules associated to the fluid particle localized in x and μ_i, $i = 1, \ldots, N(x)$ are the velocities of these molecules (Fig. 1.1).

It would be very interesting to deduce the evolution equation for the fields,
$u = u(x)$, $\rho = \rho(x)$, etc., starting from the Newton equation which governs the
motion of the molecules. To give a measure of the difficulty of this program
we note that the macroscopic observables u, ρ, T, etc., give us a reduced
description of the physical system we are considering. Such a system is described, in much more detail, by the positions and the velocities of all the
microscopic molecules. Therefore, it is not at all obvious that we are able to
deduce some closed equations involving only the interesting observables.

Until now, a rigorous microscopic derivation of the fluid equations from
the Newton laws is not known. For some discussion on this point we address
the reader to Section 1.5, which is devoted to comments and bibliographical
notes. In the absence of this deduction we limit ourselves to fixing the mathematical model of a fluid by heuristic considerations only, without taking into
account its microscopic structure. We will deduce the basic equation, called
the Euler equation, by the use of reasonable assumptions on the motion of
the fluid parlicles. In the following sections, our study will be essentially
deductive, starting from the Euler equation, which constitutes our mathematical model. Obviously, we will not neglect the physical interpretation which
is important to verify the validity of the model itself and the relevance of the
results.

The rest of the present section is devoted to the derivation of the Euler
equation.

Let $D \subset \mathbb{R}^3$, an open and bounded set of the physical space with a regular
boundary ∂D. D contains a fluid represented as a continuum of particles
localized in any point $x \in D$.

An *incompressible displacement* of the fluid is a transformation $s: D \to D$
such that the following properties hold:

(a) s is invertible and $s(D) = D$;
(b) s, $s^{-1} \in C^1(D)$; and
(c) s preserves the Lebesgue measure.

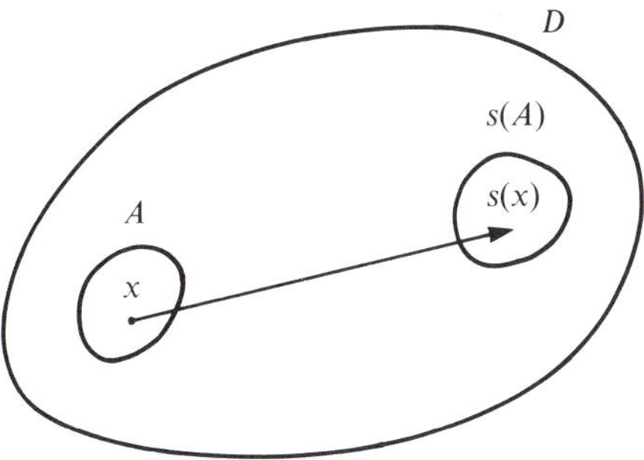

Figure 1.2

The property (c) means that, for any measurable set A, $A \subset D$, denoted by

$$s(A) = \{x \in D \,|\, s^{-1}(x) \in A\}, \tag{1.2}$$

we have

$$|s(A)| = |A|, \tag{1.3}$$

where $|A| = \text{meas } A$ denotes the Lebesgue measure of A (Fig. 1.2). We denote by S the set of all the incompressible displacements. It is evident that S has a group structure with respect to the law of natural composition

$$s_1 \circ s_2(x) = s_2(s_1(x)).$$

An incompressible motion is, by definition, a function s, $t \in R^1 \to \Phi_{s,t} \in S$ such that:

(1) $\Phi_{s,t}(\Phi_{t,r}(x)) = \Phi_{s,r}(x)$;
(2) $\Phi_{t,s}(\Phi_{s,t}(x)) = \Phi_{t,t}(x) = x$; and
(3) $\Phi_{t,s}(x)$ is continuously differentiable in t and s.

Here $\Phi_{t,s}$ denotes the position at time t of the particle of fluid that at time s was in x. We will denote by M, in the sequel, the family of incompressible motions.

We note that these conditions are reasonable properties of regularity. The requirement that the transformation be invertible means also that two different particles of fluid cannot occupy the same position. Moreover, the definition of Φ itself gives the conservation of the Lebesgue measure during the motion.

These conditions make it very easy to study the time evolution of the *density field* $\rho = \rho(x, t)$. We denote by $\rho(x, t)\,dx$ the mass of fluid contained in the element of volume dx at time t, and we assume that $\rho \in C^1(D)$. By the law of conservation of mass we have

$$\frac{d}{dt} \int_{V_t} \rho(x, t)\,dx = 0, \tag{1.4}$$

where

$$V_t = \{\Phi_t(x) | x \in V_0\} \tag{1.5}$$

is the region moving along the trajectories of an incompressible motion and $\Phi_t(x) = \Phi_{t,0}(x)$.

Let

$$u(\Phi_t(x), t) = \frac{d}{dt}\Phi_t(x) \tag{1.6}$$

be the velocity field associated with this motion.

By (1.4) we have

$$\frac{d}{dt}\int_{V_t} \rho(x, t)\, dx = \frac{d}{dt}\int_{V_0} \rho(\Phi_t(x), t)J_t(x)\, dx$$

$$= \frac{d}{dt}\int_{V_0} \rho(\Phi_t(x), t)\, dx = 0, \tag{1.7}$$

where $J_t(x)$ is the Jacobian of the transformation $x \to \Phi_t(x)$. The incompressibility condition (together with the continuity of the transformation) implies that it is one.

Hence, by the arbitrariness of V_0, we have

$$\frac{d}{dt}\rho(\Phi_t(x), t) = (\partial_t + u \cdot \nabla)\rho(\Phi_t(x), t) = 0. \tag{1.8}$$

From a physical point of view there are interesting situations in which the density is initially (and hence by (1.8) for all times) not constant in space. We will provide an example in Chapter 6. However, in most of the physically relevant cases, in which the model of incompressible fluid applies, the density can be assumed to be essentially constant. In the present book we will assume the density to be always constant (for simplicity $\rho = 1$), unless explicitly mentioned otherwise.

The condition of incompressibility is equivalent, by a well-known theorem on differential equations (the Liouville Theorem, see Appendix 1.1), to the condition

$$\operatorname{div} u(x, t) \equiv \nabla \cdot u(x, t) = 0, \qquad \forall x \in D, \quad t \in \mathbb{R}. \tag{1.9}$$

Equation (1.9) is usually called the continuity equation for incompressible flows.

From this point on, in this section, we are assuming $u \in C^1(D \times \mathbb{R}^1)$. Moreover, for any t, $u(x, t)$ is assumed continuous in $x \in \overline{D} \equiv D \cup \partial D$. This allows us to define the velocity $u(x, t)$ on the boundary ∂D as a limit.

We will now establish the boundary conditions. In general, for partial differential equations describing physical systems, the boundary conditions are a mathematical expression of the interaction of the system with the boundary. In our case, we must assume the most general and natural assumption which can be deduced from kinematic considerations only: the fluid particles cannot pass through the boundary so that

$$u(x,t)\cdot n = v(x)\cdot n \quad \text{on } \partial D, \tag{1.10}$$

where $v(x)$ is the velocity of the boundary at the point x. Most of the time, later in this book, we will comsider the container D at rest so that $v(x) = 0$ for all $x \in \partial D$.

Once the velocity field u is known, the trajectories $\Phi_t(x)$ can be uniquely built by solving the initial value problem (1.6) for the unknown quantity $\Phi_t(x)$ with initial value x at time $t = 0$.

We now want to state the equations of motion of an incompressible fluid. To determine the motion of the fluid particles we must specify the interactions among the particles themselves. We consider the only interaction produced by the incompressibility. This means that each particle tries to move freely, the only constraint being that it cannot occupy the site in which there is another particle. Later on we will be more precise. This model of an incompressible fluid is called *ideal* (or *perfect*) and it is the simplest model we can conceive.

To find the equations of motion it is convenient to consider the *Principle of Stationary Action* as suggested by the classical mechanics of systems with a finite number of the degrees of freedom.

The kinetic energy (and also the Lagrangian) of the system is given by the following expression:

$$E = \frac{1}{2}\int_D dx \left[\frac{d}{dt}\Phi_t(x)\right]^2. \tag{1.11}$$

So the action is defined as

$$A(\Phi; t_1, t_2) = \frac{1}{2}\int_{t_1}^{t_2} dt \int_D dx \left[\frac{d}{dt}\Phi_t(x)\right]^2. \tag{1.12}$$

Then $\Phi \to A(\Phi; t_1, t_2)$ is a functional defined on M, the space of incompressible motions. We have not added an interaction energy since the motion we have in mind is the same as the free motion, on a given manifold, of a finite particle system. In our case the "manifold" is given by the incompressibility constraint. Therefore, as in the mechanical analogue where the variation is chosen in accord with the constraint, here we will consider variations in the class M. Hence, to determine the physical motion Φ, we ask that the action be stationary for variations, $\Phi \to \Phi + \delta\Phi$, which are compatible with the constraint of incompressibility, and to satisfy $\delta\Phi_{t_1}(x) = \delta\Phi_{t_2}(x) = 0$ for all $x \in D$ (Fig. 1.3). Moreover, the variation must also satisfy the boundary conditions

$$\frac{d}{dt}\Phi_t^\varepsilon(x)\cdot n = 0, \qquad x \in \partial D.$$

We denote by Φ^ε, $\varepsilon \in [0, \varepsilon_0]$, a family of varied motions, tangent to the boundary ∂D, such that

$$\Phi^0 = \Phi, \qquad \Phi_{t_1}^\varepsilon = \Phi_{t_1}, \qquad \Phi_{t_2}^\varepsilon = \Phi_{t_2}, \qquad \forall \varepsilon \in [0, \varepsilon_0].$$

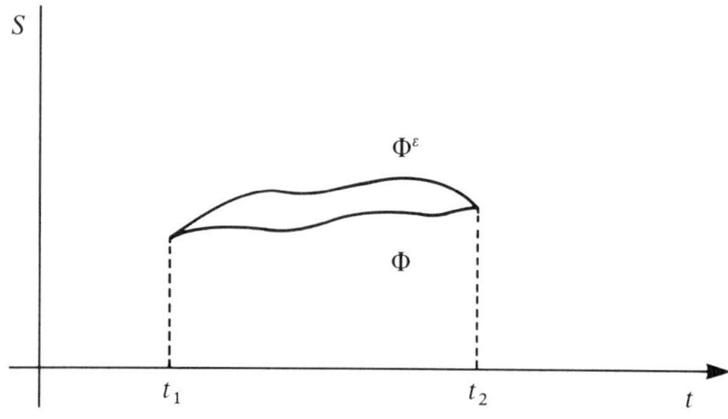

Figure 1.3

We impose that the action A be stationary on Φ, namely,

$$\frac{d}{d\varepsilon} A(\Phi^\varepsilon; t_1, t_2)|_{\varepsilon=0} = 0. \tag{1.13}$$

From (1.13) we easily obtain

$$\int_{t_1}^{t_2} dt \int_D dx \frac{d}{dt}\Phi_t(x) \cdot \frac{d}{dt}\gamma_t(\Phi_t(x)) = 0, \tag{1.14}$$

where $\gamma_t = \gamma_t^0$ and γ_t^ε is defined by

$$\gamma_t^\varepsilon(\Phi_t(x)) = \frac{d}{d\varepsilon}\Phi_t^\varepsilon(x). \tag{1.15}$$

γ_t^ε is the vector field transversal to the motion that generates a flow pa-

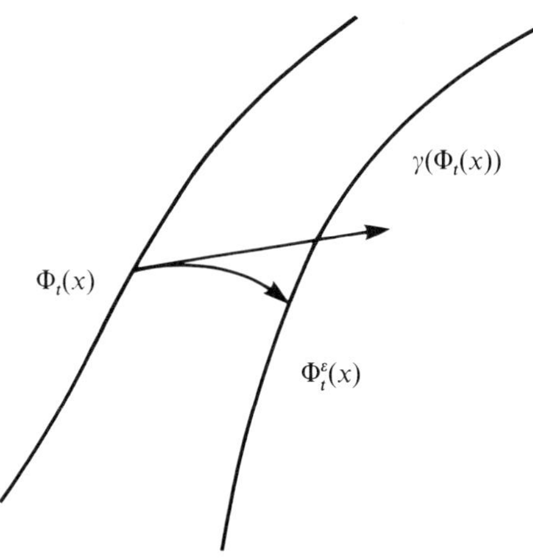

Figure 1.4

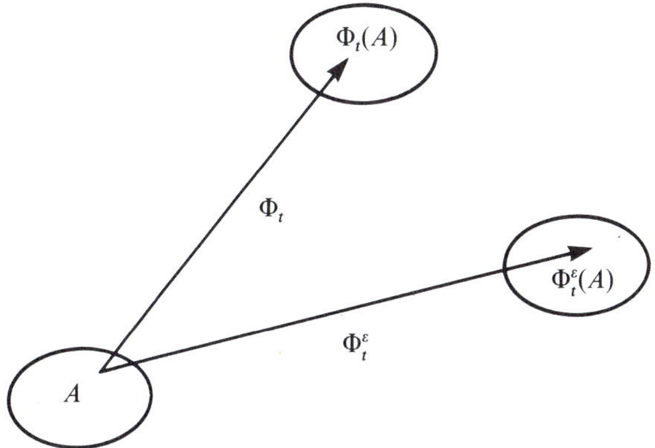

Figure 1.5

rametrized by ε (Fig. 1.4)

$$\Phi_t(x) \to \Phi_t^\varepsilon(x). \tag{1.16}$$

Obviously, such flow preserves the Lebesgue measure (Fig. 1.5)

$$(|\Phi_t^\varepsilon(A)| = |A| = |\Phi_t(A)|), \tag{1.17}$$

and hence, by the Liouville theorem,

$$\operatorname{div} \gamma_t = 0. \tag{1.18}$$

Moreover, it follows easily from definition (1.15) that

$$\gamma_t \cdot n = 0 \qquad \text{for} \quad x \in \partial D.$$

Coming back to (1.14), we obtain, by integration by parts,

$$\int_{t_1}^{t_2} dt \int_D dx \left\{ \frac{d^2}{dt^2} [\Phi_t(x)] \cdot \gamma_t(\Phi_t(x)) \right\} = 0. \tag{1.19}$$

Moreover,

$$\frac{d^2}{dt^2} \Phi_t(x) = \frac{d}{dt} u(\Phi_t(x), t) = D_t u(\Phi_t(x), t). \tag{1.20}$$

Here we used the notation

$$D_t f \equiv \partial_t f + (u \cdot \nabla) f = \partial_t f + \sum_{i=1}^{3} u_i \frac{\partial}{\partial x_i} f \tag{1.21}$$

for the derivative of a function f along the trajectories $\Phi_t(x)$ (D_t is sometimes also called the *material* or *substantial* or *molecular* derivative).

We insert (1.20) in (1.19). Since the Jacobian of the time transformation is one, by virtue of the arbitrariness of the times t_1 and t_2, we obtain

$$\int_D (D_t u)(x) \cdot \gamma_t(x) \, dx = 0. \tag{1.22}$$

From (1.22) it follows that $D_t u$ is orthogonal (in the sense of $L_2(D)$) to all divergence-free vector fields tangent to the border. (The arbitrariness of γ follows from the arbitrariness of Φ^ε). By virtue of a classical lemma (see Appendix 1.2), which states that a vector field, which is orthogonal to all the divergence-free fields tangent to the boundary, is the gradient of a scalar function, we can conclude that

$$D_t u = -\nabla p \tag{1.23}$$

for some function $p\colon R \times D \to R$. We observe that the minus sign in (1.23) is purely conventional.

Equation (1.23), together with the equations,

$$\nabla \cdot u = 0, \tag{1.24}$$

$$u \cdot n = 0 \quad \text{on } \partial D, \tag{1.25}$$

form the *Euler equation* for an ideal (or perfect) incompressible fluid.

The physical meaning of these equations is transparent: $D_t u$, the acceleration of a fluid particle, is equal to a force $-\nabla p$ to be determined on the basis of the principle of the incompressibility. $-\nabla p$ plays the same role as the constraint force for a free particle system constrained to move on a manifold. It is easy to verify (see Exercise 4) that a completely free motion in general violates the incompressibility condition. The scalar field $p = p(x, t)$ is called *pressure*.

An interesting class of solutions of the Euler equation are the *steady or stationary flows* which are the solutions, $u = u(x)$, not explicitly depending on time. For such flows the material derivative $D_t u$ consists only of the term $(u \cdot \nabla)u$, so that the stationary flows are those divergence-free fields for which $(u \cdot \nabla)u$ is the gradient of a scalar field. In this case, the integral lines of the velocity field are constant in time and they coincide with the trajectories of the particles of the fluid.

Equations (1.23), (1.24), (1.25) form a system of partial differential equations that we rewrite explicitly

$$\partial_t u_j(x, t) + \sum_{i=1}^{3} [u_i \cdot \partial_i] u_j(x, t) = -\partial_j p(x, t),$$

$$\sum_{i=1}^{3} \partial_{x_j} u_j(x, t) = 0, \tag{1.26}$$

$$\sum_{i=1}^{3} u_i \cdot n_i(x) = 0.$$

This system of equations, in spite of the simplicity of the physical model from which they have been deduced, gives rise to a rather complicated mathematical problem, as we will see in detail in the next chapter. Here we want to outline only that the main problem of fluid dynamics consists in determining the velocity field, $u = u(x, t)$, at time t once known at time zero. When the velocity field is determined, the trajectories of the fluid particles are the

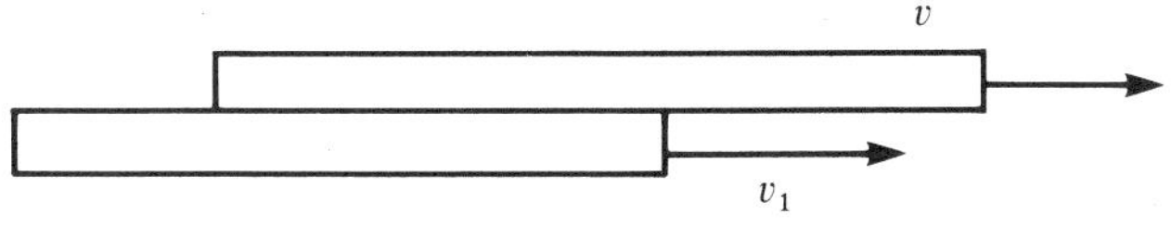

Figure 1.6

solutions of the ordinary differential problem (1.6) which has a unique solution for $t \in [0, T]$ if the field $u(x, t)$ (obtained as the solution of (1.23), (1, 24), (1.25)) is $C^1(D, [0, T])$.

We conclude with some physical considerations on the model we have introduced. As we have often outlined, the only interaction we have supposed among the particles of fluid arises from the constraint of incompressibility. As a consequence, two layers of fluid with different velocities cannot transfer kinetic momentum and so they move independently (Fig. 1.6). In particular, the fluid cannot produce rotation by itself and there is no mechanism to dissipate energy. This property gives rise to the conservation laws which will be discussed in Section 1.3.

For simplicity, we have deduced the Euler equation in a bounded domain and in the absence of external forces acting on the fluid.

When the domain D is unbounded, the equations of motion, which have a local character, remain valid. However, in this case, we must specify not only the boundary conditions ($u \cdot n = 0$ on ∂D) but also the asymptotic behavior of the velocity field $u(x)$ when $|x| \to \infty$.

When a field of external forces per unity volume, $f = f(x, t)$ is present ($f \, dx$ is the external force that acts on the particle of fluid in x) the Euler equation is modified as

$$D_t u = -\nabla p + f, \qquad \nabla \cdot u = 0. \tag{1.27}$$

When f is a potential force ($f = -\nabla U$ for some scalar field U) we have only a modification of the pressure: (1.27) equals (1.13) with p replaced by $p + U$ so that the nature of the equation does not change appreciably.

Later on we will consider (1.27) in the absence of external forces unless explicitly mentioned. A more conventional derivation of the Euler equation will be delivered in Section 1.5 in the more general case of the incompressible fluid. In this framework, the equation for incompressible fluids is obtained in a suitable asymptotic regime.

1.2. Vorticity and Stream Function

In this section we will develop some considerations of a kinematic nature on the motion of an incompressible fluid and establish some notions that will be very useful later on.

First of all, we note that in the previous section we have assumed, in deducing the equations of motion, two different points of view. In writing the action (1.12) we have studied the motion of the fluid by following the evolution of a single particle (the Lagrangian point of view). On the contrary, in the Euler equation, the velocity field $u = u(x, t)$ is the unknown quantity. This means that we fix a point x and follow the time evolution of the particle that at time t passes through x (the Eulerian point of view).

Of course, the two points of view are strictly related. If we know all the trajectories of the fluid particles, it is possible to find the velocity field by a simple differentiation. More complicated is the inverse problem. In fact, as we already seen, knowing $u = u(x, t)$, we can find the motion of each particle of the fluid by solving the differential equation

$$\frac{d}{dt}\Phi_t(x) = u(\Phi_t(x), t),$$

$$\Phi_0(x) = x. \tag{2.1}$$

In general, it is not easy to find explicitly the solutions of (2.1). In most cases we consider a dynamical problem of an incompressible fluid solved whenever we know the velocity field, $u = u(x)$, which is the real unknown quantity in our problem. The details of a trajectory are, in general, not so important (of course, this is not true in some specific problems, for instance, for diffusion and pollution problems). However, the physical trajectories $\Phi_t(x)$ of the fluid particles will often be considered later on and they will play an important role in the study of some dynamical properties of the fluid.

The lines that are tangent in any point to the velocity field, $u = u(x)$, are called *stream lines* or *flow lines*. Of course, they vary in time and they are constant in time for steady motions only. In this case the streamlines coincide with the trajectories of the particles. These last are sometimes called *path lines*.

As an example, let us consider the motion of a rigid body with an angular velocity Ω. As a consequence of the rigidity constraint, the mutual distance of the fluid particles does not change during the motion and so the motion is incompressible. Moreover, it is well known from elementary courses of mechanics that the velocities of the two points O and P are related by the expession

$$v_P = v_O + \Omega \wedge OP. \tag{2.2}$$

It is also well known that every infinitesimal rigid motion is a superposition of a rotation with an angular velocity Ω and a translation along the direction Ω. In the case of a pure rotation, the streamlines are concentric circumferences. In the case of pure translation, they are straight lines. In general, they are cylindrical helixes (of course, varying in time). On the other hand, the pathlines are quite arbitrary.

We now introduce a fundamental concept of our analysis, the *vorticity field* $\omega(x)$. By definition

$$\omega \equiv \operatorname{curl} u \equiv \nabla \wedge u = (\partial_{x_2}u_3 - \partial_{x_3}u_2, \partial_{x_3}u_1 - \partial_{x_1}u_3, \partial_{x_1}u_2 - \partial_{x_2}u_1). \tag{2.3}$$

The vorticity field $\omega(x)$ gives a measure of how the fluid is rotating. To understand more deeply the meaning of ω, we prove the following formula:

$$u(y) = u(x) + D \cdot h + \tfrac{1}{2}\omega(x) \wedge h + O(h^2), \qquad (2.4)$$

where $y = x + h$, D is a matrix defined as

$$D = \tfrac{1}{2}(\nabla u + (\nabla u)^{\mathrm{T}}), \qquad (\cdot^{\mathrm{T}})) = \text{transposed matrix}, \qquad (2.5)$$

and ∇u is the matrix

$$(\nabla u)_{ij} = \partial_{x_j} u_i. \qquad (2.6)$$

Equation (2.4) easily follows from the Taylor theorem

$$u(y) = u(x) + (\nabla \cdot u)(x) \cdot h + O(h^2) \qquad (2.7)$$

and definitions (2.5) and (2.3).

We now explain formula (2.4). By a comparison of (2.4) with (2.2) it follows, for a rigid motion, that

$$D = 0, \qquad \Omega = \tfrac{1}{2}\omega. \qquad (2.8)$$

This justifies for D the name *deformation tensor*. Then from (2.4) we obtain that the velocity of a point y "near" x is the sum of three terms: a translation, a rotation with angular velocity $\tfrac{1}{2}\omega$, and a deformation that gives a measure of the fact that the motion is not rigid.

The vorticity field is an important tool in studying the behavior of fluids so that it is natural to pose the following problem. Supposing the vorticity field ω to be known, we deduce the velocity field u generating ω. In other words, we want to solve the following equations in the unknown quantity u:

$$\nabla \wedge u = \omega, \qquad \omega \in C(D),$$
$$\nabla \cdot u = 0. \qquad (2.9)$$

We will see that the solution of this problem (absolutely general and not necessarily related to fluid dynamics) is essential for the construction of the solutions of the Euler equation in two dimensions.

We start with the two-dimensional case. Let $u = u(x)$ be a vector field, $u: D \subset \mathbb{R}^2 \to \mathbb{R}^2$, which is the unknown quantity of the problem. Equations (2.9) can be rewritten in two dimensions as

$$\partial_{x_1} u_2 - \partial_{x_2} u_1 = \text{curl } u = \omega,$$
$$\partial_{x_1} u_1 + \partial_{x_2} u_2 = \nabla \cdot u = 0, \qquad (2.10)$$

where $\omega = \omega(x)$ is a given field. It is a scalar field because, in the presence of a two-dimensional symmetry, only the third component of curl does not vanish. We denote (with a notation perhaps not completely correct) this third component as curl u.

It is evident that the system of equations (2.10) cannot have, in general, a unique solution. In fact, let u' be a solution of (2.10) then also let $u = u' + \nabla\varphi$, where φ is an harmonic function, is a solution. To obtain a unique solution we must add to (2.10) at least the boundary conditions (for instance, $u \cdot n = 0$ in ∂D).

Let us consider different cases.

(1) $D \subset \mathbb{R}^2$, D *simply connected and bounded*

In this case, the condition $\nabla \cdot u = 0$ allows us to introduce a function Ψ, called the *stream function*, such that

$$u = \nabla^\perp \Psi, \tag{2.11}$$

where

$$\nabla^\perp = (\partial_{x_2}, -\partial_{x_1}). \tag{2.12}$$

By curl $u = \omega$ we easily obtain

$$\Delta \Psi = -\omega, \tag{2.13}$$

that is, the well-known Poisson equation.

From the condition $u \cdot n = 0$ on ∂D it follows that Ψ must be a constant on ∂D. Since we are interested in determining Ψ modulo a constant, we can put

$$\Psi|_{\partial D} = 0. \tag{2.14}$$

It is known from potential theory that under the reasonable hypotheses of regularity on ω, we have a unique solution of the problem (2.13), (2.14). On the other hand, the uniqueness of the solution we have found via (2.11) can be proved quickly by the following few steps. We denote by u' another solution of the problem (2.10). Then $v = u - u'$ must satisfy the relations

$$\nabla \cdot v = 0, \qquad \text{curl } v = 0. \tag{2.15}$$

From the second equation of (2.15), because D is simply connected, we have

$$v = \nabla \varphi, \qquad v \cdot n = 0 \quad \text{on } \partial D. \tag{2.16}$$

Taking the divergence of (2.16), we obtain the Neumann problem

$$\Delta \varphi = 0,$$
$$\frac{\partial}{\partial n} \varphi = 0 \quad \text{on } \partial D, \tag{2.17}$$

which has only the trivial solution $\varphi = \text{const}$. Thus, $v = 0$.

(2) $D = \mathbb{R}^2$

Proceeding as in the previous paragraph we introduce the stream function and study (2.13). This equation can be solved by the method of the Green function.

Let $G = G(x, x')$, $x, x' \in \mathbb{R}^2$, be the fundamental solution of the Poisson equation

$$\Delta_x G(x, x') = -\delta(x - x'), \tag{2.18}$$

where $\delta(x - x')$ is the Dirac measure.

A solution of (2.18) is explicitly known

$$G(x, x') = -\frac{1}{2\pi} \ln|x - x'|. \tag{2.19}$$

By (2.18) we have

$$\Psi(x) = \int G(x, x')\omega(x')\, dx', \tag{2.20}$$

$$\nabla^{\perp}\Psi(x) = u(x) = \int K(x, x')\omega(x')\, dx', \tag{2.21}$$

where

$$K(x, x') = -\frac{1}{2\pi} \frac{(x - x')^{\perp}}{(x - x')^2} \tag{2.22}$$

with the natural notation $x^{\perp} = (x_2, -x_1)$ if $x = (x_1, x_2)$. The physical meaning of (2.22) is obvious: $K(x, x')$ is the velocity field (in x) generated by a point charge of intensity one fixed in x' (Fig. 1.7).

A sufficient condition for which expression (2.21) makes sense is that $\omega \in L_1 \cap L_{\infty}(\mathbb{R}^2)$. Is this solution unique? Certainly not if we do not specify the asymptotic behavior of the velocity field when $|x| \to \infty$. As in case 1, we observe that the difference v of the two solutions of (2.10) is a potential field, $v = \nabla\varphi$, where φ is an harmonic function. The requirement $u(x) \to u_{\infty} \in \mathbb{R}^2$ implies that the unique harmonic function φ, such that $\nabla\varphi \to u_{\infty}$, has the form $\varphi = u_{\infty} \cdot x + \text{const.}$ (Here we have made use of the Liouville theorem for harmonic functions.) Hence the solution of our problem is given by

$$u(x) = \nabla^{\perp}\Psi(x) + u_{\infty}. \tag{2.23}$$

(3) $D = [-\pi, \pi]^2$: *Flat torus in two dimensions*

In this case we want to solve (2.10) in the ambit of 2π-periodic functions or, which is the same, we consider a region of the form $D = \mathbb{R}^2/\mathbb{Z}^2$. D is a rectangle without boundaries in which we identify the points $(x_1, 0)$ with

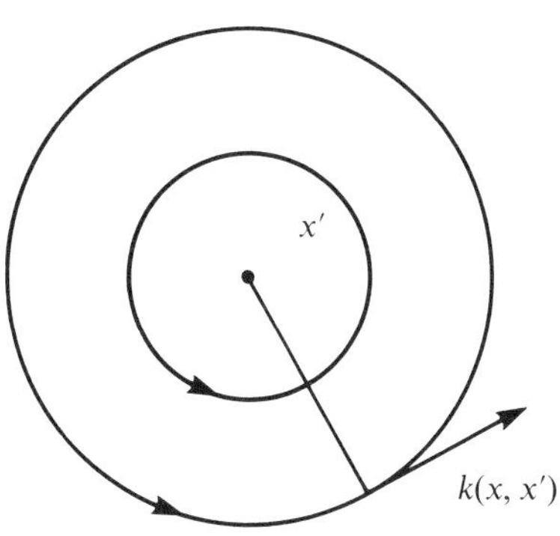

Figure 1.7

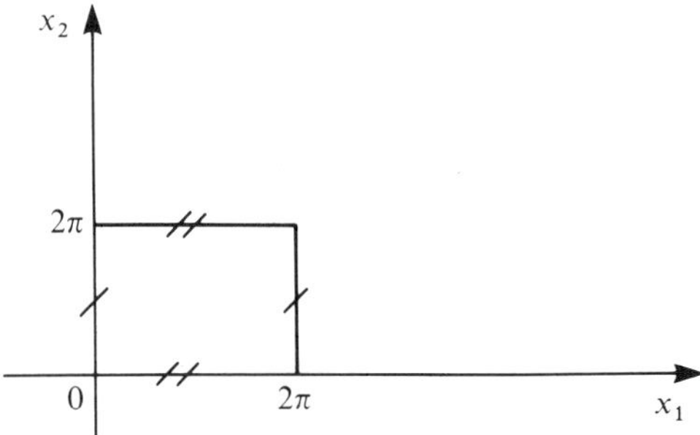

Figure 1.8

$(x_1, 2\pi)$ and $(0, x_2)$ with $(2\pi, x_2)$ (Fig. 1.8). A natural way to solve (2.10) is to introduce, once again, the stream function Ψ. The Poisson equation (2.13) can be written in terms of the Fourier transform as

$$k^2 \Psi(k)^\wedge = \omega(k)^\wedge, \qquad k \in Z^2, \tag{2.24}$$

where we denote by $f^\wedge$ the Fourier transform of f.

We note that, by virtue of the circulation theorem,

$$\omega(0)^\wedge = \frac{1}{2\pi} \int_D \omega \, dx = 0. \tag{2.25}$$

This property makes (2.24) solvable. The velocity field u is given by

$$u(x) = \frac{i}{2\pi} \sum_{k \in Z^2; k \neq 0} \exp\{ik \cdot x\} \omega(k)^\wedge \frac{k^\perp}{k^2}. \tag{2.26}$$

We can prove that $u(x)$ is real (see Exercise 5). It is easy to verify that the property $\operatorname{curl} u = \omega$ holds. Of course (2.26) is not the unique solution of (2.10) since we can add an arbitrary potential field. In this case all the poten-

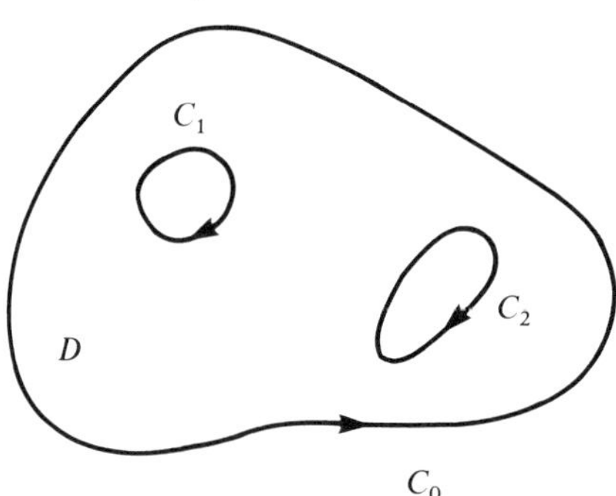

Figure 1.9

tial fields are constant. The series expressed by (2.26) is absolutely convergent if $\omega(k)^\wedge$ decays fast enough for $|k| \to \infty$.

(4) *Bounded, not simply connected D*

We consider a domain D surrounded by regular closed curves C_0, C_1, C_2, ..., C_N (see Fig. 1.9). We start our analysis by studying irrotational fields in this domain. We suppose that we know the *circulations*

$$\Gamma_i = \oint_{C_i} u \cdot dl, \qquad i = 0, \dots, N. \tag{2.27}$$

Then the following theorem holds:

Theorem 2.1. *There exists a unique irrotational divergence-free field, tangent to ∂D, with given circulations Γ_i, $i = 1, \dots, N$ ($\Gamma_0 = \sum_{i=1}^{N} \Gamma_i$ is automatically determined by the Stokes theorem).*

PROOF. *Existence.* We consider N points x_i, $i = 1, \dots, N$, each one inside the domain D_i surrounded by the border C_i. Consider the velocity field

$$u_0(x) = \sum_{i=1}^{N} K(x, x_i)\Gamma_i, \tag{2.28}$$

which has the following properties:

$$\operatorname{curl} u_0 = \operatorname{div} u_0 = 0 \quad \text{in } D, \tag{2.29}$$

$$\oint_{C_i} u_0 \cdot dl = \Gamma_i, \qquad i = 0, \dots, N. \tag{2.30}$$

Equations (2.29) and (2.30) easily follow by definition and the Gauss–Green theorem. Obviously, u_0 does not verify the boundary conditions $u_0 \cdot n = 0$ on ∂D. Consider now an harmonic function φ and the new field

$$u = u_0 + \nabla\varphi. \tag{2.31}$$

u verifies (2.29), (2.30). We choose φ in a way to satisfy the boundary conditions. We have

$$-u_0 \cdot n = \frac{\partial \varphi}{\partial n} \quad \text{on } \partial D, \tag{2.32}$$

and so it is enough to choose φ as a solution of the Laplace equation, $\Delta\varphi = 0$, with Neumann boundary conditions.

Uniqueness. Let u be a solution of the problem. For a fixed $x \in D$ we define a multivalued function V (called potential) by

$$V(y) - V(x) = \int_{C_{xy}} u \cdot dl \tag{2.33}$$

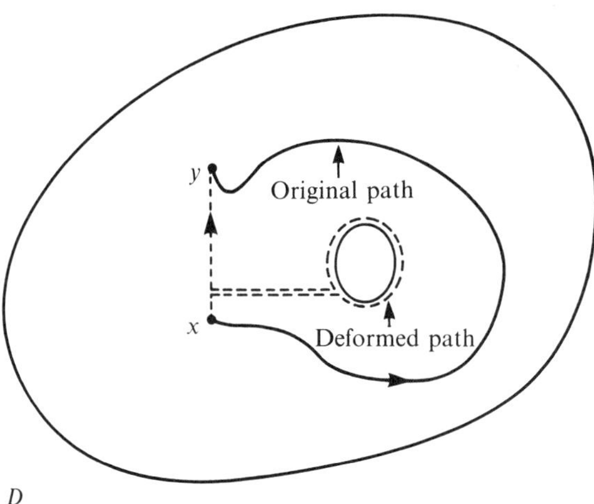

Figure 1.10

where C_{xy} is a path connecting the points x and y. The value of V in y depends on the number of revolutions that C_{xy} performs around the holes of D and the reference value $V(x)$. In fact, it is well known that the value of the integral in the right-hand side of (2.33) does not change for a continuous deformation of C_{xy}. So it can be decomposed in some paths turning around the holes of the boundary (which gives the value of the circulation) plus a path without loops (see Fig. 1.10).

The knowledge of Γ_i allows the construction of V. When we have two irrotational fields with the same circulations and the same boundary conditions, the difference field is a potential field to which is associated a single-valued harmonic potential V. V has a vanishing normal derivative on ∂D, so it is a constant. $\qquad\qquad\square$

Exactly in the same way, substituting u_0 by

$$\sum_{i=1}^{N} K(x, x_i)\Gamma_i + \int_D K(x, y)\omega(y)\, dy \qquad (2.34)$$

we can prove the following result:

Theorem 2.2. *There exists a unique solution to the problem* (2.10), *given the circulations* Γ_i. $i = 1, \ldots, N$. (Γ_0 *is automatically determined by the Stokes theorem:* $\Gamma_0 - \sum_{i=1}^{N} \Gamma_i = \int_D \omega\, dx$.)

Now we want to show that the solution obtained in the previous theorem can be expressed in terms of a stream function. For simplicity, we consider the case $N = 1$. We define

$$\Psi(x) = -\int_{C_{x_0 x}} u^\perp \cdot dl, \qquad u^\perp = (u_2, -u_1), \qquad (2.35)$$

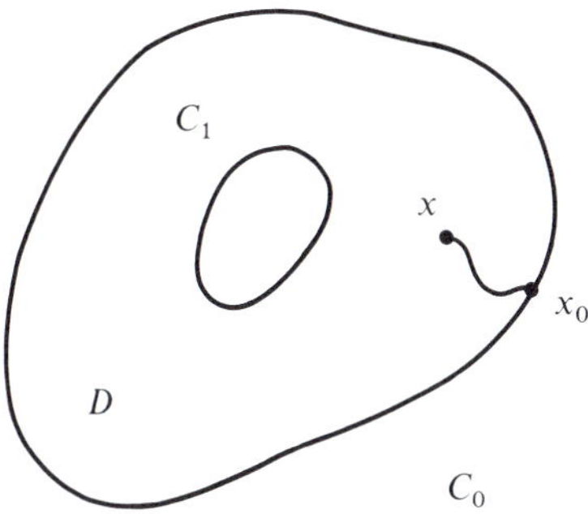

Figure 1.11

where $C_{x_0,x}$ a path connecting a point of the external boundary with the internal point x (Fig. 1.11).

We want to show that (2.35) unambiguously defines a function. This is a consequence of the identity

$$\oint_C u^\perp \cdot dl = 0, \tag{2.36}$$

where C is a closed path containing the curve C_1. In fact, by virtue of the Gauss–Green theorem

$$\oint_C u^\perp \cdot dl = \oint_C u \cdot n\, dl = \int_V \operatorname{div} u + \oint_{C_1} u \cdot n\, dl = 0, \tag{2.37}$$

where V is the domain enclosed between the two curves C and C_1.

We must mention that, although a stream function Ψ can be unambiguously defined as a solution of the Poisson problem (2.13) in a nonsimply connected domain also, however, the value it assumes at the boundary is only piecewise constant. If we fix the reference value zero at the external boundary C_0, the (constant) value at C_1 is determined by the value of the circulation around C_1. This concludes our analysis.

(5) *External nonsimply connected domain*

This case can be treated as the previous one, taking into explicit account the asymptotic behavior of the velocity field (Fig. 1.12):

$$\lim_{|x| \to \infty} u(x) = u_\infty. \tag{2.38}$$

In all the cases we have studied up to now, we have introduced a stream fuction Ψ solution of the Poisson equation with Dirichlet boundary conditions. If we impose on Ψ the zero boundary conditions, the Poisson problem can be solved by means of the technique of the Green function G_D as is well known from potential theory:

$$\Psi(x) = \int_D G_D(x, y)\omega(y), \tag{2.39}_1$$

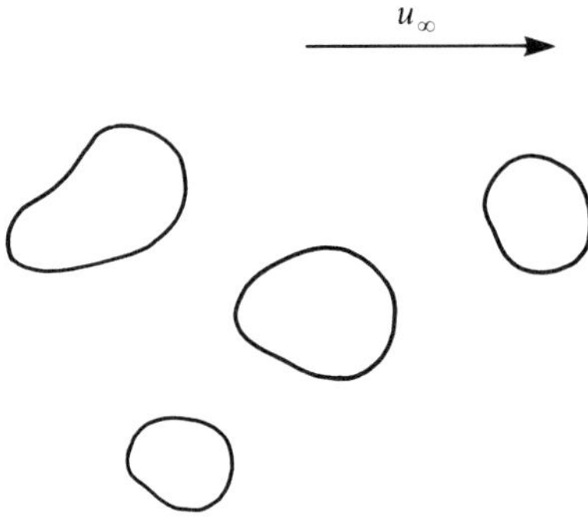

Figure 1.12

where G_D is the solution of the equation

$$\Delta_x G_D(x, y) = -\delta(x, y) \tag{2.39}_2$$

and G_D is zero when either x or y are on the boundary. Finally, $\delta(\cdot, y)$ is the Dirac measure centered in y. G_D can be written as the sum of the Green function G in $\mathbb{R}^2$ given by (2.19) and a smooth function γ which is harmonic (in x and y) in D

$$G_D(x, y) = G(x, y) + \gamma(x, y). \tag{2.40}$$

Using the relation between Ψ and u and $(2.39)_2$ we have

$$u(x) = \nabla^\perp \Psi(x) = (K * \omega)(x) = \int_D K_D(x, y)\omega(y)\, dy, \tag{2.41}$$

where

$$K_D(x, y) = \nabla^\perp G_D(x, y) \tag{2.42}$$

We notice that we use the notation of the convolution product for the expression in the right-hand side of (2.41) even if D is not translationally invariant and, consequently, K_D is not a function of the distance $|x - y|$ only.

Of course, G_D and its derivatives have the same singularities as G. In particular, the following estimates hold:

$$|G_D(x, y)| \leq C(\ln|x - y| + 1), \tag{2.43}$$

$$|K_D(x, y)| \leq C|x - y|^{-1}, \tag{2.44}$$

$$|\partial_i K_D(x, y)| \leq C|x - y|^{-2}. \tag{2.45}$$

We notice that in nonsimply connected domains the velocity field (2.41) does not solve the problem (2.10) with prescribed circulations: to this we must add an irrotational flow with suitable circulations.

In general, it is not possible to calculate explicitly the Green function. However, there are some particular domains in which this function assumes a simple analytical form. We give some examples.

$$D = [-\pi, \pi]^2, \qquad G_D(x, y) = \frac{1}{2\pi} \sum_{k \in Z^2; k \neq 0} \frac{\exp\{ik \cdot (x - y)\}}{|k|^2}. \qquad (2.46)$$

(See Exercises 7 and 9 at the end of this chapter.)

$$D = \{x \in \mathbb{R}^2 \,|\, |x| < R\}, \qquad G_D(x, y) = -\frac{1}{2\pi} \ln \frac{|x - y| R}{|y||x - \bar{y}|}, \qquad (2.47)$$

where

$$\bar{y} = \left(\frac{R^2 y_1}{y^2}, \frac{R^2 y_2}{y^2} \right), \qquad y = (y_1, y_2),$$

$$D = \{x \in \mathbb{R}^2 \,|\, x_2 > 0\}, \qquad G_D(x, y) = G(x, y) - G(x, \bar{y}),$$

$$(2.48)$$

where

$$y = (y_1, y_2) \qquad \text{and} \qquad \bar{y} = (y_1, -y_2).$$

The meaning of (2.48) is transparent: the Green function in a half-plane is the sum of the Green function in the whole plane and the Green function generated by a negative charge situated in the symmetric position. This technique is called the "mirror charge method" (Fig. 1.13). The introduction of fictitious charges is also used to deduce (2.47) and other explicit forms for the Green function (see (2.49) below).

$$D = \mathbb{R} \times (0, a)$$

$$G_D(x, y) = \frac{1}{\pi a} \sum_{n=1}^{\infty} \int_{-\infty}^{\infty} dp \, \frac{\exp(-ip|x - x'|)}{p^2 + n^2} \sin \frac{n\pi}{a} y \sin \frac{n\pi}{a} y'$$

$$= \frac{1}{a} \sum_{n=1}^{\infty} \frac{1}{n} \exp(-n|x - x'|) \sin \frac{n\pi}{a} y \sin \frac{n\pi}{a} y'$$

$$= \frac{1}{4\pi} \ln \frac{\cosh(\pi/a)(x - x') - \cos(\pi/a)(y + y')}{\cosh(\pi/a)(x - x') - \cos(\pi/a)(y - y')}. \qquad (2.49)$$

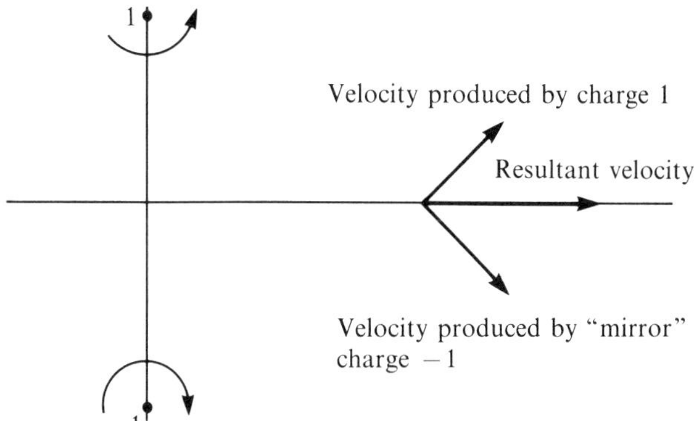

Figure 1.13

Another useful tool to solve these problems is the so-called "conformal transformation." We do not analyze here this technique which is widely discussed in the framework of potential theory.

Now we approach the three-dimensional problem. In this situation we cannot define a stream function but, as in the previous case, we can introduce a potential vector A defined by the relations

$$u = \operatorname{curl} A, \qquad \operatorname{div} A = 0 \qquad (A \in \mathbb{R}^3). \tag{2.50}$$

For reasons we will explain in a moment, we confine our analysis to the case $D = \mathbb{R}^3$ or $D = [-\pi, \pi]^3$ (the three-dimensional flat torus).

Using the vector identity

$$\operatorname{curl} \operatorname{curl} A = -\Delta A + \nabla \cdot (\nabla \cdot A) \tag{2.51}$$

we obtain the (vector) Poisson equation

$$\Delta A = -\omega. \tag{2.52}$$

We consider the case of domains without boundary only because in three dimensions there are no simple boundary conditions for A (like $A = 0$), hence the interest in introducing the potential vector A is limited to the cases discussed here.

For $D = \mathbb{R}^3$ by (2.52) we have

$$A(x) = \frac{1}{4\pi} \int \frac{\omega(x)}{|x - y|} \, dy, \tag{2.53}$$

and hence

$$u(x) = \frac{1}{4\pi} \int \frac{(x - y) \wedge \omega(y)}{|x - y|^3} \, dy \equiv \mathbf{K}\omega(x), \tag{2.54}$$

where $\mathbf{K}$ in an integral operator of the form

$$u_i(x) = (\mathbf{K}\omega)_i(x) = \sum_{j=1}^{3} \int dy \, \mathbf{K}_{ij}(x - y)\omega_j(y) \tag{2.55}$$

and the matrix kernel $\mathbf{K}(\cdot)$ is given by

$$\mathbf{K}(x)\xi = \frac{1}{4\pi} \frac{x}{|x|^3} \wedge \xi. \tag{2.56}$$

In an analogous way it is possible to find an explicit expression of $\mathbf{K}(\cdot)$ (by means of a series) that makes (2.55) valid in the periodic case. We note that relation (2.55), given the vorticity, allows us to build the velocity field, modulo potential fields. We have considered cases where the potential field is a constant, and so (2.55) holds when $u \to 0$ as $|x| \to \infty$.

It might help, at this point, to give a representation formula for a general divergence-free smooth vector field $u = u(x)$ in a regular domain $D \subset \mathbb{R}^d$,

$d = 2, 3$, satisfying the boundary conditions $u \cdot n = 0$ on ∂D. We have

$$u(x) = \text{curl}\left\{ \int_D G(x, y)\omega(y)\, dy - \int_{\partial D} G(x, y)n(y) \wedge u(y)\, d\sigma(y) \right\}, \quad (2.57)$$

where $n(y)$ is the outer normal in $y \in \partial D$ and

$$G(x, y) = -\frac{1}{2\pi} \log |x - y|, \qquad d = 2, \qquad (2.58)_1$$

$$G(x, y) = \frac{1}{4\pi |x - y|}, \qquad d = 3. \qquad (2.58)_2$$

Equation (2.57) is classical and can be easily derived (see Exercise 10). Notice that this formula does not directly solve problem (2.9) because the right-hand side of (2.57) depends on u itself.

The Euler equation can be expressed in terms of vorticity which will allow us to underline important features of the Euler flow discussed later on. The following vector identity is easily verified:

$$\tfrac{1}{2}\nabla u^2 = u \wedge \text{curl } u + (u \cdot \nabla)u. \qquad (2.59)$$

The Euler equation can be written as (recall that $\omega = \text{curl } u$)

$$\partial_t u + \tfrac{1}{2}\nabla u^2 - u \wedge \omega = -\nabla p. \qquad (2.60)$$

Taking the curl of both sides

$$\partial_t \omega - \text{curl}(u \wedge \omega) = 0. \qquad (2.61)$$

Since

$$\text{curl}(u \wedge \omega) = (\omega \cdot \nabla)u - \omega(\nabla \cdot u) - (u \cdot \nabla)\omega + u(\nabla \cdot \omega), \qquad (2.62)$$

we finally obtain

$$D_t \omega = (\omega \cdot \nabla)u. \qquad (2.63)$$

We note that (2.63) has the advantage of having eliminated the pressure. However, to study the Euler equation in form (2.63), it is necessary to reconstruct the velocity field u from the vorticity ω. As we have seen this is an easy task in two dimensions and in three dimensions for particular domains.

In two dimensions (2.63) becomes much simpler. Namely, in the presence of a planar symmetry

$$u = (u_1, u_2, 0), \qquad u_i = u_i(x_1, x_2) \qquad (2.64)$$

(notice that the Euler equation conserves this symmetry), only the third component of the vorticity $\omega_3 = \omega$ is present and the right-hand side of (2.63) vanishes. Therefore, the Euler equation for the vorticity in two dimensions becomes

$$D_t \omega = 0. \qquad (2.65)$$

22   1. General Considerations on the Euler Equation

Notice that (2.65) implies the conservation of the vorticity along the trajectories (see later for details). This fact will turn out to be of great relevance in the analysis of the two-dimensional flow. As we will see later on, the study of (2.65) is considerably simpler than the general equation (2.63), valid in three dimensions. From a physical point of view the two-dimensional case is also realized when the domain D is really two dimensional (a thin film of fluid).

We conclude this section observing that, in the case of cylindrical symmetry,

$$u_z = u_z(r, z),$$

$$u_r = u_r(r, z), \tag{2.66}$$

$$u_\theta = 0$$

(where r, z, θ are the cylindrical coordinates), (2.63) reduces to (2.65) in the half-plane $r > 0$, with the boundary conditions $u_r = 0$ for $r = 0$. Actually, expressing the gradient in cylindrical coordinates, it is easy to obtain the Euler equation which is similar to that valid in a half-plane

$$\partial_t u_r + (u_r \partial_r + u_z \partial_z) u_r = -\partial_r p,$$
$$\partial_t u_z + (u_r \partial_r + u_z \partial_z) u_z = -\partial_z p. \tag{2.67$_1$}$$

However, there is an important difference. In fact, the continuity equation becomes

$$\partial_z u_z + \frac{1}{r} \partial_r(r u_r) = 0. \tag{2.67$_2$}$$

Thus the problem reduces to the half-plane case with the incompressibility property is replaced by condition (2.67)$_2$.

Let us now discuss the equation in terms of the vorticity. Putting

$$\omega = \omega_\theta = \partial_z u_r - \partial_r u_z \tag{2.68}$$

as a direct consequence of (2.67), we obtain

$$\partial_t \omega + (u \cdot \nabla)\omega - \frac{u_r \omega}{r} = 0, \tag{2.69}$$

where

$$\nabla = \left(\partial_z, \partial_r, \frac{1}{r}\partial_\theta\right), \tag{2.70}$$

that is,

$$D_t \frac{\omega}{r} = 0. \tag{2.71}$$

With this symmetry the potential vector A assumes a simple form

$$A = (0, 0, A_\theta). \tag{2.72}$$

We define

$$A_\theta = \frac{\Psi_1(z, r)}{r}. \tag{2.73}$$

Because, in general, in cylindrical coordinates

$$\operatorname{curl} A = \left(\frac{1}{r} \partial_r (r A_\theta) - \frac{1}{r} \partial_\theta A_r, \frac{1}{r} \partial_\theta A_z - \partial_z A_\theta, \partial_z A_r - \partial_r A_z \right), \quad (2.74)$$

we have in our case

$$u = \operatorname{curl} A = \left(\frac{1}{r} \partial_r \Psi_1, \frac{1}{r} \partial_z \Psi_1, 0 \right). \quad (2.75)$$

In analogy with the two-dimensional case, Ψ_1 is called the "stream function." In fact,

$$u \cdot \nabla \Psi^1 = 0. \quad (2.76)$$

1.3. Conservation Laws

In this section we study some properties of the solutions of the Euler equation. We suppose the solutions to be regular; more precisely, we assume

$$u \in C^1(D \times [0, T]). \quad (3.1)$$

The existence of the solutions satisfying (3.1) will be the subject of the analysis in the next chapter.

The energy conservation is an expected property which provides the first conservation law valid for the ideal fluids. The *energy*, defined as

$$E = \frac{1}{2} \int_D u^2 \, dx, \quad (3.2)$$

is conserved during the motion because in our mathematical model there is no mechanism of dissipation: the fluid has neither internal friction nor friction with the boundaries.

Theorem 3.1. *Let $D \subset \mathbb{R}^3$ be a bounded domain and let u be a solution of the Euler equation with conservative external forces*

$$D_t u = -\nabla(p + V), \quad (3.3)$$

where $V = V(x, t)$ is a known function. Then

$$\frac{d}{dt} E = 0. \quad (3.4)$$

PROOF.

$$\frac{d}{dt} E = \int_D u \cdot \partial_t u = -\int_D \{ u \cdot (u \cdot \nabla) u + u \cdot \nabla P \}, \quad (3.5)$$

where we have put $P = p + V$.

We introduce here a vector identity that will be frequently used later on. Let f and v be scalar and vector fields of class C^1, respectively. Then

$$\text{div}(fv) = (\nabla f) \cdot v + f \nabla \cdot v. \tag{3.6}$$

Using (3.6) and the incompressibility condition $\nabla \cdot u = 0$, we have

$$\int_D u \cdot \nabla P = \int_{\partial D} (Pu) \cdot n. \tag{3.7}$$

The right-hand side of (3.7) vanishes by virtue of the boundary conditions $u \cdot n = 0$ on ∂D. Finally, using the boundary conditions, incompressibility, and (3.6)

$$\int_D u \cdot (u \cdot \nabla) u = \sum_i \int_D u_i u \cdot \nabla u_i$$

$$= -\sum_i \frac{1}{2} \int_D u_i^2 (\nabla \cdot u) + \sum_i \frac{1}{2} \int_{\partial D} u_i^2 u \cdot n = 0. \tag{3.8}$$

$\square$

The energy conservation law can be extended easily to unbounded domains. In this case, E is finite if u decays at infinity fast enough. When dealing with a two-dimensional symmetry (where D is a cylindrical domain in $\mathbb{R}^3$ and the velocity field is horizontal and it does not depend on the altitude), the conserved quantity is the energy density with respect to the vertical coordinate (Fig. 1.14).

For a stationary flow the energy conservation assumes a very significant form.

Theorem 3.2 (Bernoulli). *In an ideal fluid in stationary motion under the action of conservative force with potential V independent of time, the quantity*

$$\mathscr{E} = \tfrac{1}{2} u^2 + (p + V) \tag{3.9}$$

is constant along the flow lines.

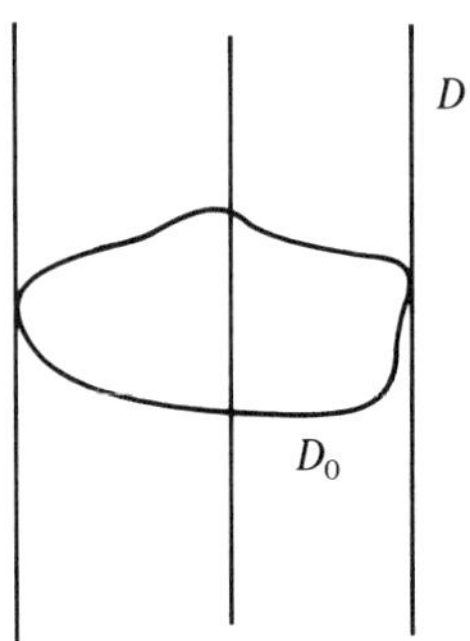

Figure 1.14

PROOF. In Section 1.1 we have seen that the Euler equation is the Newton equation of a particle of unitary mass moving under the action of the force $-\nabla(p + V)$. Such force, in general, depends on time and so the energy is not conserved. But in a stationary motion the pressure does not depend on time and therefore the expression (3.9) is conserved. $\qquad\square$

Theorem 3.2 says that E remains constant along a stream line, but in general varies when we pass from one stream line to another. On the contrary, when the velocity field is irrotational the value of E does not depend on the choice of the stream line, as follows from:

Theorem 3.3 (Bernoulli). *Consider an ideal fluid in a stationary irrotational motion in a domain D, under the action of a conservative force with potential V independent of time. Then the quantity*

$$\mathscr{E} = \tfrac{1}{2}u^2 + p + V \tag{3.10}$$

is constant.

PROOF. Using the identity (2.59), the Euler equation and the hypotheses of irrotationality and stationarity of the velocity field, we have

$$\nabla(\tfrac{1}{2}u^2 + p + V) = 0. \tag{3.11}$$

Thus the theorem is proved. $\qquad\square$

The Bernoulli theorems give us immediately some general information on stationary motions. For instance, they tell us that in the absence of external forces the pressure is greatest when the velocity is smallest and vice-versa. In particular, in a narrowing pipeline (see Fig. 1.15) the continuity equation implies that, when the pipeline has a smaller section, the velocity must be greater. The Bernoulli theorem ensures that the pressure is smaller

In the simplest case of a fluid at rest, in the presence of a gravitational field

$$V = -gx_3. \tag{3.12}$$

Theorem 3.3 give us the Stevino law:

$$p = gx_3. \tag{3.13}$$

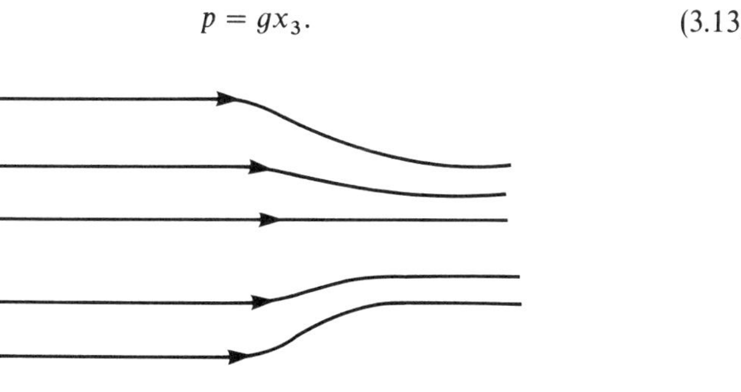

Figure 1.15

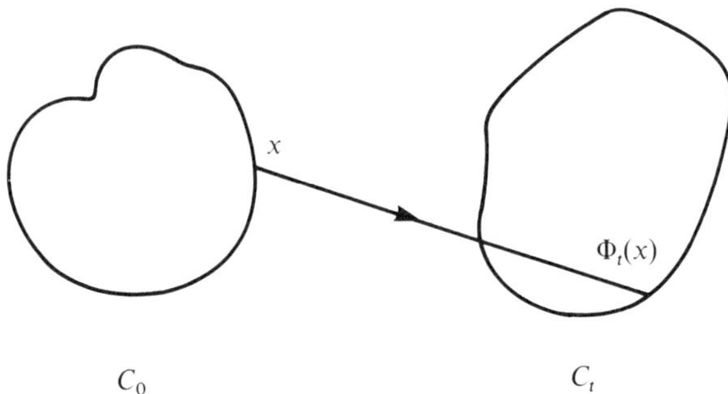

Figure 1.16

We proceed now to analyze some conservation laws involving the vorticity. We have already remarked that in a fluid in motion according to the Euler equation, different layers of the fluid cannot interact between themselves via friction forces. So it is not possible to give rise to or to change the rotation of an ideal fluid. This fact must be reflected in a conservation law involving the vorticity field. This law is expressed by the following Kelvin theorem.

Let C_t be a closed curve evolving in time according to the pathlines (Fig. 1.16)

$$C_t = \Phi_t(C_0). \tag{3.14}$$

Then, under the hypotheses of regularity of the solution to the Euler equation we have assumed, it is easy to see that regular closed curves are transported by the flow in regular closed curves.

We consider the circulation

$$\Gamma(C_t) = \oint_{C_t} u(t)\, dl, \tag{3.15}$$

where dl is the infinitesimal element of line in C_t. Then

Theorem 3.4 (Kelvin).

$$\frac{d}{dt}\Gamma(C_t) = 0. \tag{3.16}$$

PROOF. First of all we prove that

$$\frac{d}{dt}\oint_{C_t} u \cdot dl = \oint_{C_t} D_t u \cdot dl. \tag{3.17}$$

Let $[0, 1] \ni s \to x(s) \in C_0$ be a parametrization of the initial curve C_0. Then $[0, 1] \ni s \to \Phi_t(x(s))$ is a parametrization of C_t.

We have

$$\frac{d}{dt} \oint_{C_t} u \cdot dl = \frac{d}{dt} \int_0^1 u(\Phi_t(x(s)), t) \cdot \frac{\partial}{\partial s} \Phi_t(x(s))\, ds$$

$$= \int_0^1 D_t u(\Phi_t(x(s)), t) \cdot \frac{\partial}{\partial s} \Phi_t(x(s))\, ds$$

$$+ \int_0^1 u(\Phi_t(x(s)), t) \cdot \frac{\partial}{\partial t} \frac{\partial}{\partial s} \Phi_t(x(s))\, ds. \qquad (3.18)$$

The last integral vanishes. In fact, it is equal to

$$\int_0^1 u(\Phi_t(x(s)), t) \cdot \frac{\partial}{\partial s} u(\Phi_t(x(s)), t)\, ds = \frac{1}{2} \int_0^1 \frac{\partial}{\partial s} u^2(\Phi_t(x(s)), t)\, ds$$

$$= 0. \qquad (3.19)$$

So (3.17) is proved.

Using the Euler equation

$$\frac{d}{dt} \Gamma(C_t) = - \oint_{C_t} \nabla(p + V)\, ds = 0 \qquad (3.20)$$

and this achieves the proof. $\qquad\qquad\qquad\qquad\qquad\qquad\qquad\qquad\square$

In terms of the vorticity field the Kelvin theorem has an immediate consequence: the vorticity flux through a surface Σ_t moving with the fluid

$$\int_{\Sigma_t} \omega \cdot n\, d\sigma \qquad (3.21)$$

(n denotes the normal to the surface (Fig. 1.17)) remains constant in time. This follows by the Stokes theorem.

We define the *vorticity line* as a curve tangent in every point to the vorticity field (Fig. 1.18). We consider a close curve not tangent in any point to a

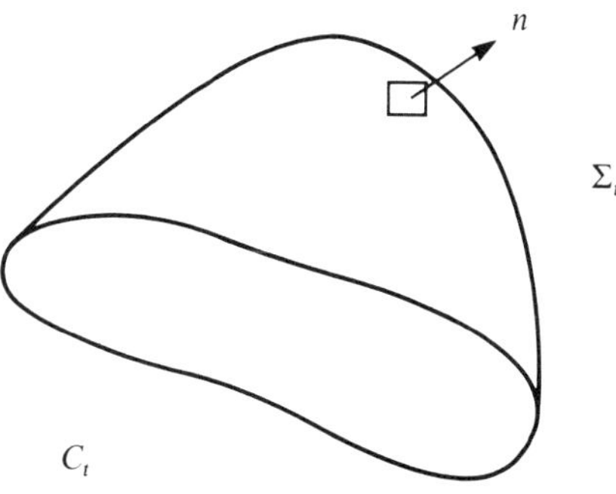

Figure 1.17

Figure 1.18

vorticity line. Consider the set of all the vorticity lines which cross (transversally) in each of its points. This set is called the *vortex tube* (or the tube of vorticity).

As we will see, the lines and tubes of vorticity are transported by the path lines in the lines and tubes of vorticity. Before proving this fact, we state the following theorem:

Theorem 3.5 (Helmholtz). *Let C_1 and C_2 be two arbitrary curves encircling the same vorticity tube. Then*

$$\Gamma = \oint_{C_1} u \cdot dl = \oint_{C_2} u \cdot dl. \tag{3.22}$$

Γ *is called the* strength *of the vortex tube. By the Kelvin theorem it remains constant during the motion.*

PROOF. This is an easy application of the Gauss and Stokes theorems. Let C_1 and C_2 be two curves oriented as in Fig. 1.19, S being the surface of the vortex tube enclosed by C_1 and C_2, and S_1 and S_2 being two surfaces whose borders are given by C_1 and C_2, respectively. Denote by Σ the region whose boundary is given by $\Sigma = S \cup S_1 \cup S_2$.

First we note that, in general,

$$\nabla \cdot \omega = 0. \tag{3.23}$$

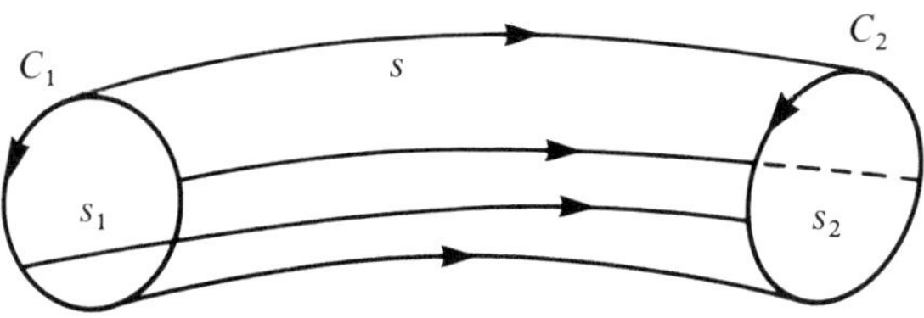

Figure 1.19

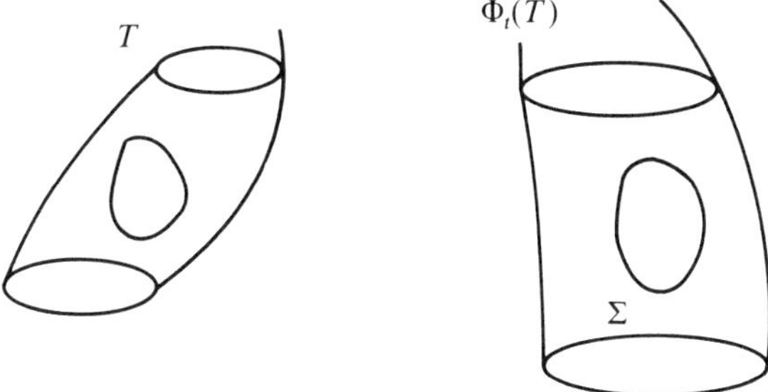

Figure 1.20

Hence

$$0 = \int_V \nabla \cdot \omega \, dx = \int_\Sigma \omega \cdot n \, d\sigma$$

$$= \int_{S_1 \cup S_2} \omega \cdot n \, d\sigma + \int_S \omega \cdot n \, d\sigma. \tag{3.24}$$

The last term vanishes because ω is tangent to S. By the use of the Stokes theorem we achieve the proof. $\qquad\square$

We see now that the Kelvin theorem implies that the motion evolves vortex tubes into vortex tubes. Let T be a vortex tube at time zero and $\Phi_t(T)$ its evolution (Fig. 1.20). We want to prove that $\Phi_t(T)$ is also a vortex tube. Let Σ be a part of the external surface of $\Phi_t(T)$. Let S be defined by $\Sigma = \Phi_t(S)$. Then S is part of the vortex tube T, and so the flux of ω through S vanishes. As a consequence of the Kelvin theorem this flux through Σ also vanishes, and so ω is tangent to Σ. The conclusion follows by the arbitrariness of Σ. As a consequence, the vortex lines, which are intersections of vortex tubes, evolve into vortex lines.

As we will see in more detail in the next chapter, Theorems 3.4 and 3.5 are not sufficient to give us an a priori bound on the growth of ω. In fact, a vortex tube during the motion can stretch without violating the previous conservation laws. When the tube becomes narrower the intensity of ω increases so that the flux remains constant. This lack of an a priori bound on ω is the main obstacle in extending to arbitrary times a local existence theorem for the solutions. The situation is completely different when we are in the presence of a two-dimensional symmetry. In this case, the vorticity is orthogonal to the plane of the motion, and so the vortex tube is a cylinder with basis, denoted by C_t, in the plane of the motion. This curve envelops a surface of area which is conserved (because of the incompressibility) during the motion.

The arbitrariness of this curve implies that ω is a constant along the path

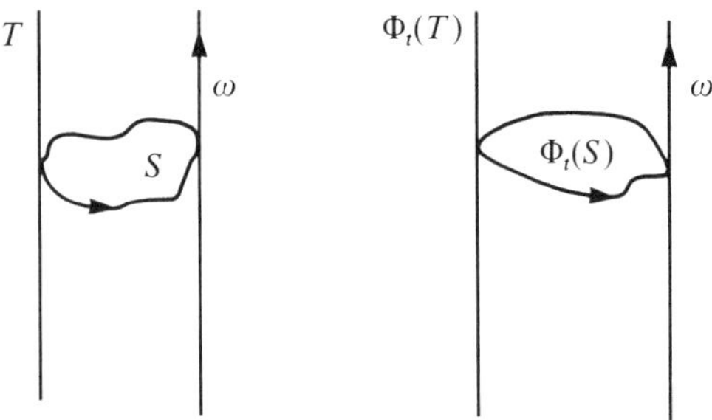

Figure 1.21

lines (Fig. 1.21). On the other hand, this property could be deduced directly observing that the Euler equation for the vorticity reduces in the two-dimensional case to

$$D_t\omega(x, t) = 0. \tag{3.25}$$

By integration of (3.25) along the path lines, we obtain the conservation of the vorticity along the trajectories

$$\omega(\Phi_t(x), t) = \omega_0(x). \tag{3.26}$$

We observe that in the presence of cylindrical symmetry, the quantity which is conserved along the trajectories of the fluid particles is ω/r, as follows by (2.71). We also remark that, thanks to (3.26) and the conservation of the Lebesgue measure, the quantity $\int_D dx\, \phi(\omega(x))$ (where ϕ is any bounded measurable function) is conserved in the two-dimensional evolution. When $\phi(r) = \frac{1}{2}r^2$ then the previous conserved quantity becomes $\frac{1}{2}\int_D dx\, \omega(x)^2$ and is called *enstrophy*. This object will play an important role in the sequel.

Now we want to discuss some conservation laws that hold when the domain D coincides with all the space. Let $u \in C^1 \cap L_\infty(\mathbb{R}^3 \times [0, T])$ be a solution of the Euler equation with finite energy. The finiteness of the energy implies that $u_0 \in L_2(\mathbb{R}^3)$ (and hence for the conservation of the energy $u(x, t) \in L_2(\mathbb{R}^3)$). Moreover, we suppose that $\omega \in C^1 \cap L_2(\mathbb{R}^3 \times [0, T])$. Then the quantity

$$A(t) = \int_{\mathbb{R}^3} u(x, t) \cdot \omega(x, t)\, dx \tag{3.27}$$

is defined for any $t \in [0, T])$ and is called *helicity*.

The following theorem states the conservation of A in time:

Theorem 3.6. *In an ideal fluid under the action of conservative forces*

$$\frac{d}{dt}A = 0. \tag{3.28}$$

PROOF. The statement can be proved by a direct verification

$$\frac{d}{dt} A = \int \{(\partial_t u) \cdot \omega + u \cdot (\partial_t \omega)\}$$

$$= -\int \{(u \cdot \nabla u) \cdot \omega + \nabla(p + V) \cdot \omega - u \cdot (\omega \cdot \nabla)u + u \cdot (u \cdot \nabla)\omega\}. \quad (3.29)$$

The condition $\nabla \cdot \omega = 0$ implies that the second and third terms vanish. Using the incompressibility condition and integrating by parts, we conclude that the sum of the first and fourth terms vanishes. This achieves the proof. (The proof we have given is very formal, but it can be made rigorous [see Exercise 12].) $\qquad\square$

We note that in domains D with boundary the helicity is not conserved This is due to the fact that, in general, neither ω nor its normal component vanishes on the boundary of D.

From now until the end of this section, we suppose that the external forces, if present, are conservative. The quantities

$$\Omega = \int_{\mathbb{R}^3} \omega \, dx, \quad (3.30)$$

$$M = \int_{\mathbb{R}^3} x \wedge \omega \, dx, \quad (3.31)$$

$$I = \int_{\mathbb{R}^3} |x|^2 \omega \, dx, \quad (3.32)$$

are *constants of the motion*. This can be proved directly by using the Euler equation and the continuity equation, and by integrating by parts when necessary. To give a rigorous proof of these facts, it is necessary that the solutions of the Euler equation which we are considering are smooth enough. A deeper understanding of the conservation in time of M and I follows from the fact that the Euler equation is essentially Hamiltonian. Actually, as discussed in Section 1.1, the Euler equation is the Euler–Lagrange equation associated to a variational principle, the stationary action principle, relative to a Lagrangian which is the kinetic energy of the fluid. Thus, according to the Noether theorem, we expect the existence of conservation laws associated to invariance properties of the kinetic energy with respect to suitable symmetry groups. In particular, the conservation in time of M and I is related to the translational and rotational invariance, respectively. On the other hand, these conservation laws are quite expected. In fact, at least at a formal level, $\frac{1}{2}M$ and $-\frac{1}{2}I$ are equal to the momentum and angular momentum of the fluid itself, respectively. As we remarked, the invariance of M and I is related to the invariance of $\mathbb{R}^3$ under a group of transformations (translation or rotation). In the case of particular domains, with these invariance properties, we expect that other corresponding quantities are conserved.

In the two-dimensional case, the above conservation law yields to the conservation of

$$\Omega = \int_{\mathbb{R}^2} \omega \, dx, \qquad \text{total vorticity,} \qquad (3.33)$$

$$B = \int_{\mathbb{R}^2} x\omega \, dx, \qquad \text{center of vorticity,} \qquad (3.34)$$

$$I = \int_{\mathbb{R}^2} |x|^2 \omega \, dx, \qquad \text{moment of inertia.} \qquad (3.35)$$

For domains different from $\mathbb{R}^2$, the total vorticity Ω is always conserved (Exercise 15). Moreover, the quantities $\frac{1}{2}B \cdot n$ (n is an arbitrary unitary vector) and $-\frac{1}{2}I$ are conserved in domains which are invariant for the group of translation on the direction n and for the group of rotation, respectively (Exercises 16 and 17). They have the meaning of momentum, angular momentum, and boundary terms that, in the two dimensional case, are constant.

We conclude this section by showing the existence of a quantity which is conserved along the trajectories for the motion in $\mathbb{R}^d$, $d = 2, 3$. Define

$$\gamma = u + \nabla\varphi, \qquad (3.36)$$

where φ satisfies the following equation

$$D_t\varphi = -\tfrac{1}{2}\nabla u^2 + p. \qquad (3.37)$$

Here u and p are those arising from the solution of the Euler equation. Then γ is a solution of

$$D_t\gamma_i = (\partial_i u_j)\gamma_j. \qquad (3.38)$$

This follows by direct computation. Notice that (3.36) expresses the unique decomposition of γ in its divergence-free part and gradient part (discussed in Appendix 1.2).

Finally, by using (3.38) and (2.63), we realize that the quantity

$$\omega \cdot \gamma = \operatorname{curl} \gamma \cdot \gamma, \qquad (3.39)$$

called *spirality*, is conserved along the trajectories.

We leave as an exercise, for the reader to prove that the conservation of the spirality implies the conservation of the helicity. This conservation law is similar to the conservation of the vorticity in the two-dimensional case (see (3.25)), however, this is less suitable because γ appearing in (3.39) is not explicitly known (in terms of u). On the other hand, if γ is assumed as the main unknown of the problem, we have that the spirality is a legitimate first integral and u can be recovered from γ as the projection on the divergence-free fields.

1.4. Potential and Irrotational Flows

In this section we discuss in more detail the *irrotational flows*, that is, flows in which the vorticity vanishes everywhere. A particular example is given by the so-called *potential flows*, those for which there exists a function $\varphi(x, t)$, such that

$$u(x, t) = \nabla\varphi(x, t). \tag{4.1}$$

Clearly any potential flow is also irrotational. The converse is not true: although it is possible to find, for any irrotational flow, a function φ satisfying (4.1), in general, it may be multivalued for a nonsimply connected domain since it can assume many different values depending on the number of loops around the holes.

We observe initially, that there are, in reality, situations schematizable by irrotational flows. However, they do not describe in a sufficiently accurate way the interaction between the fluid and the obstacles immersed in it. In addition, in dealing with these kinds of problems, some paradoxes arise which can be solved only by a drastic change of the mathematical model. This point will be more fully discussed later on.

The interest in studying irrotational divergence-free flows lies in the fact that they are a stationary solution of the Euler equation, and so are compatible with the mathematical model we are dealing with. To verify this statement, we assume that $u = u(x)$ is a solution of the equations

$$\nabla \cdot u = 0, \qquad \text{curl } u = 0, \tag{4.2}$$

in a domain $D \subset \mathbb{R}^d, d = 2, 3$. From (2.59) we know that $(u \cdot \nabla)u$ is a gradient

$$(u \cdot \nabla)u = \tfrac{1}{2}\nabla u^2 \tag{4.3}$$

and so, for boundary conditions not dependent on time, we have a stationary solution of the Euler equation, with the pressure $p = -\tfrac{1}{2}u^2$.

In some problems we are interested in studing potential flows when the boundary conditions depend on time, for instance, when the wall of the region moves with a given law. The Euler equation becomes (by use of (4.3) and (4.1))

$$\nabla(\partial_t\varphi + \tfrac{1}{2}|\nabla\varphi|^2 + p) = 0; \tag{4.4}$$

hence

$$\partial_t\varphi + \tfrac{1}{2}|\nabla\varphi|^2 + p = \text{constant}. \tag{4.5}$$

We now discuss some qualitative properties of irrotational flows. First, we note that in a bounded, simply connected domain, all the irrotational flows reduce to the trivial one: $u = 0$. In fact, (4.1) and (4.2) yield $\Delta\varphi = 0$, and the boundary conditions give $\partial\varphi/\partial n = 0$ on the boundary ∂D. Thus $\varphi = \text{const}$. This fact corresponds to the following geometric property: it is impossible to extend to the interior of a bounded simply connected domain D, a vector field

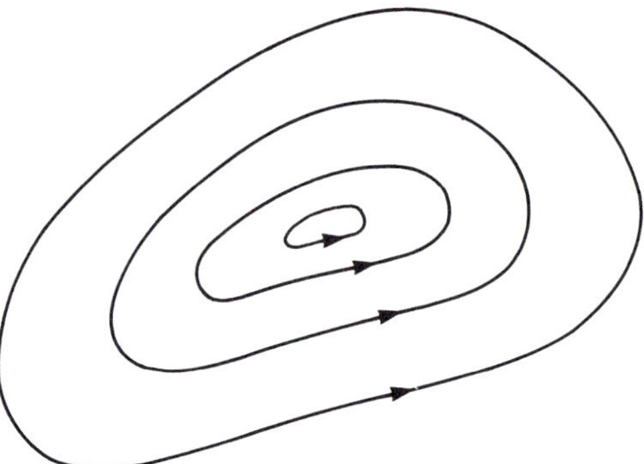

Figure 1.22

tangent to the boundary ∂D in a smooth and divergence-free way, without creating curls (Fig. 1.22).

So in order to have a nontrivial irrotational field we must consider either nonsimply connected or unbounded domains. Very important in the applications are the so-called *external domains*. They are defined as the complement of a finite union of simply connected bounded regions. A domain as in Fig. 1.23 has a different topological structure in two and in three dimensions. In fact, it is simply connected in three dimensions and so $u = \nabla \varphi$ where φ is a harmonic function satisfying the boundary conditions $\partial \varphi / \partial n = 0$ on ∂D. Specifying the asymptotic behavior

$$\lim_{|x| \to \infty} u(x) = U_\infty \tag{4.6}$$

we have a unique potential flow (following the uniqueness of the Neumann problem modulo a constant).

In two dimensions, such a domain in not simply connected and the irrotational flows are not necessarily potential flows. In Section 1.2 we saw that

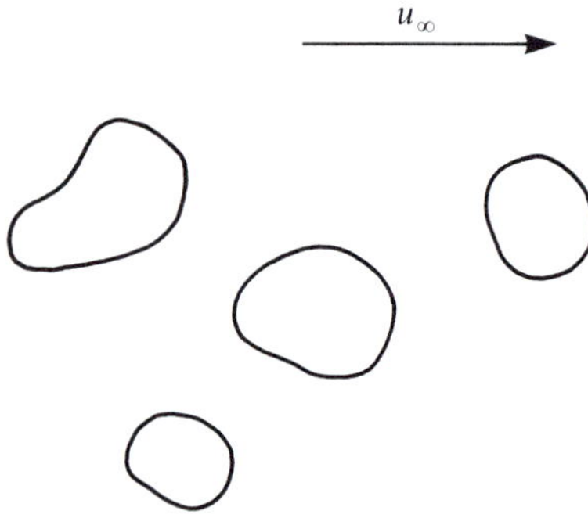

Figure 1.23

there are as many irrotational flows as there are possible values of the circulations around the obstacles (once we have fixed the asymptotic behavior (4.6)).

For example, let us consider the domain external to a circle

$$D = \{x \in \mathbb{R}^2 \mid x > 1\}. \tag{4.7}$$

The potential function φ (in terms of polar coordinates ρ, θ) is

$$\varphi(\rho, \theta) = (\rho + \rho^{-1}) \cos \theta. \tag{4.8}$$

It generates a unique potential flow

$$v_\rho = \left(1 - \frac{1}{\rho^2}\right) \cos \theta, \qquad v_\theta = -\sin \theta \left(1 + \frac{1}{\rho^2}\right), \tag{4.9}$$

with asymptotic conditions

$$u_\infty = (1, 0). \tag{4.10}$$

The circulation around the obstacle vanishes (Fig. 1.24).

In the same domain we can define an irrotational flow with a circulation α. Its stream function is given by the expression

$$\Psi(x) = -\frac{\alpha}{2\pi} \ln |x|. \tag{4.11}$$

It is exactly the velocity field produced by a charge of vorticity of intensity α situated at the origin (that point, of course, does not belong to D) (Fig. 1.25).

$$u(x) = \frac{\alpha}{2\pi} \frac{x^\perp}{|x|^2}. \tag{4.12}$$

The circulation is

$$\Gamma = \oint_{\partial D} u \cdot dl = \alpha. \tag{4.13}$$

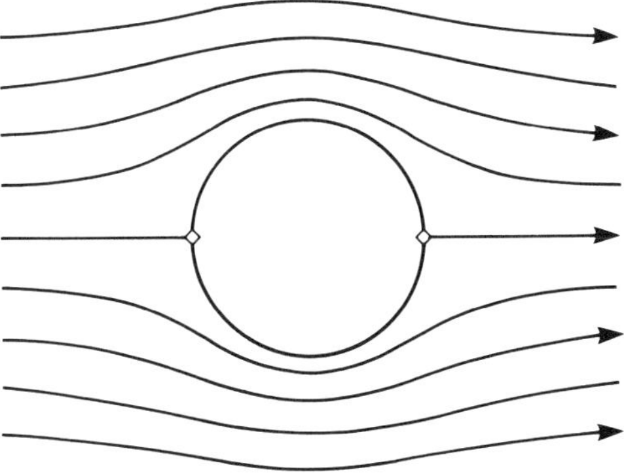

Figure 1.24

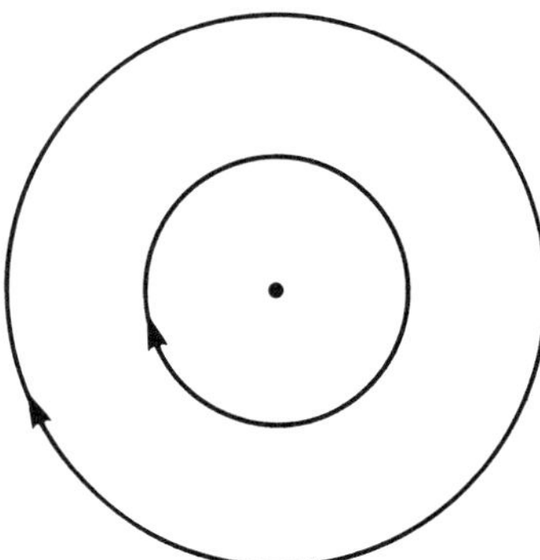

Figure 1.25

The sum of two flows (4.9) and (4.11) gives us the unique irrotational flow with circulation α and asymptotic behavior u_∞ (Fig. 1.26).

We now discuss an important property of the minimum of the potential flows. Let $\hat{u} = \hat{u}(x)$ be a solution of (4.2) in $D \subset \mathbb{R}^d$, $d = 2, 3$. We consider a simply connected domain D_0, $D_0 \subset D$. Let $\varphi = \varphi(x)$ be the potential of $\hat{u}$: $\hat{u} = \nabla\varphi$. We prove that $\hat{u}$ is a minimum of the functional energy in D_0, among all the divergence-free flows satisfying the same boundary condition of $\hat{u}$ on ∂D_0.

We consider the variation of the energy in D_0 in the passage from $\hat{u}$ to another arbitrary divergence-free profile u

$$
\begin{aligned}
E_{D_0}(u) - E_{D_0}(\hat{u}) &= \frac{1}{2} \int_{D_0} (u^2 - \hat{u}^2) \\
&= \frac{1}{2} \int_{D_0} (u + \hat{u})(u - \hat{u}) = \tfrac{1}{2}\|u - \hat{u}\|_2^2 + \int_{D_0} \nabla\varphi \cdot (u - \hat{u}) \\
&= \tfrac{1}{2}\|u - \hat{u}\|_2^2 + \oint_{\partial D_0} \varphi(u - \hat{u}) \cdot n = \tfrac{1}{2}\|u - \hat{u}\|_2^2 \geq 0. \quad (4.14)
\end{aligned}
$$

From the inequality (4.14) we conclude that the following theorem holds:

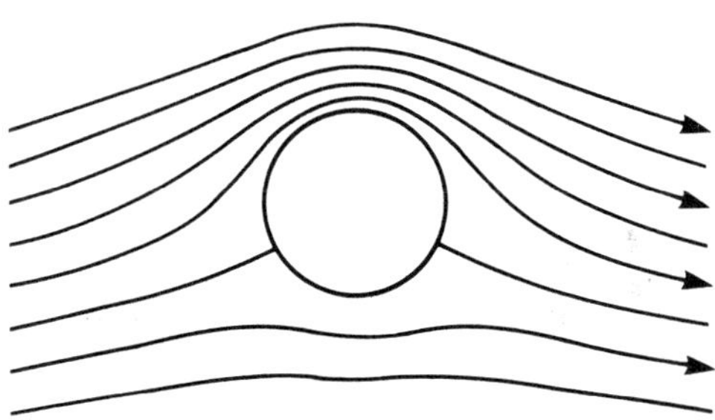

Figure 1.26

Theorem 4.1. *Let $\hat{u}$ be a divergence-free potential flow in a regular and bounded domain, $D_0 \subset \mathbb{R}^d$ ($d = 2, 3$). Then $\hat{u}$ minimizes the energy*

$$E = \frac{1}{2} \int_{D_0} u^2$$

in the class of all the divergence-free flows which satisfy the same boundary conditions of $\hat{u}$

$$u \cdot n = \hat{u} \cdot n \quad \text{on } \partial D_0. \tag{4.15}$$

Now we want to study the interaction between an irrotational flow and an obstacle B immersed in it. The flow is defined in all the space external to B, and we will assume that its asymptotic behavior is given by the constant vector $u_\infty = (\hat{u}_1, \hat{u}_2)$. We want to calculate the force F acting on B. We will obtain paradoxical results.

Theorem 4.2 (Kutta–Joukowski). *In dimension two we have*

$$F = -\Gamma |u_\infty| n, \tag{4.16}$$

where Γ is the circulation around B and n is a unitary vector orthogonal to u_∞ (see Fig. 1.27).

As a consequence of Theorem 4.2 the force produced on a symmetric obstacle by a symmetric flow (as in (4.9)) vanishes (Fig. 1.28). The proof of Theorem 4.2 is given in Appendix 1.3.

Theorem 4.3 (d'Alembert Paradox). *In dimension three*

$$F = 0. \tag{4.17}$$

The proof of Theorem 4.3 is given in Appendix 1.4.

Some comments on the Kutta–Joukowski and d'Alembert theorems. First, we notice that for the Galilean invariance the force remains unchanged if we consider an obstacle moving with velocity $-u_\infty$ and the fluid at rest at

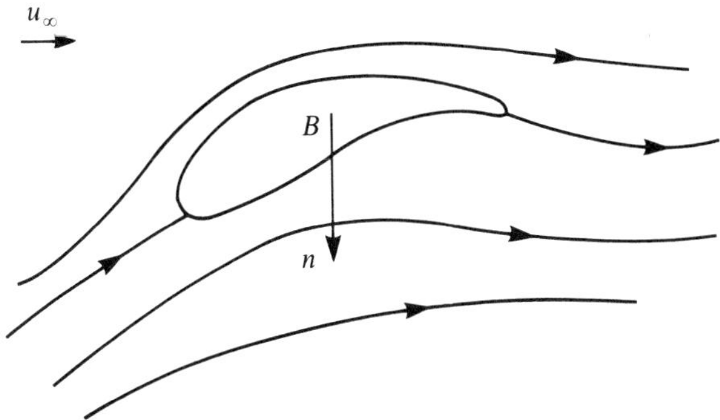

Figure 1.27

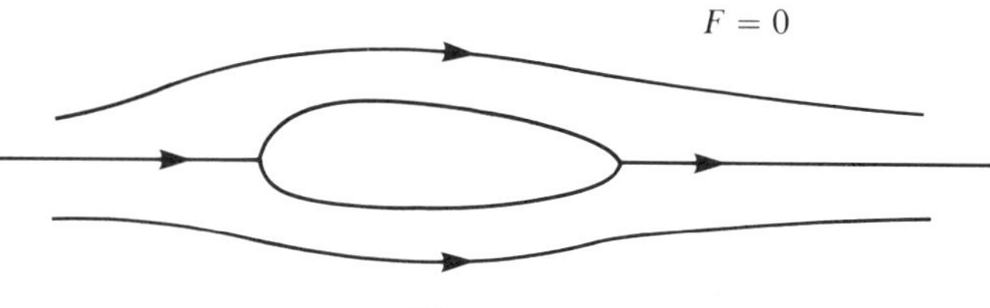

Figure 1.28

infinity. The Kutta–Joukowski theorem tell us that the drag (i.e., the component of the force in the direction opposite to the velocity) produced by the fluid on an obstacle (for instance, an infinitely long wing) vanishes. Only the lift (which is the component orthogonal to the direction of the velocity) remains and, in general, does not vanish whenever the wing is not symmetric. This effect is largely used in aerodynamics. A more detailed analysis of these arguments constitutes an important chapter in applied fluid dynamics, however, further discussion is beyond the scope of this book. The d'Alembert theorem states that an object moving with velocity u_∞ in an irrotational field does not feel any force (neither drag nor lift).

The results we have obtained are in sharp contrast with experience. For instance, an airplane could not fly. In fact, suppose that, initially, the airplane and the fluid (air) are both at rest. Then the airplane begins to move. Since vorticity cannot be produced, the irrotational flow around the airplane cannot produce any lift, so that flight is impossible. Another example comes from the football game: a ball kicked with spin performs a curved trajectory in the direction of the rotation, and not a straight line as we could expect in the absence of force, according to the d'Alembert theorem.

Such paradoxes can be avoided if vorticity is present. However, the problem remains of understanding how vorticity can be created in the system. The conservation of vorticity in an ideal fluid, while reasonable far from the obstacle, is too drastic near the boundary. A more accurate description of the interaction among the particles of the fluid and the obstacle leads us to introduce the Navier–Stokes equation, which is a correction to the Euler equation. Such a new equation can explain the effects, such as vorticity production, which are relevant near the boundary. The Navier–Stokes equation is obtained from the Euler equation by adding a dissipative term $\nu \Delta u$ ($\nu > 0$ is called the viscosity coefficient). The addition of a term of second order in the spatial derivatives of the velocity makes it necessary to modify the boundary conditions on the velocity. For Navier–Stokes flows we suppose that the tangential component af the velocity on the boundary also vanishes. This is a physical requirement, and means a perfect adherence of the fluid to the boundary.

As an example, we consider the stationary irrotational solution of the Euler equation in a half-plane

$$u_1 = u_\infty, \qquad u_2 = 0, \tag{4.18}$$

and the solution of the Navier–Stokes equation with the same initial conditions

$$u_1(x_1, x_2, 0) = u_\infty \qquad \text{when} \quad x_2 > 0,$$

$$u_2(x_1, x_2, 0) = 0, \qquad (4.19)$$

$$\nabla p = 0,$$

with the boundary conditions $u(x_1, x_2, t) = 0$ when $x_2 = 0$.

Notice that even if the initial datum does not satisfy the boundary conditions, a solution of the Navier–Stokes problem with this initial datum can be obtained nonetheless, by virtue of the regularizing property of the parabolic term Δu. Actually, the initial datum can be thought of as satisfying the boundary conditions by means of a jump discontinuity at $x_2 = 0$ which will be mollified at any positive time.

The solution of our problem is

$$u_1(x_1, x_2, t) = \frac{2u_\infty}{\sqrt{\pi}} \int_0^\eta \exp\{-y^2\}\, dy, \qquad \eta^2 = \frac{x_2^2}{4vt}, \qquad (4.20)$$

$$u_2(x_1, x_2, t) = 0.$$

Hence

$$\omega(x_1, x_2, t) = -\frac{x_2 u_\infty}{\pi v t} \exp\left\{-\frac{x_2^2}{4vt}\right\}. \qquad (4.21)$$

This solution, that is neither stationary nor irrotational, significantly differs (on a fixed time scale) from the corresponding Euler solution only near the boundary when the distance from ∂D is of the order of $\sqrt{vt}$. This region, in which the Euler and Navier–Stokes solutions significantly differ, is called the *boundary layer* (Fig. 1.29).

Notice that the Navier–Stokes dynamics can generate vorticity. In fact, the condition $u = 0$ on ∂D is responsible for gradients of velocity so large near the boundary (which gets larger as v gets smaller) that they may be considered true discontinuities in the velocity field. These discontinuities are

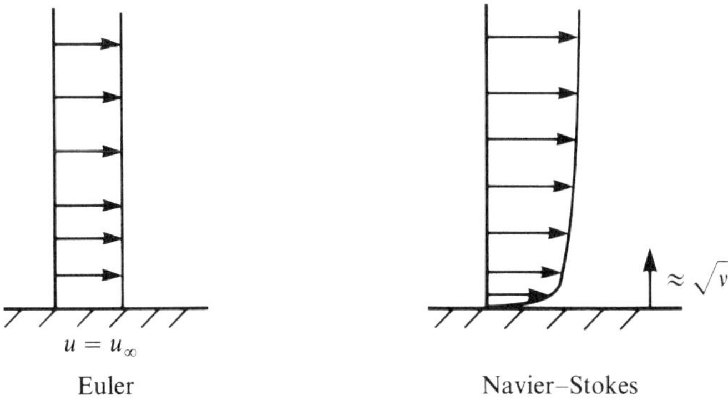

Figure 1.29

transported and diffused into the fluid by means of surfaces (or curves in two dimensions). On these lines the vorticity is very large and roughly we can consider the vorticity essentially concentrated on these surfaces. Moreover, these surfaces are very instable (instability of Kelvin–Helmholtz) as we will see in Chapter 6. In conclusion, the behavior of a fluid near an obstacle, for small viscosity v, is a very complicated phenomenon not yet completely understood.

A final remark. We have seen how the Navier–Stokes equation plays an important role in explaining the vorticity production on the boundary. However, the time evolution of the produced vorticity, at least for small v, follows essentially the Euler equation, that is, the nonlinear part of the Navier–Stokes equation. From this we conclude that the model of a nonviscous fluid has a great conceptual and practical relevance, in spite of the fact that a naïve use of this model produces the paradoxes we have just discussed.

1.5. Comments

The equations of motion of an ideal fluid were derived by Euler in 1755 [Eul]. Several mathematicians and physicists, for instance, Bernoulli, d'Alembert, Lagrange, Cauchy, Helmholtz, Kelvin, and others, have contributed to further developments of the theory. Regarding the deduction of the Euler equation via a variational principle, as was done in the first section, see the paper by Serrin [Ser 59] wherein similar topics are discussed. For a more traditional deduction of the Euler equation (not necessary for incompressible flows), consult the classical books in fluid mechanics such as [Bat 67], [ChM 79], [HuM 76], [LaL 68]₁, [Lam 32], [Mey 81], [MiT 60], [Shi 73], and [VMF 71].

The deduction is based on the following considerations. Let $u = u(x)$ and $\rho = \rho(x)$ be the velocity and density fields, respectively. The conservation of the mass and of the momentum implies

$$\frac{d}{dt} \int_{V_t} \rho(x, t)\, dx = 0, \tag{5.1}$$

$$\frac{d}{dt} \int_{V_t} \rho(x, t) u(x, t)\, dx = F, \tag{5.2}$$

where V_t is a region moving with the fluid and F is the sum of the external forces acting on the part of fluid localized in V_t.

The left-hand sides of (5.1) and (5.2) transform, by a change of variables, into

$$\frac{d}{dt} \int_{V_0} \rho(\Phi_t(x), t)\, |J_t(x)|\, dx, \tag{5.3}$$

$$\frac{d}{dt} \int_{V_0} \rho(\Phi_t(x), t) u(\Phi_t(x), t)\, |J_t(x)|\, dx, \tag{5.4}$$

where J_t is the Jacobian of the transformation $x \to \Phi_t(x)$. It is easy to verify the relation

$$\partial_t J_t(x) = J_t(x) \nabla \cdot u(\Phi_t(x), t) \tag{5.5}$$

(for the easy proof see Appendix 1.1). We remark that as a consequence of (5.5), $J_t(x) > 0$ for all t and x (making use of the fact that $J_0(x) = 1$).

Using (5.5), from (5.1) and (5.3) we obtain

$$\int_{V_0} (D_t \rho)(\Phi_t(x), t) J_t(x)\, dx + \int_{V_0} \rho(\Phi_t(x), t) \nabla \cdot u(\Phi_t(x), t) J_t(x)\, dx = 0. \tag{5.6}$$

Hence, by the arbitrariness of V_0, and supposing the integrand to be continuous, we obtain the continuity equation

$$\partial_t \rho(x, t) + \nabla \cdot [\rho(x, t) u(x, t)] = 0. \tag{5.7}$$

Moreover, by (5.2), (5.4), (5.5), and (5.7) we obtain

$$\frac{d}{dt} \int_{V_t} \rho(x, t) u(x, t)\, dx = \int_{V_t} \rho(x, t) D_t u(x, t) = F. \tag{5.8}$$

We suppose now that the forces acting on the fluid are of two kinds

$$F = \int_{V_t} \rho(x, t) f(x, t)\, dx + \int_{\partial V_t} \Phi_n(x, t)\, d\sigma, \tag{5.9}$$

where $f = f(x, t)$ is the field of the external forces per unit mass which we will suppose to be known, while the last term in (5.9) describes the surface interaction between the parts of fluid which are external and internal to V_t. More precisely $\Phi_n(x, t)\, d\sigma$ is the contact force generated by the part of fluid external to V_t, acting on the part of fluid internal to V_t through the surface $d\sigma$ of outward normal n. Φ_n is a complex object which describes the complicated (short-range) interaction between the molecules composing the fluid. In the case of an *ideal fluid*, we make the hypothesis that such interactions are purely normal, that is, there exists an unknown scalar field, $p = p(x, t)$, called *pressure*, such that

$$\Phi_n(x, t) = -p(x, t) n(x, t) \tag{5.10}$$

(the minus sign is conventional).

By the Gauss–Green lemma, putting $n_i = n \cdot c_i$, where $\{c_i\}$ is an orthonormal basis in $\mathbb{R}^3$, we have

$$\int_{\partial D_t} p n_i\, d\sigma = \int_{V_t} \nabla \cdot (p c_i)\, dx = \int_{V_t} \frac{\partial}{\partial x_i} p\, dx, \tag{5.11}$$

and then by (5.8) we obtain

$$\rho D_t u = -\nabla p + \rho f. \tag{5.12}$$

Equation (5.12) must be completed by the continuity equation (5.7). However, these equations are not sufficient to determine completely the motion of the fluid. In fact, the unknown quantities of our problem are five, u, ρ, p, while the equations we dispose of are only four.

When we introduce the hypothesis of incompressibility and homogeneity of the fluid, that is, $\rho(x, t) = $ constant, (5.7) and (5.12) reduce to the equations studied in this book. However, there are many problems in which compressibility plays an interesting role (for instance, in the shock wave problem). In general, we make the hypothesis of a constitutive relation, which depends on the nature of the fluid, connecting the pressure field to the density field

$$p = p(\rho). \tag{5.13}$$

For instance, in an ideal gas with constant temperature, we have $p = $ const. ρ. Then (5.13) must be added to (5.7), (5.12). The set of these equations is the Euler equation for compressible flows.

We consider now the very interesting problem to see in which limit the incompressible fluid model can be obtained from the compressible fluid model described by (5.7), (5.12), and (5.13). For simplicity, we assume the relation of an ideal gas

$$p = c^2\rho. \tag{5.14}$$

The constant c can be interpreted as the velocity of sound. In fact, in the linear approximation, a small perturbation in the constant density field travels according to the wave equation with a propagation velocity given by c (see, for instance, [ChM 79]). A naïve attempt would be to suppose $\rho = $ const., but we see that $p = $ const. would be implied by (5.14). and so the incompressible model cannot be obtained trivially by this assumption. What is true is that the solutions u^c, ρ^c, p^c of the Euler equation for a compressible fluid

$$D_t u^c = -\frac{\nabla p^c}{\rho^c},$$
$$D_t \rho^c + \rho^c \nabla u^c = 0, \tag{5.15}$$
$$p^c = c^2 \rho^c$$

(where we have written explicitly the dependence on the sound velocity c) approach in the limit $c^{-1} \to 0$ the solutions of an incompressible fluid. We prove it at a very formal level. We write the solution of (5.15) in the form

$$u^c = u_0 + c^{-1}u_1 + c^{-2}u_2 + O(c^{-3}), \tag{5.16}$$

$$p^c = p_0 + c^{-1}p_1 + c^{-2}p_2 + O(c^{-3}), \tag{5.17}$$

and then

$$\rho^c = c^{-2}p_0 + c^{-3}p_1 + c^{-4}p_2 + O(c^{-5}). \tag{5.18}$$

We put these expressions in the first equation (5.15) and then compare the terms of the same order in c. We obtain

(c^2) $$\frac{\nabla p_0}{p_0} = 0, \tag{5.19}$$

(c)
$$\frac{\nabla p_1}{p_0} = 0, \tag{5.20}$$

(1)
$$p_0 D_t u_0 = -\nabla p_2. \tag{5.21}$$

From (5.19) we learn that p_0 is a function of time only. The same thing holds for p_1 (and for ρ_0 and ρ_1 as a consequence). In reality, they must also be constant in time because of the conservation of the total mass.

By the continuity equation

$$D_t p^{c} = -p^{c} \nabla \cdot u^{c} \tag{5.22}$$

we obtain

$$0 = D_t p_0 = -p_0 \nabla \cdot u_0, \tag{5.23}$$

that is,

$$\nabla \cdot u_0 = 0. \tag{5.24}$$

So u_0 satisfies the Euler equation for an incompressible flow, provided the initial data satisfy the obvious relations of compatibility $\nabla \cdot u_0(x, 0) = 0$, $p_0 = \text{const}$.

The previous considerations can be formulated in a precise mathematical way including more realistic state equations than (5.14). The reader interested in these topics can consult the book by Majda [Maj 84] for a more careful analysis of the problem. Here we want to underline only that the model of an incompressible fluid is valid (starting from the model of compressible fluid) when the sound velocity is very large or, more exactly, when the adimensional quantity v/c, where v is a typical velocity of the fluid (for instance, the maximum of the velocity at time zero), is very small. The ratio $M = v/c$ is called "Mach number." These considerations explain why the model of an incompressible flow is largely used in applications in which fluids, a priori compressible, are moving in a regime in which $M \ll 1$. For instance, consider a solid body moving in air with a velocity of 10 m/s. In this case, the Mach number is almost $\cong \frac{1}{34}$, that is, sufficiently small to consider the model of an incompressible fluid as realistic.

As we have already stated in Section 1.1, there is no rigorous derivation of the equations of motion for an ideal flow starting from the dynamics of the molecules constituting the fluid. However, it is possible to convince ourselves, by heuristic arguments, that the Euler equation of a compressible fluid must be a consequence of the Newton laws. As already noted, the equations of fluid dynamics should be a macroscopic description of a system of particles subject to the laws of classical mechanics.

To recognize the fluid structure in a particle system we must make a transformation, $x \to \varepsilon^{-1} x = \xi$, of spatial coordinates that allows us to observe the fluid system, initially described on scale 1, on scale ε^{-1}, with ε very small. In other words, we are enlarging the system to investigate its internal structure. We call "macroscopic" the x-variable describing the fluid, and "microscopic" the ξ-variable, measured in the molecular length scale. Moreover,

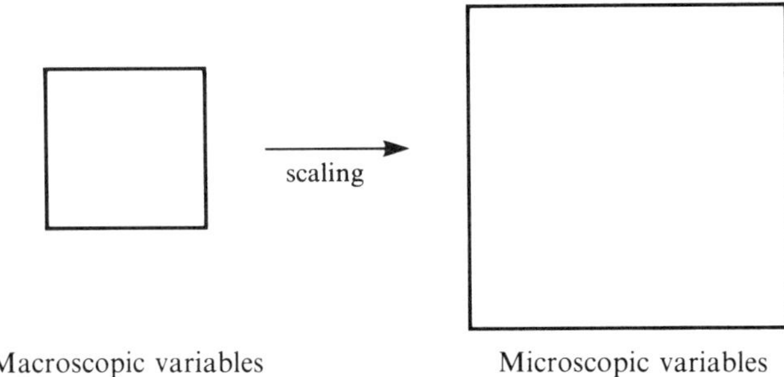

Figure 1.30

any small region around a point x contains a very large number of molecules (of the order of ε^{-3}) (Fig. 1.30).

The fluid dynamical fields $u(x)$, $\rho(x)$, $p(x)$ must be deduced from the parameters of the thermodynamical equilibrium (local because it depends on x), that we suppose the system must (locally) reach. In other words, the particle system has two scales of time: one, very fast, that allows the system to reach locally the thermodynamic equilibrium, and another one, slower, that describes the evolution of the system through the evolution of the fields u, ρ, p that follow the equations of fluid dynamics. So the microscopic time must be scaled (as the distance) by a law $t \to \varepsilon^{-1}t = \tau$. As a consequence, when we write the equations of motion of a particle system (initially described in the natural variables ξ, and τ) using the new variables x and t, we expect that the distribution function of a single particle is of the form

$$f(x, v) \approx \rho(x) \exp\{-\beta(x)[v - u(x)]^2\}[\sqrt{2\pi\beta(x)}]^{-3}, \qquad (5.25)$$

where $\beta = \beta(x)$ is a function of x proportional to the inverse of the local temperature, $\rho = \rho(x)$ is the spatial distribution of a single particle which can be interpreted as the mass density assuming $mN = 1$ (m is the mass of a molecule and N is the total number of molecules) (this assumption can always be made by a suitable choice of the measure units), and $u = u(x)$ is the mean velocity of the particles constituting the fluid. In (5.25) we are assuming that the interaction among the particles is negligible (assumption of ideal gas). The general case requires statistical mechanics concepts which are beyond the scope of the present book.

The time evolution of $f_t(x, v)$ is governed by an equation of the form

$$(\partial_t + v \cdot \nabla)f_t(x, v) = C, \qquad (5.26)$$

where C is an operator, called the "collision operator," that describes the interaction of a particle with others. It depends on the kind of intermolecular interaction we consider and on the two-body distribution if we assume binary interactions only. Equation (5.26) is the first term of a hierarchy of equations called the **BBKGY** hierarchy (from the names of its discoverers

Born, Bogoliubov, Kirkwood, Green, and Yvon). This well-known hierarchy is an easy consequence of the laws of motion (for a detailed description of these equations refer to a book on Non-Equilibrium Statistical Mechanics, for instance, [Cer 88]). For our purposes, we do not need to make explicit the collision operator C: it is enough to observe that, for a rarified gas when the ideal gas assumption is reasonable, the laws of mass, momentum and energy conservation imply

$$\int \Psi_\alpha C \, dx \, dv \simeq O, \tag{5.27}$$

where

$$\Psi_\alpha = v_\alpha \qquad \text{when} \quad \alpha = 1, 2, 3,$$
$$\Psi_4 = v^2, \tag{5.28}$$
$$\Psi_5 = 1.$$

We integrate in v the equation (5.26) and, remembering the definition

$$u(x) = \int \frac{vf(x, v)}{\rho(x)} \, dv \tag{5.29}$$

we obtain the continuity equation

$$\partial_t \rho + \nabla \cdot (\rho u) = 0. \tag{5.30}$$

Multiplying (5.26) by v_i and integrating in dv, we obtain

$$\partial_t(u_i \rho) + \nabla \cdot \int dv \, f(x, v) v_i v = 0. \tag{5.31}$$

Since, by assumption (5.25)

$$\int dv \, f v_i v = \int dv \, f \eta_i \eta + \int dv \, f u_i u, \tag{5.32}$$

where we have put $\eta = (u - v)$, we obtain

$$\partial_t(u_i \rho) + \nabla \cdot (\rho u_i u) = -\nabla \cdot \int dv \, f \eta_i \eta. \tag{5.33}$$

By virtue of (5.25) there is a function $p = p(x)$ defined as

$$\int dv \, f \eta_i \eta_j = \delta_{ij} p(x) \tag{5.34}$$

for which from (5.31), and the continuity equation, we have

$$\rho D_t u = -\nabla p \tag{5.35}$$

that expresses the momentum balance in the Euler equation. The quantity p has the interpretation of thermodynamical pressure. In fact, defining the internal energy per unit volume as

$$E = \frac{1}{\rho}\frac{1}{2}\int dv\, \eta^2 f, \tag{5.36}$$

we have

$$p = \tfrac{2}{3}\rho E \tag{5.37}$$

which is the state law of ideal gas. Observe that, in general, E is a function of space and time. We can compute the time evolution of the internal energy density to obtain a fifth equation expressing the energy balance. It is a challenging, but very difficult problem, to make rigorous the above considerations. What is necessary is a good control of the long-time behavior of Hamiltonian systems, of which very little is known. However, there are results concerning the hydrodynamical behavior of stochastic systems (see, for instance, [DeP 91]).

The potential theory is discussed in many textbooks, for instance, [Kel 53]. Other examples of Green functions not discussed in the present book are given in [CoH 37].

There are many special solutions to the Euler equations widely discussed in the literature. For instance, particular initial conditions with a cylindrical symmetry give rise to relative steady solutions, conserving the shape and moving with a constant speed in the symmetry direction. These solutions give a schematization of some physical phenomena like smoke rings or atomic mushrooms. For classical solutions see [Lam 32], for a detailed analysis see [FrB 74] and, more recently, [AmS 90].

We have seen in Section 1.3 how cylindrical symmetry makes the evolution essentially two dimensional. Another situation in which a three-dimensional problem can be reduced to a two-dimensional one is given by the initial condition

$$u(x, 0) = v(x_1, x_2, 0) + U(x), \tag{5.38}$$

where U is a potential field given by

$$U(x) = (\varphi_1(x_1, x_2), \varphi_2(x_1, x_2), -2\alpha x_3),$$
$$\nabla \cdot U = 0, \tag{5.39}$$

where $v(x_1, x_2, 0)$ is a smooth vector function contained in the x_1, x_2 plane.

We make the ansatz that the solution be of the form

$$u(x, t) = v(x_1, x_2, t) + U(x), \tag{5.40}$$

where v is contained in the x_1, x_2 plane. We look for a solution in which v vanishes at infinity. Taking the curl of the Euler equation, we have for the third component of the vorticity

$$\partial_t \omega + (v \cdot \nabla)\omega + (\varphi_1 \partial_1 + \varphi_2 \partial_2)\omega + 2\alpha\omega = 0. \tag{5.41}$$

So we have arrived at a two-dimensional problem. Notice that (5.41) differs from the Euler equation in the last two terms. We do not discuss in detail

(5.41). We only underline that, in this problem, the total vorticity is conserved while the density of vorticity is contracted or expanded. In fact, (5.41) can be rewritten in the form

$$D_t \omega = -2\alpha\omega \tag{5.42}$$

which gives

$$\omega(\Phi_t(x), t) = \omega(x, 0) \exp\{-2\alpha t\}. \tag{5.43}$$

Here $\phi_t(x)$ are the stream lines generated by the two-dimensional velocity field $v_i + \varphi_i$, $i = 1, 2$.

In general, several different situations may occur. The following simple form of φ gives rise to a quite interesting situation:

$$\varphi_1 = \alpha_1 x_1, \qquad \varphi_2 = \alpha_2 x_2, \qquad \alpha_1, \alpha_2 \in \mathbb{R}, \tag{5.44}$$

which will be discussed in some detail in Chapter 4 in the framework of the vortex theory. When $\alpha_1 = \alpha_2 = \alpha$, (5.41) can be reconstructed to the Euler equation by a change of variables

$$x' = x \exp\{-\alpha t\}, \tag{5.45}$$

$$t' = \frac{1 - \exp\{-2\alpha t\}}{2\alpha}, \tag{5.46}$$

which implies

$$v' = v \exp\{\alpha t\}, \tag{5.47}$$

$$\omega' = \omega \exp\{2\alpha t\}. \tag{5.48}$$

Equation (5.41) becomes

$$\partial_{t'}\omega'(x', t') + (v' \cdot \nabla_{x'})\omega'(x', t') = 0. \tag{5.49}$$

When we seek a solution of the form

$$u(x, t) = (u_1(x_2, t), -\gamma x_2, -\gamma x_3) \tag{5.50}$$

then the vorticity is

$$\omega(x, t) = (0, 0, -\partial_2 u_1(x_2, t)) \tag{5.51}$$

and the Euler equation in the third component yields

$$\partial_t \omega - \gamma x_2 \partial_{x_2} \omega = \gamma\omega. \tag{5.52}$$

This linear equation can be solved by transformations like (5.45) and (5.46) and the characteristic method.

A class of interesting stationary three-dimensional flows are the Beltrami flows. They are defined by the condition

$$\omega = \alpha u, \tag{5.53}$$

where α is a constant. Obviously, condition (see (2.61))

$$\omega \cdot \nabla u = u \cdot \nabla \omega \tag{5.54}$$

is verified, so that such flows are stationary. The existence of the Beltrami flows follows by solving the linear problem which is the consequence of (5.53) after taking the curl:

$$-\Delta u = \alpha^2 u. \tag{5.55}$$

Of course (5.55) must be completed by the boundary conditions on u.

There is a wide literature on the subject. See, for instance, [Dri 91] and references quoted therein.

Appendix 1.1 (Liouville Theorem)

Theorem. *Let $\Phi_t(x)$ be a flow and let $u(\Phi_t(x), t)$ be the vector field defined as*

$$u(\Phi_t(x), t) = \frac{d}{dt} \Phi_t(x). \tag{A1.1}$$

Then the following two statements are equivalent:

(i) $\Phi_t(x)$ *is incompressible*
(ii) $\nabla \cdot u = 0.$

PROOF. The theorem easily follows from the fact that $J_t(x)$ (the Jacobian of the transformation $x \to \Phi_t(x)$) satisfies the equation

$$\partial_t J_t(x) = J_t(x) \nabla \cdot u(\Phi(x), t) \tag{A1.2}$$

which we are going to prove.

Let $S(x, t)_{ij} = \partial \Phi_t(x)_i / \partial x_j$ be the Jacobian matrix. By definition $J(x, t) = \det S(x, t)$. We have

$$\frac{d}{dt} S(x, t) = \nabla u(\Phi_t(x), t) S(x, t), \tag{A1.3}$$

where $\nabla u = \partial u_i / \partial x_j$. From (A1.3) we have (neglecting the x-dependence for notational simplicity)

$$S(t + h) = S(t) + h \nabla u(\Phi_t(x), t) \cdot S(t) + O(h^2). \tag{A1.4}$$

Multiplying on the right by $S(t)^{-1}$ and taking the determinants

$$J(t + h) J(t)^{-1} = \det(1 + h \nabla u(\Phi_t(x), t) + O(h^2)). \tag{A1.5}$$

It is easy to realize that the computation on the determinant in the right-hand side of (A1.5) gives

$$1 + h \nabla \cdot u(\Phi_t(x), t) + O(h^2) \tag{A1.6}$$

which implies

$$J(t + h) - J(t) = h \nabla \cdot u(\Phi_t(x), t) J(t) + O(h^2). \tag{A1.7}$$

Dividing by h and taking the limit $h \to 0$, we obtain (A1.2). $\square$

Appendix 1.2 (A Decomposition Theorem)

Lemma. *Any vector field* $v: D \to R^3$, $v \in C^1(D) \cap C(\bar{D})$, *can be univocally decomposed in the sum*

$$v = v_1 + \nabla p, \tag{A2.1}$$

where $p \in C^1(D)$ *and* $v_1 \in C^1(D)$ *satisfy to*

$$\nabla \cdot v_1 = 0, \qquad v_1 \cdot n = 0 \quad on \ \partial D. \tag{A2.2}$$

Moreover, the fields v_1 *and* ∇p *are orthogonal in the* $L_2(D)$ *sense*

$$\int_D v_1 \cdot \nabla p \, dx = 0. \tag{A2.3}$$

PROOF. Given v, we want to construct a potential field $v_2 = \nabla p$ having the same normal component on the boundary and the same divergence as v. Therefore p must satisfy

$$(\text{div } v_2 =) \ \Delta p = \text{div } v,$$
$$\partial_n p = v \cdot n \quad \text{on } \partial D. \tag{A2.4}$$

The problem (A2.4) is a Poisson problem with Neumann boundary conditions. Therefore, it has a unique solution p modulo an additive constant.

Given v_2, let us put

$$v_1 = v - v_2. \tag{A2.5}$$

It is clear that div $v_1 = 0$ and $v_1 \cdot n = 0$ on ∂D. The orthogonality relation (A2.3) easily follows

$$\int_D v_1 \cdot v_2 \, dx = \int_D v_1 \cdot \nabla p \, dx = \int_D \text{div}(v_1 p) \, dx$$

$$- \int_D p \ \text{div } v_1 \, dx = \int_{\partial D} p v_1 \cdot n = 0. \tag{A2.6}$$

Let us now prove that the above decomposition is unique. Suppose

$$v = w_1 + w_2 = w_1 + \nabla \pi \tag{A2.7}$$

another decomposition. Then

$$v_1 - w_1 = -\nabla(p - \pi) = 0 \tag{A2.8}$$

and hence

$$\int_D |v_1 - w_1|^2 \, dx = -\int_D (v_1 - w_1) \cdot \nabla(p - \pi) \, dx. \tag{A2.9}$$

Since $\nabla(p - \pi)$ is orthogonal to all divergence-free fields satisfying the impermeability boundary conditions, we conclude that

$$\int_D |v_1 - w_1|^2 \, dx = 0. \tag{A2.10}$$

This achieves the proof because $v_1 = w_1$ and, as a consequence, $p = \pi + $ const.
$\hfill\square$

Corollary. *Suppose u to be a smooth vector field orthogonal to all $C^1(D) \cap C(\overline{D})$ divergence-free vector fields satisfing the impermeability boundary conditions. Then u is the gradient of a scalar function p.*

PROOF. Apply the lemma

$$u = u_1 + \nabla p. \tag{A2.11}$$

Since ∇p is orthogonal to all divergence-free vector fields satisfying the boundary conditions, it follows that u_1 must enjoy the same property. Hence

$$\int_D u_1^2 \, dx = 0. \tag{A2.12}$$

$\hfill\square$

Appendix 1.3 (Kutta–Joukowski Theorem and Complex Potentials)

In this appendix we prove Theorem 2.2 and introduce a tool useful in many problems arising from the study of two-dimensional irrotational incompressible flows: the complex formalism.

First, we write F by means of quantities appearing in the Euler equation. Remembering the meaning of the pressure as contact force, we have

$$F = -\int_{\partial B} pn \, d\sigma, \tag{A3.1}$$

where n is the external normal to B.

By using the connection between p and u, that in the irrotational stationary case has the form (see (4.5)),

$$p = -\tfrac{1}{2}(u_1^2 + u_2^2), \tag{A3.2}$$

$u = (u_1, u_2)$, we have

$$F = \frac{1}{2} \int_{\partial B} (u_1^2 + u_2^2)n \, d\sigma. \tag{A3.3}$$

We define *complex velocity* as

$$F = u_1 - iu_2, \tag{A3.4}$$

where i is the imaginary unity. It is useful to introduce complex variables to automatically take into account the continuity equation and irrotationality

$$\partial_1 u_1 + \partial_2 u_2 = 0, \tag{A3.5}$$

$$\partial_1 u_2 - \partial_2 u_1 = 0. \tag{A3.6}$$

F is an analytic function in $z = x_1 + ix_2$ because (A3.5) and (A3.6) are the Cauchy–Riemann equations. If F has a primitive

$$F = \frac{d}{dz} W \tag{A3.7}$$

we call W the *complex potential*. (Of course W always exists in the class of multivalues functions.)

It follows immediately from the definition that

$$W = \varphi + i\psi, \tag{A3.8}$$

where φ is the velocity potential and ψ is the stream function.

Lemma (Blasius).

$$F = -\frac{i}{2}\left(\int_{\partial B} F^2 \, dz\right)^*. \tag{A3.9}$$

Here $(\cdot)^*$ denotes complex conjugation and $F = F_1 + iF_2$ is identified with a complex number where F_i are the components of F appearing in (A3.1) as a vector.

PROOF. Let $dz = dx + i \, dy$ be an infinitesimal displacement along the curve ∂B. Then $-i \, dz = dy - i \, dx$ repesents the normal displacement. Equation (A3.1) becomes

$$F = -\int_{\partial B} p \, dy + i \int_{\partial B} p \, dx = i \int_{\partial B} p(dx + i \, dy)$$

$$= -\frac{i}{2} \int_{\partial B} (u_1^2 + u_2^2) \, dz, \tag{A3.10}$$

where in the last step we used (A3.2).

On the other hand,

$$F^2 \, dz = (u_1 - iu_2)^2 \, dz = (u_1^2 - u_2^2 - 2iu_1 u_2)(dx + i \, dy). \tag{A3.11}$$

Because u is parallel to ∂B, that is,

$$u_1 \, dy - u_2 \, dx = 0 \tag{A3.12}$$

we have

$$F^2 \, dz = (u_1^2 + u_2^2)(dx - i \, dy). \tag{A3.13}$$

Taking the complex conjugate and comparing with (A3.10) we have proved the lemma. $\square$

We return to the proof of the main theorem. The function F is analytic outside B. Then we choose the origin in B and we can expand F in Laurent series outside any circle centered in the origin and containing B. Because F at infinity is a constant, all the coefficients of positive powers in z must vanish

$$F = a_0 + \frac{a_{-1}}{z} + \frac{a_{-2}}{z^2} + \frac{a_{-3}}{z^3} + \cdots. \tag{A3.14}$$

By the condition at infinity

$$a_0 = u_{\infty,1} - iu_{\infty,2}. \tag{A3.15}$$

By the Cauchy theorem (for analytic functions)

$$\int_{\partial B} F \, dz = 2\pi i a_{-1}. \tag{A3.16}$$

Since

$$\int_{\partial B} F \, dz = \int_{\partial B} (u_1 - iu_2)(dx + i\, dy) = \int_{\partial B} (u_1 \, dx + u_2 \, dy) = \Gamma \tag{A3.17}$$

we have

$$a_{-1} = \frac{\Gamma}{2\pi i}. \tag{A3.18}$$

We square the Laurent expansion, by the previous lemma and the Cauchy theorem

$$F = -\frac{i}{2}\left(\int_{\partial B} F^2 \, dz\right)^* = -\frac{i}{2}\left(\int_{\partial B} \left(a_0^2 + \frac{2a_0 a_{-1}}{z} + \cdots\right) dz\right)^*$$

$$= -\frac{i}{2}(2a_0\Gamma)^* = \Gamma(u_{\infty,2} - iu_{\infty,1}). \tag{A3.19}$$

$$\square$$

Appendix 1.4 (d'Alembert Paradox)

We prove the following theorem due to d'Alembert:

Theorem. *Let Λ be a bounded region in $\mathbb{R}^3$, with a smooth boundary, topologically homeomorphic to a sphere. (Later on we will call Λ an obstacle.) Let u be a given potential flow in $D = \mathbb{R}^3/\Lambda$ constant at infinity ($u(x) \to u_\infty$ when $|x| \to \infty$).*

The fluid, then, does not produce any force on the obstacle, that is,

$$F = -\int_{\partial D} pn \, d\sigma = 0. \tag{A4.1}$$

Remark. This theorem can be extended to an obstacle Λ which is not simply connected.

PROOF. We denote by φ the potential associated to the velocity field u (it exists because D is simply connected)

$$u = \nabla\varphi. \tag{A4.2}$$

Then, by div $u = 0$

$$\Delta\varphi = 0 \quad \text{in } D \tag{A4.3}$$

with the boundary and asymptotic conditions

$$\frac{\partial}{\partial n}\varphi = 0 \quad \text{on } \partial D, \tag{A4.4}$$

$$\nabla\varphi = u_\infty \qquad \text{when} \quad |x| \to \infty. \tag{A4.5}$$

We look for a solution of the form

$$\varphi(x) = u_\infty \cdot x + \eta(x), \tag{A4.6}$$

where $\eta(x)$ is an harmonic function which vanishes at infinity. We write η in the form

$$\eta(x) = \frac{1}{4\pi} \int_\Lambda \frac{\rho(y)}{|x - y|} \, dy, \tag{A4.7}$$

where ρ is a suitable "charge" distribution. (It becomes obvious that (A4.7) satisfies the Laplace equation in D if we observe that

$$\frac{1}{4\pi} \frac{1}{|x - y|} \tag{A4.8}$$

is the Green function, vanishing at infinity, for the Poisson equation.)

The function ρ is determined by the condition (A4.4) which produces an integral equation in ρ. It is well known, from the study of the Laplace equation, that ρ exists and has the property

$$\int_\Lambda |\rho(y)| \, dy < \infty. \tag{A4.9}$$

We now want to study the behavior of η at infinity. For this purpose, we observe that

$$\int_\Lambda \rho(y) \, dy = 0. \tag{A4.10}$$

In fact, by the Gauss–Green lemma this integral is equal to the flux of the velocity $\hat{u} = \nabla\eta$ through ∂D. This flux vanishes by virtue of (A4.4) and because the flux of u_∞ is zero.

We denote by ρ^+ and ρ^- the positive and the negative parts of ρ. Of course

$$\int_\Lambda \rho^+ \, dy = \int_\Lambda \rho^- \, dy \tag{A4.11}$$

(the two integrals are finite by (A4.9)).

Moreover,

$$\frac{1}{4\pi} \int_\Lambda \frac{\rho^+(y)}{|x-y|} \, dy \xrightarrow[|x|\to\infty]{} \frac{a}{|x|} + O\left(\frac{1}{|x|^2}\right), \tag{A4.12}$$

$$\frac{1}{4\pi} \int_\Lambda \frac{\rho^-(y)}{|x-y|} \, dy \xrightarrow[|x|\to\infty]{} \frac{a'}{|x|} + O\left(\frac{1}{|x|^2}\right). \tag{A4.13}$$

We have $a = a'$. (To convince ourselves, we have only to consider the more unfavorable case in which the positive and negative parts are as distant as much as possible, and to use (A4.11) or consider (A4.12) and (A4.13), and apply the Gauss–Green lemma to a sphere of radius $|x| \to \infty$.)

By subtracting (A4.13) from (A4.12) we get

$$\eta(x) \xrightarrow[|x|\to\infty]{} O\left(\frac{1}{|x|^2}\right) \tag{A4.14}$$

and hence

$$u = u_\infty + O\left(\frac{1}{|x|^3}\right). \tag{A4.15}$$

The pressure also has similar behavior. Indeed,

$$p = -\frac{|u|^2}{2} = -\tfrac{1}{2}\{|u_\infty|^2 + (u - u_\infty)(u + u_\infty)\} \tag{A4.16}$$

from which

$$p = p_0 + O(|x|^{-3}). \tag{A4.17}$$

The force on the obstacle Λ is

$$F = -\int_{\partial D} pn \, d\sigma. \tag{A4.18}$$

To evaluate it, we consider a closed surface Σ enclosing Λ and we compute the quantity

$$\int_D \{(u \cdot \nabla)u + \nabla p\}, \tag{A4.19}$$

where D' is the region enclosed between Σ and ∂D. By the Euler equation this quantity vanishes. By the Gauss–Green lemma and the boundary conditions (A4.4) we have

$$(A4.19) = \int_\Sigma \{(u \cdot n)u + pn\} \, d\sigma - \int_{\partial D} pn \, d\sigma = 0. \tag{A4.20}$$

Hence

$$F = -\int_\Sigma \{(u \cdot n)u + pn\} \, d\sigma. \tag{A4.21}$$

By choosing Σ as the surface of a sphere with radius $\to \infty$, the asymptotic behaviors (A4.15) and (A4.17) allow us to achieve the proof. □

EXERCISES

1. Prove that the action (defined in (1.12)) is minimum (besides being stationary) on the motions which satisfy the Euler equation
$$\frac{d^2}{dt^2}\Phi_t = -\nabla p(\Phi_t).$$

2. Prove that the laminar flow $u(x_1, x_2) = (f(x_2), 0)$ in an infinite channel $D = (-\infty, +\infty) \times [0, 1]$ is a stationary solution of the Euler equation (Fig. 1.31). Calculate the pressure.

3. Let $D = \{x \in \mathbb{R}^2 \,|\, |x| < r\}$. Prove that the velocity field $u_0 = u_0(u_\rho, u_\theta)$, where $u_\rho = 0$ and $u_\theta = f(\rho)$, is a stationary solution of the Euler equation (Fig. 1.32). Calculate the pressure.

4. Put $r = \infty$ in the previous exercise. Prove that the motion (no longer stationary) in the absence of the pressure term
$$\frac{d^2}{dt^2}\Phi_t(x) = 0,$$
$$\Phi_0(x) = x, \qquad \frac{d}{dt}\Phi_0(x) = u_0(x),$$
violates the condition of incompressibility.

5. Verify that $u(x)$ given by (2.26) is real if $\omega = \omega(x)$ is real. Prove that $\omega \in C^1(D)$ implies that the series (2.26) is absolutely convergent.

6. Verify (2.29) and (2.30).

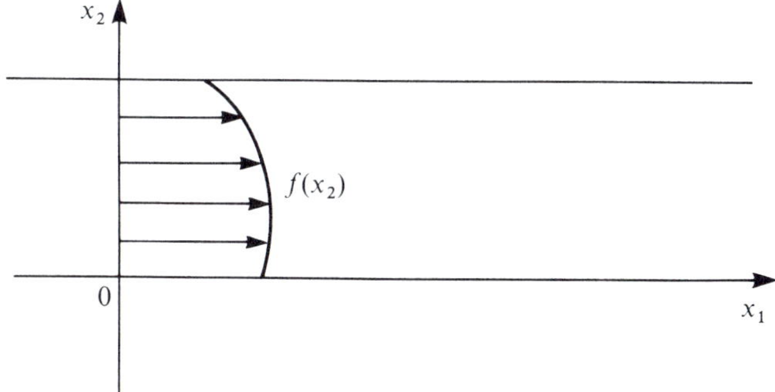

Figure 1.31

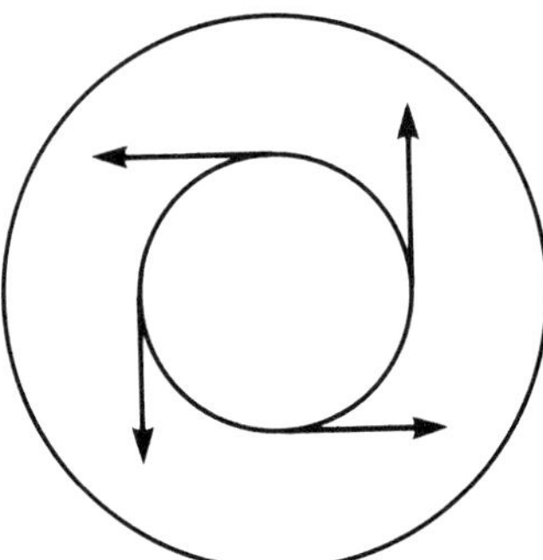

Figure 1.32

*7. Prove the convergence of the series (2.46) in the following sense: let $\Lambda_N = [-N, N]^2 \subset \mathbb{Z}^2$ be a set of regions invading $\mathbb{Z}^2$. We define

$$S_N = \sum_{k \in \Lambda_N; k \neq 0} \frac{\exp(-ik \cdot x)}{|k|^2}.$$

Then S_N converges (a proof of this statement can be found in [BaR 93]).

8. By using (2.47), (2.48) find the Green function in a semicircle and in a quarter of circle.

*9. Prove that the Green function in a flat torus $D = [-\pi, \pi]^2$ has the following formal expression:

$$G_D(x, y) = \sum_{k \in Z^2; k \neq 0} -\frac{1}{2\pi} \ln|x - y_k|, \tag{E.1}$$

where

$$y_k = (y_1 + 2k_1\pi, y_2 + 2k_2\pi), \qquad k = (k_1, k_2).$$

The meaning in which (E.1) must be understood is that

$$\Psi(x) = \sum_{k \in Z^2; k \neq 0} -\frac{1}{2\pi} \int_D \ln|x - y_k|\omega(y) \, dy, \tag{E.2}$$

where ω satisfies the relation

$$\int \omega \, dx = 0 \tag{E.3}$$

and the series in (E.2) must be understood in the meaning of Exercise 7*.

10. Prove (2.57) along the following lines. Start from the obvious identity

$$u(x) = -\Delta \int_D G(x, y)u(y) \, dy \tag{E.4}$$

and use the identities

$$\Delta = \nabla \, \mathrm{div} - \mathrm{curl} \, \mathrm{curl}, \tag{E.5}$$

$$\mathrm{curl}(Gu) = G \, \mathrm{curl} \, u - \nabla G \wedge u. \tag{E.6}$$

11. Let $\Lambda \subset D$ be a region strictly contained in a bounded domain D where the fluid is confined. Express the time variation of the kinetic energy in Λ in terms of the energy flux and the effect of the pressure in $\partial\Lambda$.

*12. Prove rigorously Theorem 3.6 under the hypothesis $u \in C^1(\mathbb{R}^3 \times [0, T]; \forall t,$ $u(\cdot, t) \in L_\infty \cap L_2(\mathbb{R}^3))$ and $\omega \in C^1([0, T]; L_2(\mathbb{R}^3))$.

13. Prove in a formal way Theorem 3.6 in $T^3 = [-\pi, \pi]^3$.

14. Consider a fluid in $\mathbb{R}^3$ (with $u \to 0$ as $|x| \to \infty$). Prove that the quantity I defined in (3.32) is the angular momentum of the fluid $I = \int (x \wedge u)\, dx$.

15. Prove that Ω, defined in (3.30), is conserved when $D = \mathbb{R}^3$ and $D = [-\pi, \pi]^3$. In the two-dimensional case, Ω defined in (3.33) is conserved in any domain.

16. Prove that the quantity

$$B_1 = \int_D x_1 \omega\, dx_1\, dx_2 \tag{E.7}$$

   is conserved for the planar Euler flow in the domain $D = (-\infty, +\infty) \times (-a, 0)$; $a > 0$.

17. Prove that the quantity

$$I = \int_D |x|^2 \omega\, dx \tag{E.8}$$

   is conserved for the planar Euler flow in the domain $D = \{(x_1, x_2) | r^2 < x_1^2 + x_2^2 < R^2; 0 < r < R < \infty\}$.

18. Let $D = (-\infty, +\infty) \times (-\infty, +\infty) \times [-A, A]$ and suppose $u \to 0$ as $|x| \to \infty$. Prove that the quantities

$$B_1 = \int_D x_2 \omega_3\, dx, \qquad B_2 = -\int_D x_1 \omega_3\, dx, \tag{E.9}$$

   are conserved during the motion and find their relation with the momentum of the fluid.

19. Let $D = (-\infty, +\infty) \times (-\infty, +\infty) \times [0, A]$ and suppose $u \to 0$ when $|x| \to \infty$. Find, if it exists, an $\alpha \in \mathbb{R}$ such that the quantity

$$\int_D \{(x \wedge \omega)_1 - \alpha\omega_2\}\, dx \tag{E.10}$$

   is conserved during the motion.

20. Let $D = (-\infty, +\infty) \times (-\infty, +\infty) \times [-A, A]$ and suppose $u \to 0$ when $|x| \to \infty$. Prove that the quantity

$$\int_D \omega\, dx \tag{E.11}$$

   is *not* conserved during the motion.

21. Find the stream function Ψ of (4.9) and write the complex potential. Observe that (4.9) gives an irrotational fluid also in $D = \{x \in \mathbb{R}^2 | x_1^2 + x_2^2 > r^2, r > 0; x_2 > 0\}$.

22. Let $D = \{x \in \mathbb{R}^3 | |x| > r, r > 0\}$. Verify that the irrotational velocity field of potential

$$\varphi = \frac{r^3}{2|x|^2} u_\infty \cdot n + x \cdot u_\infty \qquad \text{where} \quad n = \frac{x}{|x|}, \tag{E.12}$$

gives to the d'Alembert paradox.

CHAPTER 2

Construction of the Solutions

In this chapter we study the problem of the existence and uniqueness of the solutions of the Euler equation. The Euler equation, deduced and discussed in the previous chapter, is a nonlinear equation. This implies that the construction of its solutions may be a nontrivial task. In this chapter we study this problem.

2.1. General Considerations

The first problem we meet in the rigorous study of a differential equation describing physical phenomena is to establish an existence and uniqueness theorem for the solutions. This problem is of obvious interest: if a mathematical model of the real world is described by a differential equation, the proof of the existence of a large enough class of solutions is a first verification of the validity of the model. If the answer is negative the model must be forgotten or, at least, deeply modified. Moreover, once the existence of the solutions is ensured, we would like there to be only one solution having a given value at a given instant. If not, the physical state of the system at a time t, could not be uniquely determined by the differential equation itself and the knowledge of the state of the system at a previous time $t_0 < t$. In other words, we would like the Cauchy (initial value) problem associated with our differential equation to be well-posed: that is, to have a unique solution with a smooth dependence on the initial data.

Once we have positively answered this first question, thereby clarifying that there are no evident mathematical inconsistencies within the model, we must deal with the important problem of an explicit evaluation of the solu-

tions. That is, we must develop methodologies and algorithms (implementable numerically, if possible) that allow, at least in principle, the approximate
calculation of the solutions. As a consequence, among the existence and
uniqueness theorems, we will prefer those that, in the strategy of their proof,
suggest algorithms for the computation of the solutions and, when possible,
give information on the qualitative behavior of the solutions.

In the case of the Euler equation we have satisfactory answers in two
dimensions. In three dimensions the theory is, unfortunately, much more
difficult and, as we will see, it is conceivable that the solutions may develop
singularities in a finite time. Therefore we confine ourselves to an existence
and uniqueness theorem local in time only. The difficulty we encounter in
constructing a global solution of the Euler equation is a consequence of the
lack of an a priori estimate, valid for all time, for the gradient of the vorticity
field. As we will see in Section 2.5, energy conservation and the Kelvin theorems do not give sufficient control for obtaining this estimate. At the present
time, it is not known whether singularities may be created by a three-dimensional flow. The study of this challenging problem is a matter of current
research.

In this section we prove an a priori estimate on the solutions valid for
short times.

As we will see in the next section, the situation is completely different in
two dimensions. In this case, we are able to construct a solution for any time
using another conservation law related to two-dimensional symmetry.

Let us come back to the general three-dimensional case. We assume the
existence of a smooth enough solution of the Euler equation, and try to
obtain estimates on the growth in time of suitable norms of the solution itself.
As we will see, because of the nonlinearity of the equation, we are able to do
this for short times only. Using the estimates obtained here, we will construct,
in Section 2.4, the solutions as limits of regularized solutions (always for
short times).

To simplify our analysis we assume that the fluid moves in a three-dimensional torus $D = [-\pi, \pi]^3$. This hypothesis simplifies some estimates that
will be presented later. The case of a bounded domain with a smooth boundary can be recovered with a minor conceptual effort and some more technicalities.

Let f and g be two functions in $C^\infty(D)$. We introduce a scalar product,
defined for any $m = 0, 1, 2, \ldots$, as

$$(f, g)_m = \sum_{\alpha;\, 0 \le |\alpha| \le m} (D^\alpha f, D^\alpha g), \tag{1.1}$$

where

$$\alpha = (\alpha_1, \alpha_2, \alpha_3), \qquad |\alpha| = \sum_{i=1}^{3} \alpha_i, \tag{1.2}$$

and

$$D^\alpha = \frac{\partial^{|\alpha|}}{\partial^{\alpha_1} x_1 \partial^{\alpha_2} x_2 \partial^{\alpha_3} x_3}. \tag{1.3}$$

We have used the usual notation

$$(f, g) = (f, g)_0 = \int_D dx\, f(x)g(x). \tag{1.4}$$

The space $C^\infty(D)$, equipped with the scalar product (1.1), becomes a pre-Hilbert space. We denote its completion by $H_m(D)$ (or briefly by the sequel H_m), usually called the Sobolev space of index m.

Given two vector fields, $u = \{u_j\}_{j=1}^3$, $v = \{v_j\}_{j=1}^3$, we extend the above definition in an obvious way

$$(u, v)_m = \sum_{j=1}^3 (u_j, v_j)_m. \tag{1.5}$$

We denote by $\mathbb{H}_m$ the Hilbert space associated with the vector fields with components in H_m. Finally, we use the notation $|\cdot|_m$ for the associated norm, that is,

$$|u|_m = \sqrt{(u, u)_m}. \tag{1.6}$$

The $\mathbb{H}_m$ spaces are very useful in the study of the Euler equation because the second term in this equation involves the derivatives of the velocity field.

The following proposition will be proved in Appendix 2.2:

Proposition 1.1. *For $f, g \in C^\infty(D)$, the following inequalities hold:*

$$\|f\|_\infty \le C_m |f|_m \qquad \textit{if} \quad m \ge 2, \tag{1.7}$$

$$|fg|_m \le C_m |f|_m |g|_m \qquad \textit{if} \quad m \ge 2, \tag{1.8}$$

$$|fg|_0 \le C |f|_2 |g|_0, \tag{1.9}$$

$$|fg|_0 \le C |f|_1 |g|_1. \tag{1.10}$$

Later in this section we will suppose a priori that we have a solution of the Euler equation $u \in C^\infty(D \times [0, T])$. We want to estimate the norm $|u_t|_m$ of such a solution. By the Euler equation we have

$$\frac{1}{2}\frac{d}{dt}(D^\alpha u, D^\alpha u) = -(D^\alpha u, D^\alpha(u \cdot \nabla)u). \tag{1.11}$$

In the previous formula the pressure disappears because $D^\alpha u$ is a divergence-free field and so $(D^\alpha u, \nabla p) = 0$. We estimate now the right hand side of (1.11). We have

$$(D^\alpha u, D^\alpha(u \cdot \nabla)u) = \sum_\beta C_{\alpha,\beta}(D^\alpha u, [D^\beta u \cdot \nabla]D^{\alpha-\beta}u). \tag{1.12}$$

The sum in (1.12) is made over all the terms with $\beta = \{\beta_i\}_{i=1}^3$, such that $0 \le \beta_i \le \alpha_i$ and $C_{\alpha,\beta}$ are suitable positive constants. The central point in our analysis is the fact that the term $\beta = 0$ in (1.12) vanishes because (see Exercise 1)

$$(D^\alpha u, u \cdot \nabla D^\alpha u) = 0, \tag{1.13}$$

so that the degree of derivatives in the right-hand side equals the degree in the left-hand side, and so we can hope to obtain a closed equation for some norm.

Hence, by the Cauchy–Schwarz inequality

$$|(D^\alpha u, D^\alpha(u \cdot \nabla)u)| \le |D^\alpha u|_0 \sum_{0 < |\beta| \le \alpha} C_{\alpha,\beta} |(D^\beta u \cdot \nabla)D^{\alpha-\beta}u|_0. \tag{1.14}$$

By using Proposition 1.1, we have

if $|\beta| \ge 3$, by (1.9)

$$|(D^\beta u \cdot \nabla)D^{\alpha-\beta}u|_0 \le C|u|_{|\beta|}|u|_{|\alpha|-|\beta|+3} \le C(|u|_{|\alpha|})^2 \tag{1.15}$$

if $|\beta| = 1$, by (1.9)

$$|(D^\beta u \cdot \nabla)D^{\alpha-\beta}u|_0 \le C|u|_3|u|_{|\alpha|} \tag{1.16}$$

if $|\beta| = 2$, by (1.10)

$$|(D^\beta u \cdot \nabla)D^{\alpha-\beta}u|_0 \le C|u|_3|u|_{|\alpha|}. \tag{1.17}$$

For $3 \le |\alpha| \le m$, the estimates (1.15), (1.16), and (1.17) give

$$|(D^\beta u \cdot \nabla)D^{\alpha-\beta}u|_0 \le C_m(|u|_m)^2. \tag{1.18}$$

Here C and C_m denote positive constants, the latter depending only on m. By summing on all α's, such that $|\alpha| \le m$, we obtain, by (1.11) and (1.14),

$$\frac{d}{dt}(|u|_m)^2 \le C_m(|u|_m)^3, \qquad m \ge 3. \tag{1.19}$$

Using now the lemma on integral inequalities discussed in Appendix 2.1, we can conclude that

$$|u(t)|_m \le \frac{1}{1/|u_0|_m - \frac{1}{2}C_m t} \tag{1.20}$$

so that there exists a time interval $[0, T]$, $T = T(|u_0|_m)$, such that u_t, the solution of the Euler equation with initial datum $u_0 \in \mathbb{H}_m$, $m \ge 3$, lives in $\mathbb{H}_m$. T gets smaller as the $\mathbb{H}_m$ norm of the initial datum gets larger. This bound on the size of the interval of time for which we can control $|u|_m$ will allow us, in Section 2.4, to construct a local solution. The strategy will be to consider a sequence of approximating problems, whose solution is smooth and easy to construct. The a priori estimate (1.20) then applies as well to these approximations, allowing us to pass to the limit to find a solution to the Euler equation.

We conclude this section with a comment on the nature of the pressure. First, we observe that the gradient of the pressure is not an unknown quantity of the initial value problem. In Chapter 1, we saw that ∇p is the force term acting on the particles of fluid allowing them to move as freely as possible, but in a way compatible with the incompressibility constraint, $\nabla \cdot u = 0$. The equation, $D_t u = 0$, which describes the free motion of the particles of fluid (u is conserved along the trajectory of the particles), admits solutions violating the condition $\nabla \cdot u$ at $t > 0$, even if the divergence is vanishing at time zero. So the pressure gradient plays the same role that the constraint forces play in classical mechanics when we study the motion of a particle in a manifold. In our case, the constraint is given by the incompressibility condition . As in mechanics, where the constraint force is determined a posteriori once the motion of the particle is found, in our context the pressure can be determined when we have found the velocity field, which is the solution of the Euler equation. In fact, taking the divergence of this equation we obtain

$$\nabla \cdot (u \cdot \nabla)u = \Delta p. \tag{1.21}$$

From (1.21), knowing u, we can find p by solving an elliptic problem.

On the torus $D = [-\pi, \pi]^3$ this problem certainly has a solution. In fact, the equation

$$\Delta p = g \tag{1.22}$$

has a unique solution given in terms of the Fourier transform by

$$-|k|^2 p^\wedge(k) = g^\wedge(k), \qquad k \in \mathbb{Z}^3, \tag{1.23}$$

provided that the compatibility condition

$$g^\wedge(0) = \int dx\, g(x) = 0 \tag{1.24}$$

is verified. It is easy to see that this condition is verified by the function $\nabla \cdot (u \cdot \nabla)u = 0$.

In domains with a boundary, the Euler equation projected on the outward normal gives the condition

$$(u \cdot \nabla)u \cdot n = \frac{\partial p}{\partial n} \tag{1.25}$$

and so problem (1.21) is completed by the Neumann boundary conditions (1.25).

2.2. Lagrangian Representation of the Vorticity

In this section we establish some representation formulas for the time evolution of the vorticity field. Sometimes we will use, throughout this chapter, the notation ω_t and u_t for $\omega(\cdot, t)$ and $u(\cdot, t)$, respectively.

We have seen in Chapter 1 that the Euler equation, in terms of vorticity, reads

$$D_t \omega = (\omega \cdot \nabla)u, \tag{2.1$_1$}$$

$$\nabla \cdot u = 0, \tag{2.1$_2$}$$

$$\omega = \operatorname{curl} u. \tag{2.1$_3$}$$

This equation becomes much simpler in two dimensions. Since $(\omega \cdot \nabla)u = 0$, denoting by $\omega = \omega_3$ the third component of the vorticity field, we obtain

$$D_t \omega = (\partial_t + u \cdot \nabla)\omega = 0. \tag{2.2}$$

This equation implies that the derivative along the particle paths vanishes

$$\frac{d}{dt}\omega_t(\Phi_t(x)) = 0. \tag{2.3}$$

Finally, integrating with respect to time, we obtain

$$\omega_t(\Phi_t(x)) = \omega_0(x), \tag{2.4}$$

where $\Phi_t(x)$ denotes (as already seen in Chapter 1) the trajectory of the particle starting from x at time zero

$$\frac{d}{dt}\Phi_t(x) = u(\Phi_t(x), t),$$
$$\Phi_0(x) = x. \tag{2.5}$$

According to (2.4), the vorticity (in two dimensions) is conserved along the particle paths, making the fluid dynamics in two dimensions much simpler than in three dimensions.

As we will see in the next section, (2.4) will be the basic tool for the construction of the solutions in two dimensions. From formula (2.4), the quantity

$$\int_D F(\omega(x))\, dx, \tag{2.6}$$

$F: \mathbb{R} \to \mathbb{R}$ measurable (D is the domain in which the fluid moves), is a constant of the motion (sec Exercise 2). So we have infinitely many first integrals of motion in two dimensions that, however, tell us nothing more than the conservation of the vorticity along the trajectory. In particular, the L_p norms of ω are conserved

$$\|\omega_t\|_p = \|\omega_0\|_p, \qquad p = 1, 2, \ldots, \infty. \tag{2.7}$$

Equations (2.7) with $p < +\infty$ are a consequence of (2.6) . The case $p = +\infty$ follows directly from (2.4).

We consider now the problem of finding a representation formula (if any) that generalizes (2.4) to the three-dimensional flow. We hypothesize the existence of a matrix field

$$A = A(x, t), \tag{2.8}$$

taking into account the term $(\omega \cdot \nabla)u$ in the Euler equation (2.1), so that

$$\omega_t(\Phi_t(x)) = A \cdot \omega_0(x). \tag{2.9}$$

We try to determine the form of A assuming the existence of a smooth solution of the Euler equation. Taking the time derivative of the expression

(2.9) we obtain

$$D_t \omega_t(\Phi_t(x)) = \partial_t A \cdot \omega_0(x). \tag{2.10}$$

By using the Euler equation

$$(\omega_t \cdot \nabla) u_t(\Phi_t(x)) = [(A \cdot \omega_0) \cdot \nabla] u_t(\Phi_t(x)) = \partial_t A \cdot \omega_0(x). \tag{2.11}$$

By virtue of the arbitrariness of ω_0, (2.11) can be written as

$$\frac{\partial}{\partial t} A_{ik}(x, t) = \sum_j \partial_i u_j(\Phi_t(x), t) A_{jk}(x, t)). \tag{2.12}$$

This linear equation has a unique solution (if $u(x, t)$ is sufficiently regular) of the form

$$(\nabla \Phi_t(x))_{ji} = \partial_i \Phi_t(x)_j, \tag{2.13}$$

which is the Jacobian matrix of the group of transformations induced by the path lines. In fact, differentiating (2.5) with respect to the initial data, we obtain

$$\partial_i \partial_t \Phi_t(x)_j = \frac{\partial}{\partial x_i} u_j(\Phi_t(x), t) = \sum_k \partial_k u_j(\Phi_t(x), t) \partial_i \Phi_t(x)_k. \tag{2.14}$$

By a comparison of (2.12) with (2.14) we obtain $A = \nabla \Phi_t$. Thus we have proved the following theorem:

Theorem 2.1. *Let $u = u(x, t)$, $u \in C^1(D, [0, T])$, be a solution of the Euler equation and $\omega_t = \text{curl } u_t$. Then*

$$\omega_t(\Phi_t(x)) = \nabla \Phi_t(x) \cdot \omega_0(x). \tag{2.15}$$

The geometrical interpretation of formula (2.15) is the following: the vorticity field is transported along the path lines and deformed under the action of matrix $\nabla \Phi_t$ (Fig. 2.1).

The transport action is described by the term $(u \cdot \nabla)\omega$ (also present in two dimensions, see (2.4)). The deformation action on the field ω is described by the term $(\omega \cdot \nabla)u$ which is present only in the absence of a two-dimensional

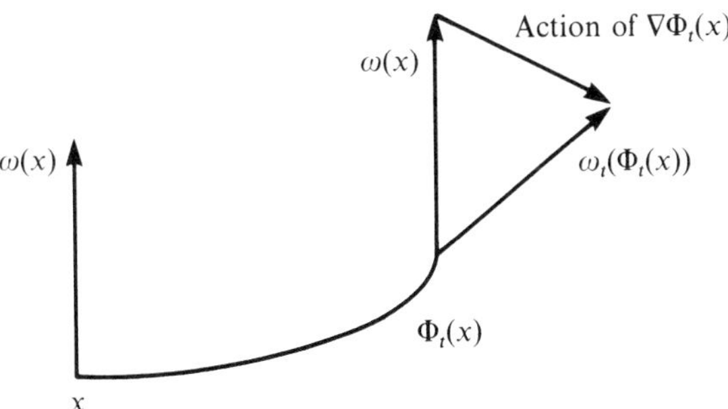

Figure 2.1

symmetry (in this case, $\nabla\Phi_t$ equals the identity matrix). The action of this term will be further discussed in Section 2.6.

2.3. Global Existence and Uniqueness in Two Dimensions

In this section we will show an existence and uniqueness theorem for the solutions of the initial value problem (ivp) associated to the Euler equation in two dimensions. To obtain this result, it is essential to use the representation formula (2.4) given in the previous section.

We will study this problem in the simplest case: we suppose the fluid to be confined in an open regular bounded domain $D \subset \mathbb{R}^2$. The Cauchy problem for the Euler equation in terms of vorticity is

$$D_t\omega = (\partial_t + u \cdot \nabla)\omega = 0, \tag{3.1}$$

$$\omega = \operatorname{curl} u, \qquad \nabla \cdot u = 0, \tag{3.2}$$

$$u \cdot n = 0 \quad \text{on } \partial D, \tag{3.3}$$

$$u(x, 0) = u_0(x). \tag{3.4}$$

As already discussed in Chapter 1, see (2.41), we have

$$u_t(x) = K_D * \omega_t(x) \equiv \int_D K_D(x, y)\omega(y, t)\, dy, \tag{3.5}$$

where

$$K_D(x, y) = \nabla^\perp G_D(x, y) \tag{3.6}$$

and G_D is the fundamental solution of the Poisson equation with the Dirichelet condition on ∂D (see Chapter 1, Section 2).

On the basis of the considerations of the previous section we observe that, knowing the velocity field, $u = u_t(x)$, solution of the Euler equation (which is just what we want to find), denoting by $\Phi_t(x)$ the path lines starting from $x \in D$, that is, the solutions of the ordinary differential equation

$$\frac{d}{dt}\Phi_t(x) = u(\Phi_t(x), t),$$

$$\Phi_0(x) = x, \tag{3.7}$$

we have

$$\omega(x, t) = \omega_0(\Phi_{-t}(x)). \tag{3.8}$$

Then the following problem is strictly related to the Cauchy problem (3.1)–(3.4), previously established.

Problem 3.1. Having fixed ω_0, find a (possibly unique) triple (Φ_t, u_t, ω_t), such that (3.7), (3.8), and (3.5) are simultaneously satisfied.

While a solution of the ivp (3.1)–(3.4) also provides a solution to Problem 3.1, a solution (Φ_t, u_t, ω_t) to Problem 3.1, on the contrary, does not necessarily solve the differential problem (3.1)–(3.4), ω_t not being differentiable with respect to x and t in order to satisfy the Euler equation in a "classical" sense. In this way Problem 3.1 has the advantage of not requiring any property of differentiability on ω_t, and so it extends the original problem. A solution to Problem 3.1 then is a weak solution to the Euler problem (3.1)–(3.4).

Before approaching Problem 3.1, we first pose the problem of finding what is the natural space in which it lives. In the previous section, we have seen that all the L_p norms of the vorticity are conserved, and this elementary remark gives us an a priori estimate on the vorticity. In a bounded domain the norm L_∞ gives the maximal control, and so it is natural to look for solutions $\omega_t \in L_\infty$.

However, another question arises: Is the velocity field generated by $\omega \in L_\infty$ regular enough to give sense to the ordinary differential problem (3.7)? The following two lemmas give a positive answer to this question.

Lemma 3.1. *The following two inequalities hold* $(|D| = $ *volume of* $D)$:

$$|u(x)| = |K_D * \omega|_\infty \le C\|\omega\|_\infty, \tag{3.9}$$

$$\int_D |K_D(x, y) - K_D(x', y)|\, dy \le C(1 + |D|)\varphi(|x - x'|), \tag{3.10}$$

where

$$\varphi(r) = \begin{cases} r(1 - \ln r) & \text{if } r < 1, \\ 1 & \text{if } r \ge 1. \end{cases} \tag{3.11}$$

The proof will be given in Appendix 2.3.

By Lemma 3.1 it follows that if $\omega \in L_\infty$, then $u = K * \omega$ is a vector field uniformly bounded (by (3.9)) continuous in x, but does not satisfy the Lipschitz condition. However, from (3.10)

$$|K_D * \omega(x) - K_D * \omega(y)| \le C\|\omega\|_\infty \varphi(|x - y|). \tag{3.12}$$

Condition (3.12) will be called a quasi-Lipschitz condition, and corrects the usual Lipschitz condition by a logarithmic diverging factor. Obviously, it is weaker than the Lipschitz condition, but is sufficient to prove the existence and uniqueness for the solutions of the ordinary differential problem (3.7) we are dealing with. In fact, the following lemma holds:

Lemma 3.2. *Let us consider the Cauchy problem in* $\mathbb{R}^n$

$$\frac{d}{dt}x = b(x, t),$$

$$x(0) = x_0, \tag{3.13}$$

with $b \in C(\mathbb{R}^n \times [0, \infty))$ uniformly bounded and satisfying the condition

$$|b(x, t) - b(y, t)| \le C\varphi(|x - y|) \tag{3.14}$$

for some positive constant C independent of t.

Then the problem (3.13) has a unique solution.

PROOF. We make use of the classical iterative method. We define

$$x_n(t) = \int_0^t b(x_{n-1}(s), s)\, ds + x_0, \tag{3.15}$$

$$x_0(t) = x_0. \tag{3.16}$$

We have

$$|x_n(t) - x_{n-1}(t)| \le \int_0^t ds\, |b(x_{n-1}(s), s) - b(x_{n-2}(s), s)|$$

$$\le C \int_0^t ds\, \varphi(|x_{n-1}(s) - x_{n-2}(s)|)$$

$$\le C \int_0^t ds\, L_\varepsilon |x_{n-1}(s) - x_{n-2}(s)| + Ct\varepsilon, \tag{3.17}$$

where

$$L_\varepsilon = -\ln \varepsilon, \qquad \varepsilon < 1. \tag{3.18}$$

In (3.17) we have used the inequality (Fig. 2.2)

$$\varphi(r) \le -(\ln \varepsilon)r + \varepsilon. \tag{3.19}$$

Inequality (3.17) can be iterated and we obtain (for any $0 \le t \le T$ and $n \ge 2$)

$$|x_n(t) - x_{n-1}(t)| \le CT\varepsilon \sum_{k=0}^{n-2} \frac{C^k L_\varepsilon^k t^k}{k!} + \frac{t^{n-1} C^{n-1} L_\varepsilon^{n-1}}{(n-1)!} \sup_{t \le T} |x_1(t) - x_0|. \tag{3.20}$$

Because b is bounded we have

$$|x_1(t) - x_0| \le CT.$$

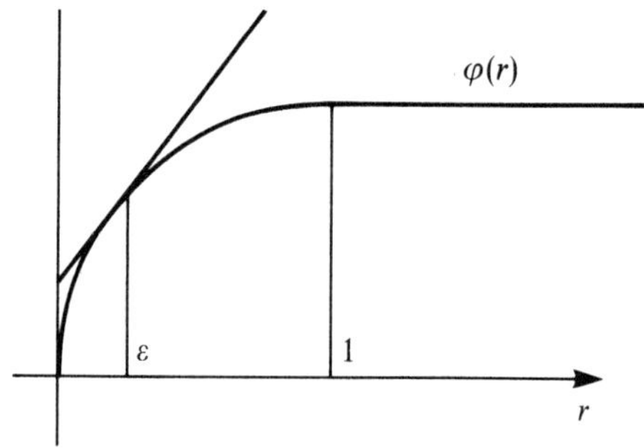

Figure 2.2

Hence

$$|x_n(t) - x_{n-1}(t)| \le CT\varepsilon \exp\{-CT \ln \varepsilon\} + C^n \frac{T^n L_\varepsilon^{n-1}}{(n-1)!}. \tag{3.21}$$

We choose $\varepsilon = \exp\{-n\}$ and T sufficiently small such that $1 - CT > \frac{1}{2}$. We have

$$|x_n(t) - x_{n-1}(t)| \le CT \exp\left\{-\frac{n}{2}\right\} + C^n \frac{T^n n^{n-1}}{(n-1)!}. \tag{3.22}$$

This estimate proves that, if T is sufficiently small (independently of x_0), $|x_n(t) - x_{n-1}(t)|$ is bounded by the terms of a convergent geometrical series, and thus it is exponentially small (remember the Stirling formula $n^n \le n! \exp\{Cn\}$). Then $\{x_n(t)\}_{n=1\ldots\infty}$ for $0 < t < T$ (T small) is a uniformly convergent Cauchy sequence. By denoting its limit with $x(t)$, the uniform convergence implies that $x(t)$ satisfies (3.13) in integral form because, by (3.14), $b(x_n(t), t)$ is also uniformly convergent to $b(x(t), t)$. Finally, by the continuity of $t \to b(x(t), t)$, it follows that $x(t)$ satisfies (3.13) in differential form.

Because T does not depend on x_0, the procedure can be iterated until arriving at arbitrary large times. Uniqueness can be proved by the same method. $\qquad\square$

Remark. It is not necessary to assume b to be uniformly bounded: inequality (3.14) is enough to prove the result established in Lemma 3.2.

We return now to Problem 3.1, by introducing a sequence of approximations defined as follows:

$$\frac{d}{dt}\Phi_t^n(x) = u_t^n(\Phi_t^n(x)), \tag{3.23}$$

$$u_t^n(x) = K_D * \omega_t^{n-1}, \tag{3.24}$$

$$\omega_t^n(x) = \omega_0(\Phi_{-t}^n(x)). \tag{3.25}$$

Initially, we choose

$$\omega_t^0(x) = \omega_0(x), \qquad \omega_0 \in L_\infty. \tag{3.26}$$

In terms of differential equations such approximated problems correspond, whenever ω_t is smooth enough, to the sequence of linear problems

$$(\partial_t + u_t^n \cdot \nabla)\omega_t^n = 0,$$
$$u_t^n = K_D * \omega_t^{n-1}. \tag{3.27}$$

The approximate problems we have introduced make sense on the basis of Lemma 3.2 and inequality (3.10). In fact, $u^n(x, t)$ satisfies the same hypotheses as $b(x, t)$ in Lemma 3.2 (see Exercise 5). Moreover, there is no problem with the boundary, since u^n are tangent to the border, the trajectories cannot leave the domain D (see Exercise 6).

We now want to investigate the convergence of the triple (ϕ^n, u^n, ω^n) to find a solution to Problem 3.1. For any $x \in D$, we have

$$\Phi_t^n(x) - \Phi_t^{n-1}(x) = \int_0^t ds\{u^n(\Phi_s^n(x), s) - u^n(\Phi_s^{n-1}(x), s)\}$$

$$+ \int_0^t ds\{u^n(\Phi_s^{n-1}(x), s) - u^{n-1}(\Phi_s^{n-1}(x), s)\}. \quad (3.28)$$

On the other hand, by virtue of the Liouville theorem and (3.25)

$$\int_D |u^n(x, t) - u^{n-1}(x, t)| \, dx$$

$$= \int_D dx \left| \int_D dy \, K_D(x, y)[\omega^{n-1}(y, t) - \omega^{n-2}(y, t)] \right|$$

$$= \int_D dx \left| \int_D dy \{K_D(x, \Phi_t^{n-1}(y)) - K_D(x, \Phi_t^{n-2}(y))\} \omega_0(y) \right|$$

$$\leq \|\omega_0\|_\infty \int_D dx \int_D dy \, |K_D(x, \Phi_t^{n-1}(y)) - K_D(x, \Phi_t^{n-2}(y))|$$

$$\leq C\|\omega_0\|_\infty (1 + |D|) \int_D \varphi(\Phi_t^{n-1}(y), \Phi_t^{n-2}(y)) \, dy, \quad (3.29)$$

where, in the last step, we have used (3.10).

We define

$$\delta^n(t) = \frac{1}{|D|} \int_D dx \, |\Phi_t^n(x) - \Phi_t^{n-1}(x)|. \quad (3.30)$$

By (3.28) and (3.29) we obtain

$$\delta^n(t) \leq C \int_0^t ds \, \varphi(\delta^n(s)) + C \int_0^t ds \, \varphi(\delta^{n-1}(s)), \quad (3.31)$$

where the constant C which appears in (3.31) depends on $\|\omega_0\|_\infty$ and on $|D|$. We observe that we have used the Jensen inequality: for any positive f, by the concavity of φ, we have

$$\frac{1}{|D|} \int_D \varphi(f(x)) \, dx \leq \varphi\left(\frac{1}{|D|} \int_D f(x) \, dx\right). \quad (3.32)$$

We define

$$\rho^N(t) = \sup_{n > N} \delta^n(t). \quad (3.33)$$

By (3.31) we have, for $n \geq N$,

$$\delta^n(t) \leq C \int_0^t dx \, \varphi(\rho^{N-1}(s)) \quad (3.34)$$

and hence

$$\rho^N(t) \leq C \int_0^t ds \, \varphi(\rho^{N-1}(s)). \quad (3.35)$$

So we have obtained an inequality like the second formula in (3.17). Proceeding as in Lemma 3.2, we obtain

$$\lim_{N \to \infty} \rho^N(t) \to 0 \tag{3.36}$$

uniformly in $t \in [0, T]$, for T sufficiently small. The size of T depends only on $\|\omega_0\|_\infty$ and on $|D|$, and so the procedure can be extended to any time since $\|\omega_t\|_\infty = \|\omega_0\|_\infty$.

Thus we have proved that there exists $\Phi \in C([0, T], M(D))$, where $M(D)$ is the family of measurable transformations of D in itself such that

$$\lim_{n \to \infty} \int_D dx \, |\Phi_t(x) - \Phi_t^n(x)| = 0. \tag{3.37}$$

It is obvious that this application can be defined as well for negative t. Putting

$$\omega_t(x) = \omega_0(\Phi_{-t}(x)), \tag{3.38}$$

it is easy to realize that ω_t is the weak limit of ω_t^n. In fact, for $f \in D^1(D)$,

$$\left| \int_D f(x) [\omega_t^n(x) - \omega_t(x)] \, dx \right|$$

$$= \int_D [f(\Phi_t^n(x)) - f(\Phi_t(x))] \omega_0(x) \, dx|$$

$$\leq \|\nabla f\|_\infty \|\omega_0\|_\infty \int_D |\Phi_t^n(x) - \Phi_t(x)| \, dx \xrightarrow[n \to \infty]{} 0. \tag{3.39}$$

Using the usual density argument we can prove that the same limit holds for $f \in C(D)$.

Finally, defined

$$u_t(x) = (K_D * \omega_t)(x) \tag{3.40}$$

we obtain

$$\int_D |u_t(x) - u_t^n(x)| \, dx$$

$$= \int_D \int_D dy \{K_D(x, \Phi_t(y)) - K_D(x, \Phi_t^{n-1}(y))\} \omega_0(y)|$$

$$\leq C \|\omega_0\|_\infty (1 + |D|) \int_D dx \, \varphi(|\Phi_t(x) - \Phi_t^{n-1}(x)|)$$

$$\leq C \|\omega_0\|_\infty (1 + |D|) \left[\varepsilon + L_\varepsilon \int_D |\Phi_t(x) - \Phi_t^{n-1}(x)| \right], \tag{3.41}$$

where we have used (3.19).

Then

$$u_t(x) = \lim_{n \to \infty} u_t^n(x) \tag{3.42}$$

in $L_1(D)$. At this point it is not difficult to prove the following identity:

$$\Phi_t(x) = \int_0^t ds\, u_s(\Phi_s(x)) \tag{3.43}$$

valid for almost all $x \in D$. Moreover, u is continuous in t and it is quasi-Lipschitz in x (because $\omega_t \in L_\infty(D)$). Thus, for any $x \in D$, (3.7) holds.

In conclusion, we have proved the following theorem:

Theorem 3.1. *Let $\omega_0 \in L_\infty(D)$. There exists a unique triple (Φ_t, u_t, ω_t) solution to Problem 3.1, where $\omega \in L_\infty([0, T], L_\infty(D))$, $u = K_D * \omega$ is a quasi-Lipschitz vector field, and Φ satisfies (3.7).*

PROOF. We have already constructed the triple. By the same argument we can prove its uniqueness. The details are left to the reader. $\qquad\square$

We have seen that a triple solution to Problem 3.1 is a weak solution of the Euler equation. A further justification for this name is given by the following considerations. Let $\omega \in C^1([0, T], C^1(D))$ be a "classical" solution of the Euler equation, that is, a function that verifies (3.1) and (3.2) for any x and t. Putting

$$\omega_t(f) = \int_D dx\, \omega_t(x) f(x) \tag{3.44}$$

it is not difficult to obtain the identity

$$\frac{d}{dt}\omega_t(f) = \omega_t(u_t \cdot \nabla f) \tag{3.45}$$

valid for any $f \in C^1(D)$. Equation (3.45) follows from a simple application of the Gauss–Green lemma. Notice that in this equation no derivatives of ω_t appear and so it could be satisfied by the solutions we have found in Theorem 3.1. Actually the following theorem holds:

Theorem 3.2. *Let $f \in C^1(D)$ and let $\omega_t \in L_\infty$ be the solution constructed in Theorem 3.1. Then $\omega(f) \in C^1([0, T])$ and (3.45) holds.*

The proof is not difficult and is left to the reader.

2.4. Regularity Properties and Classical Solutions

In the previous section we have constructed a (weak) solution of the Euler equation such that $\omega_t \in L_\infty$ for all t, provided that, initially $\omega_0 \in L_\infty$. As consequence of the hyperbolic character of the equation, we do not expect the time evolution to increase the regularity of the solution, so it is hopeless (actually, not true, see Exercise 10) to find a a classical solution starting from an initial vorticity ω_0, which is essentially bounded only. However, with the hypotheses of further regularity on the initial datum, we can construct classical solutions. The following theorem shows that the Euler evolution does not decrease the regularity of the inital datum.

Theorem 4.1. *Let* $\omega_0 \in C^k(D)$ *with* $k \geq 1$. *Then* ω_t, *a solution of the Euler equation with initial datum* ω_0, *is also a classical solution in the sense that* (3.1) *is satisfied for all x and t. Moreover, for all t,* $\omega_t \in C^k(D)$.

PROOF. We first observe that the trajectories $\Phi_t(x)$ are Hölder continuous in x. In fact,

$$|\Phi_t(x) - \Phi_t(y)| \leq |x - y| + \int_0^t ds\, |u(\Phi_s(x), s) - u(\Phi_s(y), s)|$$

$$\leq |x - y| + C \int_0^t ds\, \varphi(|\Phi_s(x) - \Phi_s(y)|). \qquad (4.1)$$

Because the ordinary differential equation

$$\frac{d}{dt} z = C\varphi(z),$$
$$z(0) = z_0, \qquad\qquad (4.2)$$

has a unique solution, given by

$$z(t) = \begin{cases} z_0^{\exp\{-Ct\}} e^{1 - \exp\{Ct\}} & \text{if } z < 1, \\ 1 + C(t - t_0) & \text{if } z \geq 1, \end{cases} \qquad (4.3)$$

where t_0 is the time in which $z(t_0) = 1$. Using the lemma given in Appendix 2.1 to the integral inequality (4.1), we obtain the existence of a function α: $t \to \alpha(t)$ such that

$$|\Phi_t(x) - \Phi_t(y)| \leq C|x - y|^{\alpha(t)}. \qquad (4.4)$$

Then, if $\omega_0 \in C^\alpha(D)$ (if $\alpha < 1$, it means that ω_0 is Hölder continuous with exponent α)

$$|\omega_t(x) - \omega_t(y)| \leq |\omega_0(\Phi_{-t}(x)) - \omega_0(\Phi_{-t}y))| \leq C(\omega_0)|x - y|^{\alpha\alpha(t)}.$$

$$(4.5)$$

Then $\omega_t \in C^{\alpha\alpha(t)}(D)$.

In Appendix 2.4 it is proved that $\omega \in C^{k+\beta}(D)$, for some $\beta \in (0, 1)$, implies $u = K_D * \omega \in C^{k+1+\beta'}$ with $\beta' < \beta$ and $k = 0, 1, \ldots$.

The Hölder continuity of ω implies the differentiability of u from which it follows that Φ_t is also differentiable with respect to the initial data, and the following identity holds:

$$\nabla \Phi_t(x) = \mathbb{1} + \int_0^t ds\ \nabla u_s(\Phi_s(x)) \cdot \nabla \Phi_s(x), \tag{4.6}$$

where $(\nabla \Phi_t(x))_{ij} = \partial \Phi_t(x)_i / \partial x_j$ is the Jacobian matrix and $\mathbb{1}_{ij} = \delta_{ij}$ is the unit matrix. Because $u \in C(D \times [0, T]) \cap L_\infty([0, T]; C^1(D))$

$$\|\nabla \Phi_t\|_\infty \leq \exp\{Ct\}. \tag{4.7}$$

Finally, if $\omega_0(C^1(D),$

$$\frac{\partial}{\partial x_i} \omega_t(x) = \sum_j \frac{\partial}{\partial x_j} \omega_0(\Phi_{-t}(x)) \frac{\partial}{\partial x_i} \Phi_{-t}(x)_j, \tag{4.8}$$

from which it follows that $\omega_t \in C^1(D)$. From the existence and the time continuity of the term $u \cdot \nabla \omega$, it follows that ω is differentiable with respect to time. Therefore, it is also a classical solution.

To investigate the further properties of regularity we observe that $\omega_t \in C^{1+\beta}(D)$. In fact, the Hölder continuity of ∇u implies the Hölder continuity of $\nabla \Phi_t$, which in turn implies the Hölder continuity of $\nabla \omega$. Therefore $u_t \in C^{2+\beta'}(D)$, and then the second derivatives $(\partial^2 / \partial x_{ij}) \Phi_t(x)_k$ exist. Moreover, $\omega_0 \in C^2(D)$ implies $\omega_t \in C^2(D)$. The procedure can be iterated as long as we want, up to the maximum order of derivatives of the initial datum. This concludes the proof of Theorem 4.1. $\square$

The analysis of the problems connected to the existence and uniqueness of the Euler flow in two dimensions in bounded domains is thus achieved. This analysis can be extended to unbounded domains if we assume suitable properties of decay of the vorticity field (see Exercise 9). More singular distributions of the vorticity or unbounded domains with infinite initial energy elude our present purposes. Further discussions and references will be given in Section 2.7.

2.5. Local Existence and Uniqueness in Three Dimensions

In this section we will prove an existence and uniqueness theorem for the solution of the Euler equation in three dimensions for short times. The shortness of the time in which the solution is constructed depends, in an essential way, on the a priori estimates (valid for short times only) which we obtained in Section 2.1. As before, we consider $D = [-\pi, \pi]^3$. We follow this strategy: first we construct a sequence of approximated solutions and then we study their convergence. Let P_N be the orthogonal projector in the subspace of $L_2(D)$ generated by the functions $\{\exp[ik \cdot x]\}_{|k| \leq N}$ $(|k| = \max\{k_i\})$.

We consider the following initial value problem:

$$\partial_t u^N + P_N[(u^N \cdot \nabla)u^N] = -P_N \nabla p_t^N,$$
$$\nabla \cdot u^N = 0, \qquad u_0^N = P_N u_0, \qquad u_0 \in \mathbb{H}^m, \qquad m \geq 3. \tag{5.1}$$

As we will see, problem (5.1) is equivalent to an ordinary differential equation which admits a global solution. Moreover, (5.1) is similar to the initial value problem associated to the Euler equation and reduces to it, at least at a formal level, in the limit $N \to \infty$. We introduce the Fourier transform of the velocity field

$$\hat{u}_t(k) = \left(\frac{1}{2\pi}\right)^{3/2} \int dx \, \exp\{-ik \cdot x\} u_t(x). \tag{5.2}$$

The Euler equation in terms of Fourier transform is

$$\frac{d}{dt}\hat{u}_t(k) = i\left\{kp^\wedge(k) - \sum_{h \in \mathbb{Z}^3} \hat{u}_t(k-h) \cdot h\hat{u}_t(h)\right\},$$
$$k \cdot \hat{u}_t(k) = 0. \tag{5.3}$$

We remember that, as we discussed at the end of Section 2.1, the Fourier transform of the pressure $p^\wedge(k)$ must be understood as a functional of the velocity field.

Consider now the ordinary differential problem, for $|k| \leq N$,

$$\frac{d}{dt}\hat{u}_t^N(k) = i\left\{kp^{\wedge N}(k) - \sum_{h:\,|h| \leq N;\,|k-h| \leq N} \hat{u}^N(k-h) \cdot h\hat{u}^N(h)\right\},$$
$$k \cdot \hat{u}_t^N = 0, \tag{5.4}$$

which constitutes a truncation of (5.3). The above ordinary differential system is obviously solvable at least for short times. We remark that also in this finite-dimensional version the truncated pressure may be eliminated by taking the scalar product of the first equation of (5.4) against $\hat{u}_t^N$ and by using the incompressibility condition $(k \cdot \hat{u}_t^N = 0)$.

It is easy to verify that equations (5.1) and (5.4) are equivalent in the sense that, denoting by $\hat{u}^N$ the Fourier transform of the solution u^N in (5.1), it verifies (5.4). This remark gives consistency to our notation.

The truncation we have done is such that the function u_t^N is real. In fact, supposing it to be real at time zero, i.e., the condition

$$\hat{u}_0^N(k) = [\hat{u}_0^N(-k)]^* \tag{5.5}$$

is satisfied, we see immediately that the structure of (5.4) is such that the reality of the solutions (which exist uniquely, at least, for short times) is preserved.

The corresponding solution (for the moment only local in time) of (5.1) is real, and so

$$(u_t^N, P_N(u_t^N \cdot \nabla)u_t^N) = 0. \tag{5.6}$$

This implies the conservation of the (truncated) energy

$$\frac{d}{dt}(\|u_t^N\|_2)^2 = 0. \tag{5.7}$$

Having an a priori estimate on the quantity

$$\sum_k |\hat{u}_t^N(t)|^2 = \sum_k |\hat{u}_0^N(k)|^2 \tag{5.8}$$

we deduce that the solutions of (5.4) (and of (5.1)) may be extended to arbitrary large times. So for arbitrary $T > 0$, we have found a unique solution $u_t^N \in C^\infty(D)$ to the problem (5.1). Finite-dimensional trunctations of partial differential equations are called Faedo–Galerkin truncations.

It is not difficult to verify that the estimates of Section 2.1 are satisfied by the solutions of (5.1), so that there exists a time T, depending on m and $|u_0|_m$, such that

$$\sup_{0 \le t \le T} |u_t^N|_m \le M, \qquad m \ge 3, \tag{5.9}$$

with M independent of N. Making use of estimate (5.9), we can prove the convergence of the sequence u^N to $u \in C([0, T]; L_2(D))$. In fact, let $H > N$

$$\frac{1}{2}\frac{d}{dt}(\|u_t^N - u_t^H\|_2)^2$$

$$= \frac{d}{dt}(u^N, u^H)$$

$$= -(P_N(u^N \cdot \nabla)u^N, u^H) - (u^N, P_H(u^H \cdot \nabla)u^H)$$

$$= ([1 - P_N](u^N \cdot \nabla)u^N, u^H) + (u^N, ([u^N - u^H] \cdot \nabla)u^H)$$

$$= ([1 - P_N](u^N \cdot \nabla)u^N, u^H) + (u^N, [u^N - u^H] \cdot \nabla(u^H - u^N)), \tag{5.10}$$

where we have used the integration by parts and the divergence-free condition.

The second term in the right-hand side of (5.10) can be estimated by

$$\|\nabla u^N\|_\infty \|u^H - u^N\|_2^2 \le C\|u^H - u^N\|_2^2 \tag{5.11}$$

by virtue of (5.9) and (1.7) of Proposition 1.1 yielding $\|\nabla u^N\|_\infty \le |u|_3$.

Moreover

$$([u^N \cdot \nabla]u^N, [1 - P_N]u^H) \le \|(u^N \cdot \nabla)u^N\|_2 \|(1 - P_N)u^H\|_2. \tag{5.12}$$

From (1.9) of Proposition 1.1 we have

$$\|(u^N \cdot \nabla)u^N\|_2 \le C|\nabla u^N|_0 |u^N|_2 \le C. \tag{5.13}$$

Finally

$$\|(1 - P_N)u^N\|_2^2 \le \sum_{|k| > N} |\hat{u}_t^H(k)|^2 |k|^6 \frac{1}{N^6}$$

$$\le |u_t^H|_3 \frac{C}{N^6} \le \frac{MC}{N^6}. \tag{5.14}$$

Hence we have obtained that

$$\frac{1}{2}\frac{d}{dt}\|u_t^N - u_t^H\|_2^2 \le C(\|u_t^N - u_t^H\|_2)^2 + CN^{-3} \tag{5.15}$$

from which

$$\sup_{0 \le t \le T} (\|u_t^N - u_t^H\|_2)^2 \le CN^{-3}. \tag{5.16}$$

Thus we have proved the convergence (uniform on compacts) of the sequence.

Observe that as regular as the initial data are (i.e., we can make m large), the faster can be made the convergence speed in estimate (5.16). This can be seen by suitably modifying estimate (5.14) using m in place of 3 (see Exercise 13). Let $u \in C([0, T], L_2(D))$ be defined as the L_2 limit of u_t^N. Because u inherits the estimates on u^N we have $u \in L_\infty([0, T]; \mathbb{H}_m)$.

It remains to prove that u is a solution of the Euler equation. We start by observing that u satisfies the Euler equation in the following weak form:

$$(\varphi, u_t) = (\varphi, u_0) + \int_0^t ds \sum_{i;j} (\partial_j \varphi_i, u_j u_i) \tag{5.17}$$

for any divergenceless vector field $\varphi \in C^\infty(D)$. In fact, u^N satisfies the equation (notice that P_N commutes with the derivative operator)

$$(\varphi, u_t^N) = (\varphi, u_0^N) + \int_0^t ds \sum_{i;j} (P_N \partial_j \varphi_i, u_j^N u_i^N) \tag{5.18}$$

and there is no problem in going to the limit $N \to \infty$ making use of (5.16) and the a priori estimate. Actually it follows that $(P_N \partial_j \varphi_i, u_j^N u_i^N) \to (\partial_j \varphi_i, u_j u_i)$ uniformly in $t \in [0, T]$.

Furthermore, we observe that $(\partial_j \varphi_i, u_j u_i)$ is a continuous function of the time so that (φ, u_t) is differentiable. Thus

$$\frac{d}{dt}(\varphi, u_t) = -(\varphi, [u_t \cdot \nabla]u_t). \tag{5.19}$$

Finally, because $(u \cdot \nabla)u \in \mathbb{H}_2$ (see (1.8) of Proposition 1.1), we have that u is strongly differentiable in $\mathbb{H}_2$ and the identity

$$\frac{d}{dt}u_t = -(u_t \cdot \nabla)u_t - \nabla p \tag{5.20}$$

holds in $\mathbb{H}_2$ for some $p \in C([0, T]; H_1)$.

Because u is differentiable in $\mathbb{H}_2$ with respect to time, it is also pointwise differentiable (see (1.7) of Proposition 1.1) and so the Euler equation is satisfied by the solution we have found also in the classic sense. So we have proved the following theorem:

Theorem 5.1. *Let $u_0 \in \mathbb{H}_m$, $m \ge 3$, and $T \ge 0$ be sufficiently small. There then exists a unique classical solution $u_t(x)$ of the Euler equation, with initial datum*

$u_0(x)$, such that

$$\sup_{0 \leq t \leq T} |u_t|_m < \infty.$$

PROOF. The existence of the solution has already been proved. The uniqueness of the solution in $\mathbb{H}_m$ is easy and is left to the reader. $\qquad\square$

The same result discussed in this section can be obtained for bounded domans by substituting the projection on the subspace generaged by the trigonometric functions by the projection on the subspaces generated by the eigenfunctions of the Laplace operator.

2.6. Some Heuristic Considerations on the Three-Dimensional Motion

In the previous sections we have quite often underlined the fact that the motion of an incompressible ideal fluid looks very different in two and three dimensons. In this section we want to develop some considerations without any attempt at mathematical rigor, which, we hope, will provide further clarification on the behavior of three-dimensional flows.

We have seen by (2.4) that in two dimensions the vorticity field is simply transported along the path lines. On the contrary, in three dimensions the Jacobian matrix $\nabla\Phi_t$ acts linearly on the transported vorticity field, which is no longer conserved along the path lines (see (2.15)). This action is described by the term $(\omega \cdot \nabla)u$ of the Euler equation for the vorticity. Moreover, this term is responsible for the difficulties which arise in the construction of a global solution of the Euler equation. In fact, ω is the antisymmetric part of $\nabla \cdot u$ (see Section 2 of Chapter 1), and so we are tempted to conjecture that $(\omega \cdot \nabla)u$ is of the order of ω^2 and $D_t\omega \approx \omega^2$. Because the ordinary equation

$$\frac{d}{dt}Y = Y^2,$$
$$Y(0) = Y_0 > 0, \tag{6.1}$$

has solutions that blow up in finite time, it is not easy to exclude a priori that the vorticity becomes infinite, in a finite time, in some point of space. This divergence would make it difficult, if not impossible, to even give a meaning to the Euler equation in whatever weak sense.

Until now neither a global existence theorem, nor an example which shows the development of a singularity in a finite time is known. This singularity is necessarily described by the blow-up of the vorticity field: if ω_t is a solution of the Euler equation, there exists a time t^* such that

$$\lim_{t \to t^*} \|\omega_t(x)\|_\infty = \infty. \tag{6.2}$$

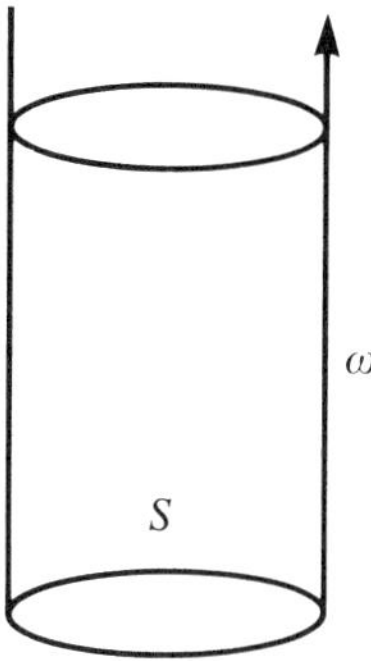

Figure 2.3

In fact we can prove (see the Comments in the next section) that, if the vorticity remains bounded, it is possible to construct smooth unique solutions to the initial value problem.

In conclusions, we are not able to obtain more than the local theorem discussed in Section 2.5. We conclude this heuristic discussion by showing that the conservation laws (energy and circulation) are not capable of preventing the development of singularities (in the sense of a blow-up of the L_∞ norm of ω) during motion. On the other hand, they outline features of an extreme complexity.

Let us suppose a vorticity field approximately constant, concentrated in a tube T of the space shown in Fig. 2.3. The circulation theorem does not prevent a large growth of the vorticity. For instance, the tube could stretch in time t conserving the circulation

$$\omega S = \Omega \sigma, \qquad \Omega \gg \omega, \quad \sigma \ll S. \tag{6.3}$$

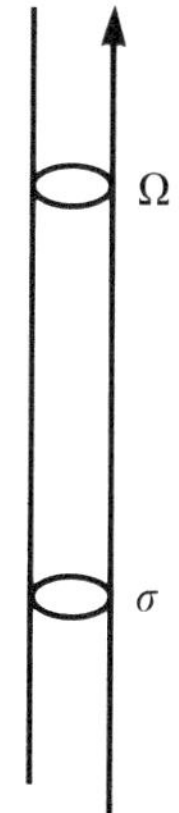

Figure 2.4

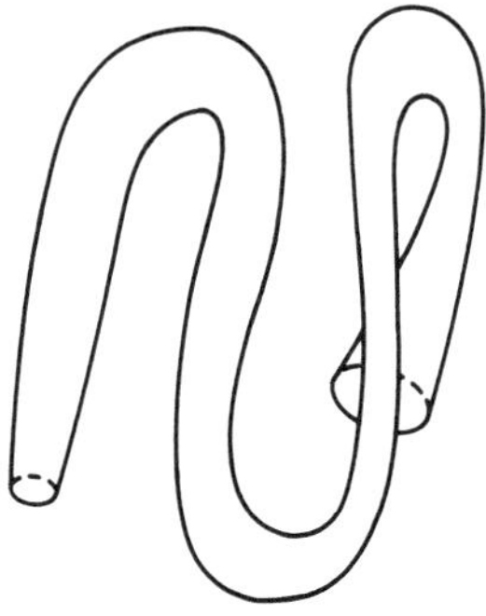

Figure 2.5

But the evolution (Fig. 2.3 → Fig. 2.4) is contrary to the energy conservation law. A simple calculation shows that the energy of the configuration described by Fig. 2.4 is much larger than that given in Fig. 2.3. This is easily confirmed by using the Biot–Savart law (2.54), Chapter 1. The energy does not increase and so, in connection with a local increase of the vorticity field produced by a stretching of the tube, the geometry of the tube and the distribution of vorticity in its interior must necessarily be combined in such a way as to produce the compensations necessary for the constancy of the energy. So the tube tends to become bent many times on itself (Fig. 2.5) in complicated geometries, and the vorticity fields may assume a very large value, in a small region of space, without any violation of the conservation laws.

Numerical simulations, actually very difficult, delicate, and far from being conclusive, seem to show this tendency.

2.7. Comments

The solution of the initial value problem associated with the Euler equation, in two dimensions for arbitrary times and in three dimensions for short times, has long been known. There is a large literature on the subject. We mention only [Wol 33], [Yud 63], [Kat 67], [Bar 72] (see also [KaP 92] in connection with this last paper) for two-dimensional existence theorems, and [EbM 70], [Kat 72], [BoB 74], [Tem 75], [Tem 76], [Tem 86], [KaP 88] for the three-dimensional case. The analysis presented here in Sections 2.3, 2.4, 2.5 is essentially due to Kato [Kat 67], [MaP 84], and [Kat 72].

We have already stated that the construction of the solutions can be easily extended to unbounded domains provided that the total energy is finite. However, the interesting physical situation is when the energy is locally finite only. We mention here that, in such a case, the solution for a two-dimensional infinite strip can be constructed. See [CaM 86].

We mention that the analysis developed in Sections 2.3 and 2.4 for a two-dimensional incompressible flow does not apply to the case of a nonconstant density. In this case, the conservation of the vorticity along the trajectories is violated, as can be seen by direct inspection. As a consequence, we can prove a result which holds locally in time only. It is not known whether there is a breakdown of the solution.

As already mentioned it is not known whether a three-dimensional Euler flow can develop a singularity in finite time. Beale, Kato, and Majda proved that a necessary condition for having a singularity at time T is that

$$\int_0^T \|\omega(t)\|_\infty \, dt = +\infty. \tag{7.1}$$

The precise statement (valid for $D = [-\pi, \pi]^2$, the same result can be obtained for all the space with minor modifications) is the following:

Theorem 7.1. *Suppose that, for a given solution to the Euler equation, we have up to time T*

$$\int_0^T \|\omega(t)\|_\infty \, dt < +\infty \tag{7.2}$$

then such a solution is classical.

PROOF. The proof is based on the following inequality, allowing the control of $\|\nabla u\|_\infty$ in terms of $|u|_3$ and $\|\omega\|_\infty$. We have [BKM 84]:

$$\|\nabla u\|_\infty \leq C(1 + \ln_+ |u|_3)\|\omega\|_\infty, \tag{7.3}$$

where $\ln_+ r = \ln r$ for $r > 1$ and $\ln_+ r = 0$ for $r \leq 1$.

Coming back to the Euler equation, applying D^α (see Section 2.1) to the equation, we obtain

$$\partial_t D^\alpha u + u \cdot \nabla D^\alpha u = -\nabla D^\alpha p - D^\alpha(u \cdot \nabla u) + u \cdot \nabla D^\alpha u. \tag{7.4}$$

The key point of this analysis is that the following inequality holds for $s \geq 3$:

$$\|D^\alpha(u \cdot \nabla u) - u \cdot \nabla D^\alpha u\|_2 \leq C |u|_s \|\nabla u\|_\infty. \tag{7.5}$$

Therefore, by taking the scalar product of (7.4) against $D^\alpha u$, we obtain

$$\frac{d}{dt}|u|_s^2 \leq C |u|_s^2 \|\nabla u\|_\infty \tag{7.6}$$

after eliminating, as usual, the pressure term, making use of the fact that $(D^\alpha u, u \cdot \nabla D^\alpha u) = 0$, and, finally, applying the Cauchy–Schwarz inequality.

We observe that inequality (7.6) constitutes an improvement of inequality (1.19) since $\|\nabla u\|_\infty \leq C |u|_s$. From (7.6)

$$|u(t)|_s \leq |u(0)|_s \exp C \int_0^t \|\nabla u(\tau)\|_\infty \, d\tau. \tag{7.7}$$

On the other hand, plugging (7.3) in (7.7) we get

$$|u(t)|_s \leq |u(0)|_s \exp C \int_0^t \|\omega(\tau)\|_\infty (1 + \ln_+ |u(\tau)|_s) \, d\tau. \tag{7.8}$$

By using the Gronwall inequality on $\ln_+ |u(t)|_s$ we finally obtain

$$|u(t)|_s \leq |u(0)|_s \exp\left\{ C \exp C \int_0^t \|\omega(\tau)\|_\infty \, d\tau \right\}. \tag{7.9}$$

In conclusion, if (7.2) holds, then $|u(t)|_s$ stays bounded in the time interval $[0, T]$. As shown in Section 2.5 this is enough to prove the existence of a unique classical solution in the same time interval. $\square$

Remark. As a corollary, we also have

$$\sup_{0 \leq t \leq T} \|\omega(t)\|_\infty \leq C \tag{7.10}$$

as follows from (7.9) and (1.7). Thus (7.2) and the fact that ω solves the Euler equation imply the stronger condition (7.10).

The behavior of a vortex tube and its capability in stretching and, eventually, in creating singularities, is related to the important problem of turbulence which will be discussed in Chapter 7. For the moment, we simply mention that numerical simulations devoted to the study of the evolution of a vortex tube, available until now, do not provide a conclusive answer to the problem of singularity formation. In fact, due to the short-distances divergence in the kernel of the Biot–Savart law, it is very difficult to conceive algorithms which describe possible singularities beyond the approximations and numerical errors.

Recently, Di Perna and Majda have investigated the following two-dimensional problem. Consider a sequence $\{u_k\}_{k=1\ldots\infty}$ of smooth solutions of the two-dimensional Euler equation in the plane. We assume that they have uniformly bounded energy

$$\int |u_k(x, t)|^2 \, dx \leq C \tag{7.11}$$

(C independent of k) and weakly convergent to u, i.e.,

$$\int_0^T dt \int dx \, u_k(x, t)\phi(x, t) \rightarrow \int_0^T dt \int dx \, u(x, t)\phi(x, t) \tag{7.12}$$

for all $\phi \in C_0^\infty$. We remark that the convergence (7.12) is a consequence, at least for suitable subsequences, of the uniform bound (7.11).

Because of the nonlinearity of the Euler equation, we cannot conclude that u solves such an equation, as would follow in the case of strong convergence, i.e.,

$$\int_0^T dt \int dx \, |u_k(x, t) - u(x, t)|^2 \rightarrow 0. \tag{7.13}$$

In fact, the product $u_i u_j$ appearing in the weak form of the Euler equation (for the velocity) is not weakly continuous. So the problem is to understand how the limit field u behaves. This problem has physical motivations: there are nonsmooth initial conditions (as a vortex sheet configuration, which will be discussed in detail in Chapter 6) whose evolution is interesting, especially in connection with the evolution of a slightly viscous flow in the presence of boundaries (see the heuristic discussion in Chapter 1). Such singular initial configurations can, at time zero, be well approximated by a sequence of smooth velocity profiles, whose evolutions are known to be smooth. The approximation can be done in a weak sense as well as in a stronger sense. In the first case, the convergence to something is (at least for subsequences) automatically ensured. However, the problem is to characterize the limit. In the second case, once the convergence is proved (see, e.g., the approach developed in Section 2.5) there is no problem in proving that the limit actually solves the right equation although the proof of the convergence is, in general, difficult.

Following the first point of view, Di Perna and Majda tried to get information on the weak limit u. To this purpose, they introduce the following object:

$$\theta(E) = \limsup \int_E dx\, dt\, |u_k(x, t) - u(x, t)|^2 \tag{7.14}$$

called the reduced defect measure. Actually θ is nontrivial only when the strong convergence fails, so it gives a measure of the lack of strong convergence. The support of θ is, roughly speaking, the subset of space–time where the Euler equation is, possibly, not satisfied by the limit u. We remark, however, that the set of solutions to the Euler equation, satisfying the bounds (7.11) and (7.15) below, could be weakly closed without the reduced defect measure being trivial.

If we assume, in addition to (7.11), the following L_1 bound for the vorticity

$$\int dx\, |\omega_k(x, t)| \leq C \tag{7.15}$$

with C independent of k, then it is possible to prove that ([DiM 87]$_1$, [DiM 87]$_2$, [DiM 88]):

(i) the reduced defect measure is concentrated on a set of Hausdorff dimension equal, at most, to one; and
(ii) if this set has Hausdorff dimension less than one, then u solves the Euler equation.

Actually, Greengard and Thomam proved ([GrT 88]) that under the hypothesis of (ii) the strong convergence is also ensured.

We remark that if $\theta(E) > 0$, then

$$\int_E |u_k(x, t)|^2\, dx \quad \text{does not converge to} \quad \int_E |u(x, t)|^2\, dx. \tag{7.16}$$

By general arguments (convexity) we get

$$\liminf \int_E |u_k(x, t)|^2 \, dx \geq \int_E |u(x, t)|^2 \, dx \tag{7.17}$$

so that the exceptional set in which θ is concentrated is also a set in which there is a loss of kinetic energy. Thus by (ii) we have an energy concentration on a one-dimensional set.

In Chapter 4, in connection with the vortex theory, we will investigate the same problem for a situation in which

$$\omega_k \to \sum_{i=1}^N \alpha_i \delta_{x_i}, \tag{7.18}$$

where δ_x is the Dirac measure concentrated on the point x, α_i are real numbers, and the above convergence must be understood in the sense of the weak convergence of measures. In this case the kinetic energy is diverging. Here we have an even more dramatic energy concentration, nevertheless, in this case the limiting behavior is described by the vortex flow, which is, in a sense, a weak solution of the Euler equation. However, the energy concentration is a phenomenon which has yet to be understood in general situations.

Coming back to the DiPerna and Majda analysis, it must be noted that if condition (7.15) is replaced by a stronger L_p control on the vorticity, for $p > 1$, then there is no loss of compactness in L_2 and, hence, the limit field does satisfy the Euler equation. However, this case does not contain interesting situations such as the vortex sheet.

At this point it is worth emphasizing the fact that interesting physical situations can elude the weak convergence and the weak solution concept for the velocity field. Actually, vortex sheet dynamics (see Chapter 6) is not very well described in terms of weak solutions for the velocity field, since we expect very many weak solutions of the Euler equation for the velocity, only one of them being physically reasonable. Further comments on this interesting problem will be given in Chapter 6.

We now want to touch on the existence of solutions of the Navier–Stokes equation that we introduced in the previous chapter. We first observe that the viscosity, in principle, makes the problem more regular. Actually, for the Navier–Stokes flow we can derive an energy inequality (along the same lines yielding the energy conservation law for the Euler flow) which gives us a $\mathbb{H}_1$ control on u. Thus we have enough compactness to obtain a global solution even in three dimensions. However, the uniqueness of such a solution is an open problem. See [Lad 69] and [Tem 84] for the existence theory of the Navier–Stokes flow.

As mentioned in Chapter 1 the Navier–Stokes flow is very different when dealing in domains with or without a boundary. In the case of the absence of a boundary, the solution for the two-dimensional Navier–Stokes initial value problem can be constructed along the same lines as those in Section 2.3. In fact, the Navier–Stokes equation in terms of vorticity reads as

$$D_t \omega = v \Delta \omega \qquad (7.19)$$

which can be interpreted as a Fokker–Plank nonlinear type of equation associated with the stochastic different equation

$$d\phi_t(x) = u(\phi_t(x))\, dt + \sqrt{2v}\, dw, \qquad (7.20)$$

where $v > 0$ denotes the viscosity coefficient, w is a standard Brownian motion, and $\phi_0(x) = x$ almost surely.

Equation (7.20) replaces the analogous equation (3.7) for the current lines, playing an important role in the construction of the solutions in two dimensions in the case of the Euler equation. It is not difficult to check that the strategy of Section 2.3 can be followed in exactly the same way to construct the solutions of the Navier–Stokes flow, by simply replacing the current lines by the stochastic analogue which is the solution of (7.20) (for more details see [MaP 82] and [MaP 84]).

Another possibility for constructing the Navier–Stokes flow is to use the splitting method. Denote by E_t and H_t the semiflows solution of the Euler equation and the heat equation, respectively. Then define an approximate solution of the Navier–Stokes equation by

$$\omega_t^n = (E_{t/n} H_{t/n})^n \omega_0. \qquad (7.21)$$

Since we know that the L_∞ norm of the vorticity is not growing in time for both of the two semiflows, we have an a priori control on $\|\omega_t^n\|_\infty$. This can be used to prove the convergence of ω^n as $n \to \infty$. We leave it as a nontrivial exercise for the reader.

When boundaries are present the problem is more difficult. We no longer control the L_∞ norm of the vorticity. Actually, the boundary can produce vorticity, which is a consequence of the interaction of the fluid with the wall. In this case other methods, based on energy inequality, can be used.

An interesting problem arising quite naturally from our discussion is the behavior of the Navier–Stokes solutions in the vanishing viscosity limit $v \to 0$. In the absence of a boundary it is easy to prove that $u^v \to u$ in L_2 when $v \to 0$. Here u^v and u denote the solutions of the Navier–Stokes problem and the Euler problem, respectively, both associated to the fixed initial value u_0. In fact, by using the equations it is easy to establish the following identity:

$$\frac{d}{dt}(u^v, u) = ((u - u^v) \cdot \nabla u, u - u^v) - v(\nabla u, \nabla u^v) \qquad (7.22)$$

from which

$$\left| \frac{d}{dt}(u^v, u) \right| \leq \|\nabla u\|_\infty \|u - u^v\|_2^2 + v\|\nabla u\|_2 \|\nabla u^v\|_2. \qquad (7.23)$$

On the other hand, in the case of the absence of boundaries, we have a uniform control (in v) of $\|\nabla u^v\|_2$, either in two dimensions for all times, or in three dimensions for short times (see the arguments in Section 2.1). Thus, by the inequality (7.23) (realizing that the time derivative of $\|u - u^v\|_2^2$ can be expressed in terms of the time derivative of (u^v, u)), we obtain the L_2 convergence.

The vanishing viscosity limit, in different contexts in two and three dimensions, is discussed in references [Gol 66], [McG 68], [Swa 72], [Bar 72], [Kat 72], [Kat 75], [MaP 84], and [EMP 88]. In the last two papers the proof is based on the convergence of stochastic processes describing the Navier–Stokes flow.

The behavior of the solutions in the vanishing viscosity limit when boundaries are present is much more complicated and not yet completely understood. A mathematical theory of the boundary layer problem is far from being achieved and an analysis of the existing results is beyond the purposes of the present book.

Appendix 2.1 (Integral Inequalities)

Proposition A.1. *Let* $u \in C([0, T]; \mathbb{R}^+)$ *and* $\varphi \in C(\mathbb{R}^+; \mathbb{R}^+)$ *be a nondecreasing function, such that*

$$u(t) \le u(0) + \int_0^t ds\, \varphi(u(s)), \qquad t \le T. \tag{A1.1}$$

Let $v = v(t)$ *be a solution of the initial value problem*

$$\frac{d}{dt} v = \varphi(v),$$

$$v(0) = u(0), \tag{A1.2}$$

that is, continuous with respect to the initial data. Then

$$u(t) \le v(t) \qquad for\ any\quad t \in [0, T]. \tag{A1.3}$$

Remark. If $\varphi(x) = kx$, $k > 0$, we have $u(t) \le u(0) \exp\{kt\}$. The proposition in this case takes the name of the Gronwall Lemma.

PROOF. Let $v_\varepsilon = v_\varepsilon(t)$ be a solution of the problem (A1.1) with initial datum $u(0) + \varepsilon$, $\varepsilon > 0$. By continuity, there exists a time $t^* > 0$ defined by the relation

$$t^* = \sup\{t \in [0, T] | u(t) < v_\varepsilon(t)\}. \tag{A1.4}$$

We want to show that $t^* = T$. In fact, if $t^* < T$, we would have

$$0 = v_\varepsilon(t^*) - u(t^*) \geq \varepsilon + \int_0^{t^*} ds[\varphi(v_\varepsilon(s)) - \varphi(u(s))] \geq \varepsilon > 0 \quad (A1.5)$$

which is absurd.

Then for every $t \in [0, T]$, $v_\varepsilon(t) > u(t)$. In the limit $\varepsilon \to 0$ we obtain relation (A1.3). $\qquad\square$

Appendix 2.2 (Some Useful Inequalities)

PROOF OF PROPOSITION 1.1. We start by proving (1.7). First, we observe that $\|f\|_\infty \leq \|f^\wedge\|_1$. Moreover, let $m \geq 2$

$$\|f^\wedge\|_1 = \sum_{k \in \mathbb{Z}^3 - 0} |f^\wedge(k)| + |f^\wedge(0)|$$

$$\leq \|f\|_1 + \sum_{k \in \mathbb{Z}^3 - 0} \frac{|f^\wedge(k)|\,|k|^m}{|k|^m}$$

$$\leq C\|f\|_2 + \left(\sum_{k \in \mathbb{Z}^3 - 0} |f^\wedge(k)|^2 |k|^{2m}\right)^{1/2} \left(\sum_{k \in \mathbb{Z}^3 - 0} \frac{1}{k^{2m}}\right)^{1/2}$$

$$\leq C(\|f\|_2 + \|D^m f\|_2 \leq C|f|_m. \quad (A2.1)$$

So (1.7) is proved.

We have

$$(D^\alpha(fg), D^\alpha(fg)) \leq \sum_k |k|^{2\alpha} \sum_{h_1;h_2} [f^\wedge(k - h_1)g^\wedge(h_1)]^*[f^\wedge(k - h_2)g^\wedge(h_2)],$$

$$(A2.2)$$

where, for notational simplicity, we denote $\alpha = |\alpha| = \sum \alpha_i$ when $|\alpha|$ appears as an exponent of $|k|$.

$$|k|^{2\alpha} = |k - h_1 + h_1|^\alpha |k - h_2 + h_2|^\alpha$$

$$\leq C_\alpha(|k - h_1|^\alpha |k - h_2|^\alpha + |h_1|^\alpha |k - h_2|^\alpha + |k - h_1|^\alpha |h_2|^\alpha + |h_1|^\alpha |h_2|^\alpha).$$

$$(A2.3)$$

Inserting the first term of this development in (A2.2) we have, for $\alpha \geq 2$,

$$\left| \sum_{k;h_1;h_2} |k - h_1|^\alpha [f^\wedge(k - h_1)]^* |k - h_2|^\alpha f^\wedge(k - h_2)[g^\wedge(h_1)]^* g^\wedge(h_2) \right|$$

$$\leq \sum_{h_1;h_2} |g^\wedge(h_1)|\,|g^\wedge(h_2)| \sum_k |k|^{2\alpha} |f^\wedge(k)|^2$$

$$\leq C(|g|_\alpha)^2 (|f|_\alpha)^2, \quad (A2.4)$$

where in the last step we have used (A2.1). The other terms can be studied in a similar way

$$\sum_{k;h_1;h_2} |h_1|^\alpha |f^\wedge(k-h_1)| |k-h_2|^\alpha |f^\wedge(k-h_2)| |g^\wedge(h_1)| |g^\wedge(h_2)|$$

$$= \sum_{k;r;h_2} |k-r|^\alpha |f^\wedge(r)| |k-h_2|^\alpha |f^\wedge(k-h_2)| |g^\wedge(k-r)| |g^\wedge(h_2)|$$

$$\leq \sum_{r;h_2} |g^\wedge(h_2)| |f^\wedge(r)| \left(\sum_k |k-r|^{2\alpha} |g^\wedge(k-r)|^2 \right)^{1/2}$$

$$\times \left(\sum_k |k-h_2|^{2\alpha} |f^\wedge(k-h_2)|^2 \right)^{1/2}$$

$$\leq C |f|_\alpha^2 |g|_\alpha^2. \tag{A2.5}$$

The term with $|k-h_1|^\alpha |h_2|^\alpha$ can be treated in exactly the same way. Finally

$$\sum_{k;h_1;h_2} |h_1|^\alpha |f^\wedge(k-h_1)| |h_2|^\alpha |f^\wedge(k-h_2)| |g^\wedge(h_1)| |g^\wedge(h_2)|$$

$$= \sum_{k;r_1;r_2} |k-r_1|^\alpha |f^\wedge(r_1)| |k-r_2|^\alpha |f^\wedge(r_2)| |g^\wedge(k-r_1)| |g^\wedge(k-r_2)|$$

$$\leq \sum_{r_1;r_2} |f^\wedge(r_2)| |f^\wedge(r_1)| \left(\sum_k |k-r|^{2\alpha} |g^\wedge(k-r)|^2 \right)$$

$$\leq C |f|_\alpha^2 |g|_\alpha^2. \tag{A2.6}$$

So (1.8) is proved.

From (A2.2), with $\alpha = 0$, we get

$$(|fg|_0)^2 \leq \left(\sum_h |g^\wedge(h)| \right)^2 (|f|_0)^2 \leq C(|g|_2)^2 (|f|_0)^2, \tag{A2.7}$$

where we have used, once more, (A2.1). So (1.9) is proved.

To prove (1.10) we use (A2.2) with $\alpha = 0$. We put

$$F(k) = f^\wedge(k)(1+k^2)^{1/2}, \qquad G(k) = g^\wedge(k)(1+k^2)^{1/2}.$$

Then

$$(|fg|_0)^2 \leq \sum_{k;h_1;h_2} \{ G(h_1)G(h_2)F(k-h_1)F(k-h_2)$$

$$\times \left(\frac{1}{1+(h_1)^2} \right)^{1/2} \left(\frac{1}{1+(h_2)^2} \right)^{1/2} \left(\frac{1}{1+(k-h_1)^2} \right)^{1/2} \left(\frac{1}{1+(k-h_2)^2} \right)^{1/2} \}$$

(using the inequality $ab \leq \frac{1}{2}(a^2 + b^2)$)

$$\leq \sum_{k;h_1;h_2} \left\{ [G(h_1)]^2 [F(k-h_1)]^2 \left(\frac{1}{1+(h_2)^2} \right) \left(\frac{1}{1+(k-h_2)^2} \right) \right.$$

$$\leq C \sum_{k;h_1} \{ [G(h_1)]^2 [F(k-h_1)]^2 \leq C \|G\|_2 \|F\|_2$$

$$= C |f|_1 |g|_1.$$

This concludes the proof of Proposition 1.1. $\square$

Appendix 2.3 (Quasi-Lipschitz Estimate)

We prove Lemma 3.1.

$$|K * \omega(x)| \leq \|\omega\|_\infty \int_D dy \, |K_D(x, y)|. \tag{A3.1}$$

Moreover

$$\int_D dy \, |K_D(x, y)| \leq C \int_D \frac{1}{|x - y|} \, dy < +\infty. \tag{A3.2}$$

So (3.9) is proved.

Let $r = |x - x'| < 1$ (If $r \geq 1$ then (3.10) is a consequence of (A3.2)) and $A = \{y \in D \,|\, |y - x| \leq 2r\}$. Then

$$\int_D |K_D(x, y) - K_D(x', y)| \, dy = \int_{D \cap A} \cdot + \int_{D \cap A^c} \tag{A3.3}$$

We consider the first integral. It is bounded by

$$C \int_{D \cap A} \left[\frac{1}{|x - y|} + \frac{1}{|x' - y|} \right] dy \leq C \int_{|x-y| \leq 3r} \frac{dy}{|x - y|} \leq Cr. \tag{A3.4}$$

To estimate the second integral we choose a convenient point x'' and observe that, if x'' belongs to the segment x, x', we have for $y \in A^c$, $|x'' - y| \geq \frac{1}{2}|x - y|$ and the second integral is bounded by

$$Cr \int_{D \cap A^c} |\nabla K_D(x'', y)| \, dy \leq Cr \int_{D \cap A^c} \frac{dy}{|x'' - y|^2}$$

$$\leq Cr \left\{ \int_{2r < |x-y| < 2} \frac{dy}{|x - y|^2} + \int dy \, \chi_D(y) \right\}. \tag{A3.5}$$

The proof is achieved by performing the integral. $\qquad\qquad\square$

Appendix 2.4 (Regularity Estimates)

Let D be a bounded open set with smooth boundary. Let $\omega \in L_\infty(D)$ be a given vorticity profile which is Hölder continuous of exponent $\beta > 0$. Then $u = K_D * \omega$ is continuously differentiable. Moreover, the derivatives are Hölder continuous of exponent $\beta' < \beta$.

To see the differentiability we use the following representation formula for the derivatives of u, which is valid when $\omega \in C^1(D)$,

$$(\partial_i u)(x) = -\omega(x) \int_{\partial D} n_i(y) K_D(x, y) \, dy$$

$$+ \int_D \{\omega(y) - \omega(x)\} (\partial_i K_D)(x, y) \, dy. \tag{A4.1}$$

The above formula is an immediate consequence of the following formula (see [CoH, Vol. II, p. 247]):

$$(\partial_i u)(x) = \int_{\partial D} n_i(y) K_D(x, y)\omega(y) - \int_D K_D(x, y)(\partial_i \omega)(y)\, dy \qquad (A4.2)$$

and a partial integration. Here $n_i(y)$ denotes the projection on the i direction of the outward normal n.

To prove that $\partial_i y$ is Hölder continuous we consider the difference

$$(\partial_i u)(x) - (\partial_i u)(x'). \qquad (A4.3)$$

The difference of the boundary terms in the right-hand side of (A4.1) is manifestly bounded by const. $|x - x'|^\beta$ (obviously the constant may diverge when x and x' approach the boundary). Let us fix $|x - x'| = r$. The contribution to the difference (A4.3) due to the integration on the domains $|y - x| < r$ and $|y - x'| < r$ can be bounded by

$$\text{const.} \int_{|x-y|<r} \frac{dy}{|x - y|^{2-\beta}} = \text{const. } r^\beta. \qquad (A4.4)$$

The rest of the difference has the following expression:

$$\int_{D \cap \{|x-y|>r\}} \{\omega(y) - \omega(x)\}(\partial_i K_D)(x, y)\, dy$$

$$- \int_{D \cap \{|x'-y|>r\}} \{\omega(y) - \omega(x')\}(\partial_i K_D)(x', y)\, dy$$

$$= \int_{D'} (\partial_i K_D)(x, y)\{\omega(y) - \omega(y - (x - x'))\}\, dy + a(r), \qquad (A4.5)$$

where $D' = D \cap \{D + (x - x')\} \cap \{|y - x| > r\}$ and $a(r) = O(r)$.

Finally, the first integral in the right-hand side of (A4.5) is estimated by (for suitable $A > 0$)

$$r^\beta \int_{D'} \frac{dy}{|x - y|^2} \leq \text{const. } r^\beta \int_r^A \frac{d\rho}{\rho} = \text{const. } r^\beta |\ln r|. \qquad (A4.6)$$

The above argument can be easily extended to prove that if $\omega \in C^{k+\beta}(D)$, k integer, and $\beta \in (0, 1)$, then $u \in C^{k+1+\beta'}(D)$ with $\beta' < \beta$. $\square$

EXERCISES

1. Prove that if u is a solution of the Euler equation in $D = [-\pi, \pi]^3$ with periodic conditions, then $\int \nabla \cdot (u \cdot \nabla)u = 0$.

2. Prove that in a two-dimensional motion, for every function F, $\int F(\omega)\, dx$ is a constant of motion.

3. In the hypothesis of Exercise 2 prove that the measure of the set $\{x \in D \,|\, \omega(x) > a\}$, $a \in \mathbb{R}$, is constant in time.

4. Consider the equation $\partial_t \omega = (\omega \cdot \nabla) u$. Write for the solution ω_t a representation formula of a form like (2.15).

5. Prove that u^n defined in (3.24) is jointly continuous in x and t. (*Hint*: By (3.10) it is enough to prove that

$$\lim_{t \to s} \sup_x |u^n(x, t) - u^n(x, s)| = 0.)$$

By considering the expression

$$\int \omega_0(y) \{ K_D(x, \phi_t^{n-1}(y)) - K_D(x, \phi_s^{n-1}(y)) \}$$

we split the integration into two parts: A and B. A is the set of all y's such that either

$$|x - \phi_s^{n-1}(y)| < \varepsilon \qquad \text{or} \qquad |x - \phi_t^{n-1}(y)| < \varepsilon.$$

B is the complementary set. For the integral on B we use the continuity (in t) $\phi_t^{n-1}(y)$ (which has been established in the previous inductive step). The integral on A is bounded by

$$\|\omega_0\|_\infty \left\{ \int_{B_s} K_D(x, y) + \int_{B_t} K_D(x, y) \right\},$$

where $B_s = \phi_s^{n-1}(B)$. Here we used the Liouville theorem. Finally, meas $B_s =$ meas $B \leq$ const. ε^2. Actually the worst possible case is when B_s is a circle around x so that the above estimate follows by direct inspection).

6. Prove that the approximate problems (3.23), (3.24), (3.25) make sense, by showing that $u^n \in C(D \times (-\infty, +\infty))$ and that the trajectories $\phi_t^n(x)$, $x \in D$, do not leave D in finite times. *Remark.* The trajectories, however, can arbitrarily approach the boundary: see the field given by (4.9) in Chapter 1.

7. By using (2.15) prove that the vorticity lines during motion go in vorticity lines.

8. Prove Theorem 3.2. (*Hint*: Prove (3.45) in integral form and then, by using the ideas in Exercise 5, prove that $t \to \omega_t(u_t \cdot \nabla f)$ is a continuous function.)

*9. Assume $\omega_0 \in L_1 \cap L_\infty$. Prove Theorem 3.1 in the case $D = \mathbb{R}^2$.

*10. Consider a two-dimensional motion and $\omega_0 \in \chi(A)$, A measurable. Prove that ω_t has the form $\omega_t = \chi(A_t)$ and characterize A_t.

*11. Assume in Theorem 3.1 that $\omega_0 \in L_p(D)$, $p \leq \infty$. For which p are we able to prove an existence and uniqueness theorem like Theorem 3.1?

12. Prove that the local solutions constructed in three dimensions are unique.

13. Prove that the convergence demonstrated in Theorem 5.1 can be made of polynomial order as large as we want. More precisely, prove that, if $\omega_0 \in C^\infty(D)$, for any $p \geq 1$

$$\sup_{0 \leq t \leq T} \|u_t - u_t^N\| \leq C_p(t)(N)^{-p}.$$

*14. Let D be a toroidal subset of $\mathbb{R}^3$ of the following type:

$$D = \{ x = (r, \theta, z) | (r, z) \in D_0 \quad 0 \leq \theta < 2\pi \},$$

where D_0 is a bounded regular domain of the plane not intersecting the z-axis. Prove the existence, for all times, of an axisymmetric flow $u = (u_z(z, r, t), u_r(z, r, t))$ satisfying the Euler equation, and discuss the uniqueness of such a solution. (*Hint*: Reduce the problem to a two-dimensional one with a new operator. Use the conservation of the quantity ωr^{-1} and a quasi-Lipschitz estimate on the Green function to mimic a proof similar to that presented in Section 2.3.)

CHAPTER 3

Stability of Stationary Solutions of the Euler Equation

In this chapter we investigate some qualitative properties of solutions of the Euler equation and, in particular, we give sufficient conditions for the stability of stationary flows and discuss some instabilities.

3.1. A Short Review of the Stability Concept

The stability theory poses a quite natural question: given an evolution equation we want to know whether a small perturbation of the initial condition produces effects which are uniformly small in time. The mere continuity of the solutions of an ordinary differential equation with respect to the initial data (this property is guaranteed under the reasonable hypotheses of regularity) ensures the smallness of the perturbation for finite times only. In general, the perturbation grows exponentially in time. We seek conditions for which this does not happen.

We begin our analysis by recalling classical results and considerations concerning ordinary differential equations, although fluids are described by partial differential equations. The reason for this is that many of the considerations we will develop extend to evolution equations in Hilbert or Banach Spaces. Moreover, the analysis of a simple problem may give useful insights in view of the study of a more complex one.

Consider an autonomous ordinary differential equation

$$\frac{d}{dt} x = F(x),$$

$$x(0) = x_0,$$

(1.1)

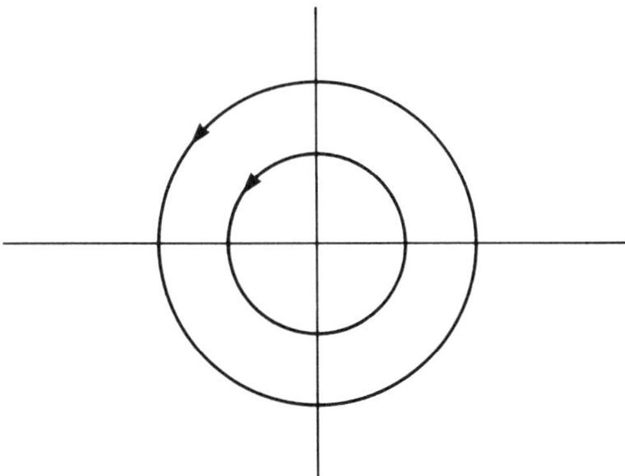

Figure 3.1

where $F: \mathbb{R}^n \to \mathbb{R}^n$ is a C^1 vector field. We will indicate by $x(t, x_0)$ the solution of (1.1). A point x^* is *stationary* (or an equilibrium point) if

$$x(t, x^*) = x^* \qquad \text{for all} \quad t > 0. \tag{1.2}$$

We say that x^* is a *critical point* of the vector field F is

$$F(x^*) = 0. \tag{1.3}$$

It is obvious that any critical point is stationary and the converse is also true.

The critical points may have a very different nature. Consider, for instance, the vector fields $F(x) = (-x_2, x_1)$ and $F(x) = (x_1, -x_2)$ drawn in Fig. 3.1 and in Fig. 3.2, respectively. Both of them have a critical point in 0. In the first case, any trajectory starting close to the origin stays indefinitely close to it. In the second case, all the trajectories, no matter how close they start to

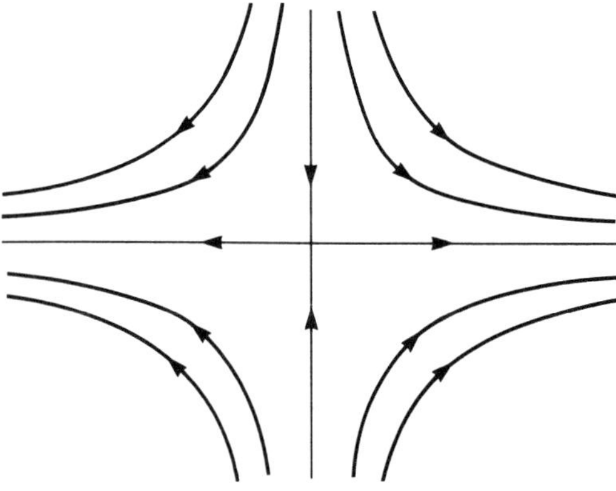

Figure 3.2

the origin, move away arbitrarily, unless they start from the manifold $x_1 = 0$. Let us suppose a physical system described by the two previous fields F.

We can easily imagine that it will turn out to be very difficult to realize, in practice, the stationary state of the system in Fig. 3.2. A small initial perturbation, due for instance to an experimental inaccuracy, will have a very large effect in the future, leading the system very far from the expected situation. On the contrary, in the first case, a small error in the realization of the initial condition will leave the system near the stationary state. These considerations should clarify how the practical realization of an equilibrium situation is related to "stability" with respect to initial perturbations of a stationary solution.

We give now a precise definition.

Definition 1.1. We say that a critical point x^* is *stable* if, for all $\varepsilon > 0$, there exists a $\delta > 0$ depending only on ε, such that the condition

$$|x_0 - x^*| < \delta \tag{1.4}$$

implies

$$\sup_{t \geq 0} |x(t, x_0) - x^*| < \varepsilon. \tag{1.5}$$

A critical point which is not stable is said to be *unstable*.

We remark that the above definition extends immediately to evolution equations in Banach spaces and this is exactly the situation in which we are interested in in fluid mechanics. There is, however, an important difference between the analysis in finite- and infinite-dimensional spaces which we want to stress now.

In R^n all the norms are equivalent to that of Definition 1.1 (i.e., $|x| = (\sum_{i=1}^{n} x^2)^{1/2}$) so that the stability notion is not dependent on the distance that we have chosen. In Banach spaces the situation is completely different. The existence of nonequivalent metrics makes Definition 1.1 norm-dependent. In fact, we have to specify a priori which kind of "vicinity" notion we are interested in. Later on we will show examples of stationary fluid flows which are stable with respect to some norms and unstable with respect to others. Obviously, the right choice of a norm in the study of a stability problem can follow from physical considerations only.

We finally notice that Definition 1.1 can be generalized in the sense that the norms appearing in (1.4) and (1.5) may be different. Actually it might be necessary, in order to have control of the solution at time t in a given norm, to express the vicinity of the initial perturbation in a stronger norm. Later on, we will give examples in which this happens.

There are possible generalizations of the stabiliy notion. For instance

Definition 1.2. Let $h: U \to R^n$ be a continuous function such that $h(x^*) = 0$ and $h(x) \geq 0$. Here U is a neighborhood of a critical point $x^* \cdot x^*$ is said

h-stable if, for all $\varepsilon > 0$, there exists $\delta > 0$ (depending only on ε) such that the condition $h(x_0) < \delta$ implies $\sup_{t \geq 0} h(x(t, x_0)) < \varepsilon$.

With this definition the origin in Fig. 3.1 is a point h-stable for $h(x) = x_2$. Therefore we have stabiliy with respect to some coordinate only. This stability concept is usually called conditional stability. We do not insist further on this point.

Let us come back to Definition 1.1. We want to find conditions ensuring the stability of a critical point x^* by an analysis of the vector field F. Let us consider the simplest case: F is a linear operator in R^2. Hence, let $F = A$ be the 2×2 matrix describing our system. Let λ_1 and λ_2 be the two eigenvalues of A (λ_1 and λ_2 are real or complex conjugate, see Exercise 2). From the analysis of the two eigenvalues we can say whether the origin is stable or not. Namely, we have the situation shown in Figs. 3.3–3.8.

(i) Real eigenvalues:

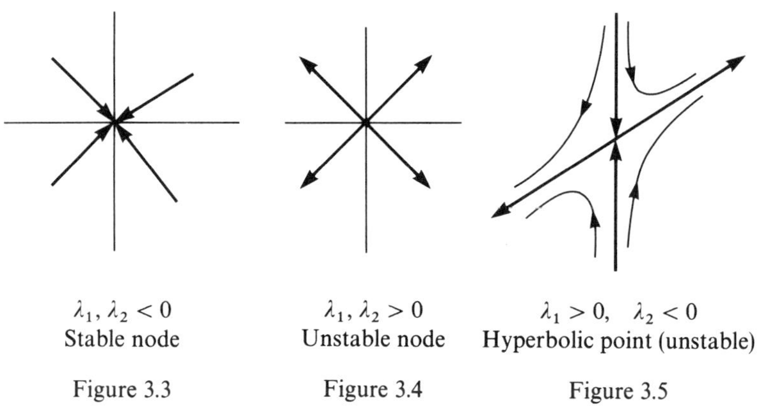

<table>
<tr><td align="center">$\lambda_1, \lambda_2 < 0$
Stable node</td><td align="center">$\lambda_1, \lambda_2 > 0$
Unstable node</td><td align="center">$\lambda_1 > 0, \quad \lambda_2 < 0$
Hyperbolic point (unstable)</td></tr>
<tr><td align="center">Figure 3.3</td><td align="center">Figure 3.4</td><td align="center">Figure 3.5</td></tr>
</table>

(ii) Complex eigenvalues: $\lambda \pm i\omega$

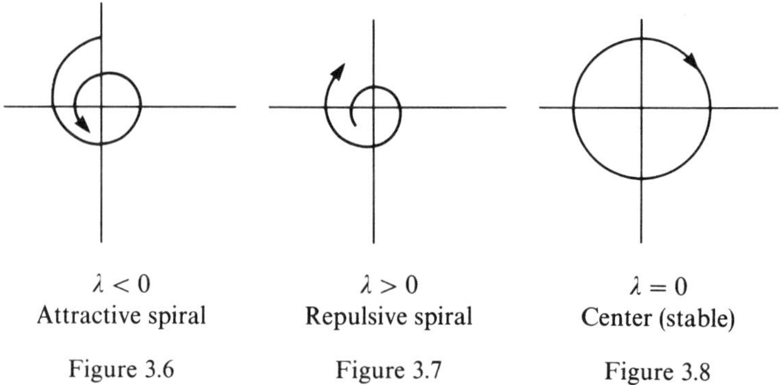

<table>
<tr><td align="center">$\lambda < 0$
Attractive spiral</td><td align="center">$\lambda > 0$
Repulsive spiral</td><td align="center">$\lambda = 0$
Center (stable)</td></tr>
<tr><td align="center">Figure 3.6</td><td align="center">Figure 3.7</td><td align="center">Figure 3.8</td></tr>
</table>

Definition 1.3. A critical point x^* is said to be *attractive* if there exists a neighborhood U of x^* such that for all $x_0 \in U$ then $\lim_{t \to \infty} x(t, x_0) = x^* \cdot U$ is called an *attraction basin*.

Definition 1.4. A critical point x^* is called *asymptotically stable* if it is stable and attractive.

In the cases considered in Fig. 3.3 the stable node and the attractive spiral are asymptotically stable, whereas the center is stable but it is not attractive. At first sight we might suspect that the attractivity is a stronger notion than the stability. This is not true: the attractivity is an asymptotic notion (for $t \to \infty$) while the stability must hold for all times (see Exercise 3).

In general, the system is not described by linear equations, i.e., F is a non-linear function. However, the analysis of the linear part of the system can give us good insights for the study of the stability problem. Without loss of generality, we can assume that the coordinate axes are fixed in such a way as to identify the origin with the critical point whose stability properties have to be investigated. In other words, we suppose that

$$F(0) = 0. \tag{1.6}$$

A natural way to approach the problem is to develop the vector field F by means of the Taylor formula, to test if the linearized equation can give us information on the full equation. We have

$$F_i(x) = \sum_j \frac{\partial F_i}{\partial x_j}(0)x_j + R_i(x). \tag{1.7}$$

Setting

$$A_{i,j} = \frac{\partial F_i}{\partial x_j}(0) \tag{1.8}$$

the evolution equation (1.1) becomes

$$\frac{dx}{dt} = Ax + R(x), \tag{1.9}$$

where

$$R(x) = O(x^2). \tag{1.10}$$

Let us consider now the linearized equation

$$\frac{dx}{dt} = Ax. \tag{1.11}$$

We want to see whether (1.11) is good approximation of (1.1). We expect such an approximation to be good when the initial point is chosen near the origin so that, at least initially, the remainder is small, and A has eigenvalues with negative real part. If so, the linear part has the tendency to bring the orbits closer and closer to the origin, while the nonlinear part will be less and less relevant during time. Therefore the linear part will be dominant forever. Such an heuristic argument is the basic content of the following theorem.

Theorem 1.1. *Let us consider problem (1.9) and suppose that all the eigenvalues of A have a negative real part. Then the origin is asymptotically stable.*

PROOF. From the hypotheses of Theorem 1.1 it follows that there exist two constants C and μ, such that (see Exercise 4)

$$|e^{At}x| \leq Ce^{-\mu t}|x|. \tag{1.12}$$

Then, from the integral equation

$$x(t) = e^{At}x_0 + \int_0^t e^{A(t-s)}R(x(s))\, ds \tag{1.13}$$

it follows, for $|x(t)| < 1$, that

$$|x(t)| \leq C\left(e^{-\mu t}|x_0| + \int_0^t e^{-\mu(t-s)}|R(x(s))|\, ds\right)$$

$$\leq Ce^{-\mu t}|x_0| + C_1 \int_0^t e^{-\mu(t-s)}|(x(s))|^2\, ds, \tag{1.14}$$

where

$$C_1 = CM \tag{1.15}$$

and M is such that, for $|x| < 1$,

$$|R(x)| \leq M|x|^2. \tag{1.16}$$

Let now T be defined by

$$T = \sup\{t\,|\,|x(t)| \leq \alpha\} \tag{1.17}$$

with $\alpha < 1$ to be determined later. We set

$$a = \sup_{0 \leq t \leq T} e^{(\mu/2)t}|x(t)|. \tag{1.18}$$

From (1.14) for $t \leq T$, we have

$$a \leq C|x_0| + \sup_{0 \leq t \leq T} C_1 \alpha a \int_0^t e^{-(\mu/2)(t-s)}\, ds$$

$$\leq C|x_0| + \frac{2C_1 \alpha a}{\mu}. \tag{1.19}$$

We set

$$\alpha = \frac{\mu}{4C_1} \tag{1.20}$$

and hence we obtain

$$a \leq 2C|x_0| \leq \alpha/2 \tag{1.21}$$

provided that

$$|x_0| \leq \frac{\alpha}{4C}. \tag{1.22}$$

From (1.21) it follows that $T = +\infty$ and

$$|x(t)| \leq Ce^{-(\mu/2)t}\frac{\alpha}{2} \tag{1.23}$$

for $|x_0|$ sufficiently small. Hence the theorem is proved. $\qquad\square$

On the contrary:

Theorem 1.2. *Suppose that A has an eigenvalue with a positive real part. Then the origin is unstable.*

PROOF. We start by considering the particular case when A is a diagonal matrix. Then $\mathbb{R}^n$ splits into the direct sum of two orthogonal subspaces E_1 and E_2, $\mathbb{R}^n = E_1 \oplus E_2$, the first corresponding to the positive eigenvalues and the other corresponding to the negative or null eigenvalues. We denote by $x = (\xi, \eta)$, $\xi \in E_1$, $\eta \in E_2$, the decomposition of the generic vector x according to this splitting, and by A_1 and A_2 the restrictions of the matrix A to these two subspaces.

We assume that the origin is stable, that is, for all positive ε, we can find $\delta > 0$ such that the condition $|x(0)| < \delta$ implies $|x(t)| < \varepsilon$ for all $t > 0$. We choose ε later on. Denoting by C the cone $C = \{x \in \mathbb{R}^n | |\xi| > |\eta|\}$ and by U the neighborhood $\{|x(0)| < \delta\}$, we see that $x(t)$ cannot leave C provided that $x(0) \in U$, and ε is small enough. In fact, setting

$$f(x) = |\xi|^2 - |\eta|^2, \tag{1.24}$$

we have

$$\frac{d}{dt}f(x) = 2\left(\frac{d}{dt}\xi, \xi\right) - 2\left(\frac{d}{dt}\eta, \eta\right)$$

$$= 2(A_1\xi, \xi) - 2(\eta, A_2\eta) + 2(R_1(\xi, \eta), \xi) - 2(\eta, R_2(\xi, \eta))$$

$$\geq 2a|\xi|^2 - 2C\varepsilon(|\xi|^2 + |\eta|^2)$$

$$\geq 2(a - 2C\varepsilon)|\xi|^2, \tag{1.25}$$

where R_i, $i = 1, 2$, are the restrictions of the remainder R to the two subspaces E_i, and a denotes the minimum positive eigenvalue. The lower bound (1.25) follows by the stability hypothesis.

On the other hand, we have:

$$\frac{1}{2}\frac{d}{dt}|x|^2 = (x, Ax + R(x)) \geq a|\xi|^2 - \varepsilon|x|^2 \geq \left(\frac{a}{2} - \varepsilon\right)|x|^2. \tag{1.26}$$

The last two inequalities are due to the stability hypothesis and the fact that $x(t) \in C$. The differential inequality (1.26) imposes that

$$|x(t)| \geq |x(0)| \exp\left\{\left(\frac{a}{2} - \varepsilon\right)t\right\} \tag{1.27}$$

from which we argue that $|x(t)|$ becomes arbitrarily large if $\varepsilon < a/2$. This contradicts the stability hypothesis and hence the origin is unstable.

The general case may be recovered by using a lemma of linear algebra, called the "canonical form theorem." It says that, given a matrix A, $\mathbb{R}^n$ can be decomposed as a direct sum $\mathbb{R}^n = E_1 \oplus E_2$ where E_1 and E_2 are two orthogonal subspaces, each of them invariant under the action of A. Denoting by A_1 and A_2 the restrictions of A to these two subspaces, we see that A_1 has eigenvalues with a positive real part and A_2 has eigenvalues with a negative or null real part. Moreover, denoting by a any number smaller than the positive real parts of the eigenvalues, we can find Euclidean metrics in E_1 and E_2 for which

$$(\xi, A_1 \xi) \geq a|\xi|^2. \tag{1.28}$$

In addition, for any positive b, we can find an Euclidean metric in E_2 such that

$$(\eta, A_2 \eta) < b|\eta|^2. \tag{1.29}$$

We choose $0 < b < a$, and the previous proof also applies to the general case with minor modifications. $\qquad\square$

The stability analysis we have done up to now by means of the linearized equation, leaves out the case in which all the eigenvalues have a vanishing real part. In this case, the nonlinear terms may play a relevant role in the study of the stability, and determine the nature af the critical point. Although the feature of having all eigenvalues with a zero real part seems very particular, there is a very important class of systems, the Hamiltonian systems, for which stability is achieved only if this circumstance is realized. We briefly recall that the Hamiltonian systems are defined through a function H, called Hamiltonian, for which the evolution is of the form

$$\frac{dp_i}{dt} = -\frac{\partial H}{\partial q_i}, \qquad \frac{dq_i}{dt} = \frac{\partial H}{\partial p_i}, \tag{1.30}$$

where $x = \{q_1, p_1, \ldots, q_N, p_N\} \in R^{2N}$. From (1.30) it follows that the divergence of the vector field is zero. This implies that the trace of the Jacobian matrix A is also zero, so that the sum of all the eigenvalues is vanishing. As a consequence, either all the eigenvalues have a zero real part, or there are necessarily eigenvalues with a positive real part. For these systems an analysis of the linear part can, at most, give us information about instability: the origin is unstable whenever there is an eigenvalue with a positive real part. If all the eigenvalues are imaginary we cannot say anything. For example, consider the system

$$\frac{dx}{dt} = \frac{\partial H}{\partial y} = x^2, \qquad \frac{dy}{dt} = -\frac{\partial H}{\partial x} = -2xy, \tag{1.31}$$

where $(x, y) \in R^2$ and $H = x^2 y$.

The linear part of the vector field, given by (1.31), is zero and therefore the origin is stable for the linearized dynamics. However, the system (1.31) may be easily integrated and we get

$$x(t) = \frac{1}{x(0)^{-1} - t}, \qquad y(t) = (1 - tx(0))^2 y(0), \tag{1.32}$$

from which we conclude, by direct inspection, that the origin is unstable.

On the other hand, the system given in Fig. 1.1 is Hamiltonian and also stable, but not asymptotically stable. In general, any Hamiltonian system cannot be asymptotically stable as follows by the Liouville theorem. We leave it as an exercise to give the details of the proof of this statement.

We finally remark that the proof of Theorem 1.1 extends, with some care, to Hilbert spaces. In this case, if the spectrum of the operator linear part of F lies in the negative half-plane without accumulating at the origin, then we can conclude that (1.12) holds and the proof of Theorem 1.1 carries through with minor modifications.

We conclude this study by showing that, even in the absence of stability, the linearized problem is still a good approximation of the full problem, under certain closeness assumptions at time zero, even on an arbitrary (but fixed a priori) interval of time.

Theorem 1.3. *Suppose* $y = y(t)$ *to be the solution of the linearized problem* (1.11) *and* $x = x(t)$ *to be the solution of the full problem* (1.9), *both with initial condition* x_0.

Given $T > 0$ *and* $\eta \in (0, 1)$, *there exists* $\delta \in \delta(\eta, T)$ *such that if* $|x_0| < \delta$, *then* $x(t)$ *is defined in* $[0, T]$ *and*

$$|x(t) - y(t)| < \delta^{1-\eta}. \tag{1.33}$$

PROOF. Set $|x_0| < \frac{1}{2}$. By continuity, there exists $T_1 = \inf\{t > 0 \,||\, |x(t)| > 1\}$. Furthermore, there exists $\delta_1 > 0$ such that, if $|x_0| < \delta_1$, $T_1 > T$. In fact, for $t \in (0, T_1)$, $|x(t)| < 1$, and from

$$x(t) = x_0 + \int_0^t [Ax(s) + R(x(s))] \, ds \tag{1.34}$$

it follows that:

$$|x(t)| \leq |x_0| + (C + M) \int_0^t |x(s)| \, ds, \tag{1.35}$$

where $C = \|A\|$ and M is defined by (1.16). From (1.35) we have

$$|x(t)| \le |x_0| e^{t(C+M)} \tag{1.36}$$

and hence, choosing $\delta_1 < \exp -\{(C + M)T\}$, we can conclude that $T < T_1$.
    Moreover

$$x(t) - y(t) = \int_0^t [A(x(s) - y(s)) + R(x(s))]\, ds \tag{1.37}$$

from which we obtain the following inequality:

$$|x(t) - y(t)| \le C \int_0^t |(x(s) - y(s)|\, ds + MT \sup_{s \le T} |x(s)|^2. \tag{1.38}$$

By using inequality (1.36) and the Gronwall lemma, we finally obtain

$$|x(t) - y(t)| \le 2MT|x_0|^2 \exp(CT + 2(C + M)T) \quad \text{for} \quad t \le T. \tag{1.39}$$

Therefore, given $\eta \in (0, 1)$, the theorem is proved by choosing

$$\delta\eta = \min\left(\frac{e^{-3(C+M)T}}{MT}, \delta_1\eta\right). \tag{1.40}$$

$\square$

We conclude this section by showing an intrinsically nonlinear stability criterion based on the existence of a particular function, called the Liapunov function, which plays the role of controlling the motion.

Definition 1.5. We call the Liapunov function any continuous positive function $L: \mathbb{R}^n \to \mathbb{R}$ vanishing only at the critical point x^* such that it is non-increasing along the trajectories, i.e.,

$$L(x(t_2)) \le L(x(t_1)) \quad \text{for} \quad t_2 > t_1. \tag{1.41}$$

Theorem 1.4. *Let us suppose that a Liapunov function exists. Then x^* is stable.*

PROOF. The proof is almost trivial and can be done by "reductio to absurdum." Suppose x^* unstable. Then we can find $\varepsilon > 0$ and $t^\wedge$ such that

$$|x(t^\wedge) - x^*| = \varepsilon \tag{1.42}$$

no matter how small δ is for which

$$|x_0 - x^*| \le \delta. \tag{1.43}$$

On the other hand, in the domain $|x - x^*| = \varepsilon$, L which is continuous, assumes an absolute minimum, $m > 0$. It cannot be zero because L vanishes only at the point x^*.

    By continuity we can find δ so small for that $L(x) < m$ for $|x - x^*| < \delta$, since $L(x^*) = 0$. In conclusion, L cannot be nondecreasing along the trajectories and the proof is achieved. $\square$

As a corollary of the above theorem we have the well-known Dirichlet theorem.

Theorem 1.5. *For a mechanical system an equilibrium point x^*, which is a strict local minimum of the potential energy, is stable.*

Remark. A mechanical system is described by first-order equations in the phase space (the set of all pair positions and velocities), so that the notion of stability to which we refer in the above theorem is the following. The trajectory $\{x(t), v(t)\}$ is arbitrarily close to the stationary solution $\{x^*, 0\}$ provided that $|x(0) - x^*|$ and $v(0)$ are sufficiently small.

PROOF OF THEOREM 1.5. As a Liapunov function we take the total energy which is a constant of motion. Moreover, assuming that the potential energy vanishes at x^*, the total energy is zero in a neighborhood of $\{x^*, 0\}$, only in this point. $\qquad\square$

Theorem 1.4 is not directly applicable to the case of infinite-dimensional Banach spaces because, in this case, the surface $|x - x^*| = \varepsilon$ is no longer compact, so that the Liapunov function may not have a minimum. This difficulty can be overcome by modifying the definition of the Liapunov function in such a way that Theorem 1.4 still works with minor modifications. We give an example.

Theorem 1.6. *Consider the differential system* (1.1) *in a Banach space B, and assume that 0 is a critical point of F. Suppose that there exists a Liapunov function L satisfying, in addition, the following inequalities:*

$$a(\|x\|) \leq L(x) \leq b(\|x\|), \qquad x \in B, \quad \|\cdot\| \text{ norm in } B, \qquad (1.44)$$

where a, b are two continuous nondecreasing functions vanishing only at the origin. Then 0 is stable. Moreover if, for a suitable positive constant c,

$$\frac{d}{dt} L(x(t)) \leq cL(x(t), \qquad (1.45)$$

then 0 is also asymptotically stable.

PROOF. The initial size of L is controlled by b, while the growth of $\|x(t)\|$ is controlled by a. Moreover (1.45) implies an exponential decrease of L. Combining these observations, it is easy to achieve a complete proof, details of which we leave to the reader. $\qquad\square$

We notice that Theorem 2.2 below, whose demonstration is explicitly given, is an application of the present theorem.

Obviously, in practical situations the difficult point is finding a Liapunov function for which (1.45) holds. However, the technique of the Liapunov function is widely used in practice. The efforts of researchers in the field are devoted to find, by means of experience and fantasy, functions which decrease along the trajectories, with the hope that they might eventually imply some stability conditions which are useful for the problem at hand. The general strategy of finding ad hoc approaches, case by case, is probably the most effective in situations in which the nonequivalence of the norms makes the stability problem not univocally posed. As we will show in the next sections, the construction of suitable Liapunov functions will allow us to prove the stability of some stationary flows. It is remarkable how a direct nonlinear approach is easier and more powerful than any procedure based on linearization. In addition, we want to stress that the linearization method cannot give conclusive results for the stability of the solutions of the Euler equation due to the Hamiltonian structure of the system (see the consideration developed in Chapter I, Section 1).

3.2. Sufficient Conditions for the Stability of Stationary Solutions: The Arnold Theorems

In this section we will give sufficient conditions for the stability of some stationary two-dimensional flows. Our analysis reduces to two-dimensional flows for two reasons. The first is that we do not know of the existence of three-dimensional solutions starting, at time zero, close to stationary flows. The second is that, even assuming the existence and regularity of such solutions, the three-dimensional motion is so complicated that general techniques devoted to understanding the qualitative behavior of the motion are not known. Therefore we confine ourselves mainly to the study of two-dimensional flows.

As we have already remarked at the end of the previous section, the study of the linearized system can give, at most, information on the instability of the system. In fact, the Euler equation defines an infinite-dimensional Hamiltonian system. We could make this statement more precise, however, an analysis of this type, of an essentially geometric nature, is beyond the scope of this book. For the moment it is enough to realize that, in the study of stationary Euler flows, the nature of the nonlinear terms is absolutely essential in providing conditions ensuring the stability of such flows. Therefore our approach will be based on the method of the Liapunov function.

Let us begin our analysis by considering the particularly simple case of irrotational flows in three dimensions. In a bounded domain D we know that there exists only one flow with such a property, the trivial one: $u = 0$. By the energy conservation theorem we obtain the stability of this solution in the L_2 norm of the velocity field. Such a result, however, is not very meaningful. The theory we have developed in Chapter 2 does not guarantee the existence of a solution in $L_2(D)$. Even possessing such a solution, the control we get from the energy conservation law is not very much. We could imagine solutions u_t with very small energy, but with very large gradients so that they are very different from the trivial solution. In two dimensions the situation changes. The conservation laws for the vorticity field allow us to prove the following theorem.

Theorem 2.1. *Let D be a regular domain in $\mathbb{R}^2$ and let $u^* = u^*(x)$ be an irrotational flow. Then u^* is stable with respect to the $L_p(D)$ norm of the vorticity field.*

PROOF. The proof is immediate. It follows by the conservation of the $L_p(D)$ norm of the vorticity field. $\qquad\qquad\square$

We remark that we have not considered the energy in Theorem 2.1 because, for unbounded domains, the irrotational flows (the only nontrivial flows which can be constructed) have infinite energy. For bounded domains we could add the energy to the $L_p(D)$ norm of the vorticity field.

We do not insist further on the irrotational flows, which are trivial from the point of view of stability problems. We begin to deal with the more difficult and interesting problem of the stability of rotational flows by considering a classical problem: the flow in a channel D with a periodic boundary condition in one direction. The domain is

$$D = [0, L] \times [-A, A], \qquad (x_2, 0) = (x_2, L). \qquad (2.1)$$

On the boundary $x_2 = \pm A$ the usual impermeability conditions are assumed: $u_2 = 0$ (Fig. 3.9).

We consider a stationary solution of the form

$$u^* = (u_1, 0), \qquad u_1 = u_1(x_2), \qquad (2.2)$$

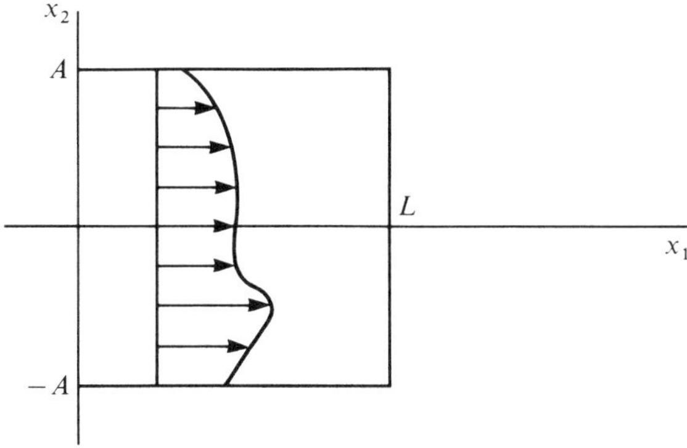

Figure 3.9

to which is associated a vorticity profile

$$\omega^* = -\partial_2 u_1. \tag{2.3}$$

We want to know under what conditions solution (2.2) is stable.

The study of this problem is old and has been approached, in the last century, by means of the linearization technique. In particular, as we will see later, this approach implies a necessary condition for the instability (called the Rayleigh condition, see Section 3.4 below) of the associated linearized system, imposing on the profile u_1, at least an inflexion point. As a consequence, it is often improperly claimed that if u_1'' is always different from zero (the vorticity profile is monotone), then the solution under consideration is stable. We say improperly because the linearization techniques give us the Rayleigh condition only when the perturbation grows exponentially in time, while, in principle, we could have polynomial instability, without satisfying the Rayleigh condition. As a consequence, as regards stability, we can only say that if the vorticity profile is monotone, then either the system is stable, or it is unstable with at most a polynomial growth in the perturbation. The second possibility is, however, excluded by the nonlinear analysis which follows. Our purpose is to provide a real proof of this fact and, at the same time, to find sufficient conditions for the stability of stationary flows which hold in more general situations. Obviously, it would also be interesting to find conditions necessary for the stability of such flows, in view of a complete characterization. This could be done by using the linear theory which could give, at least in principle, sufficient conditions for the instability and hence the conditions necessary for the stability. However the situation is not so simple. As we will see in Section 3.4, there are very few cases in which the instability analysis can be carried out, so that, until now, there are many cases still unsolved.

We will now give an important criterion due to Arnold

Theorem 2.2. *Let D be a bounded regular domain with smooth boundaries $\{\Gamma_i\}_{i=0\ldots n}$ in $\mathbb{R}^2$ (Fig. 3.10). Let $u^*: D \to R^2$, $u^* \in C^3(D)$, be a stationary solution to the Euler equation. Suppose that there exist two positive constants c_1 and c_2 such that*

$$c_1 \leq -\frac{u^*}{\nabla^\perp \omega^*} \leq c_2, \tag{2.4}$$

where $\omega^* = \mathrm{curl}\, u^*$. Then u^* is stable in the norm

$$\|u\|_2 + \|\mathrm{curl}\, u\|_2. \tag{2.5}$$

Remark 1. The ratio appearing in (2.4) makes sense because, by virtue of the stationarity, the vector field u^* and $\nabla^\perp \omega^*$ are collinear since u^* and $\nabla \omega^*$ are orthogonal.

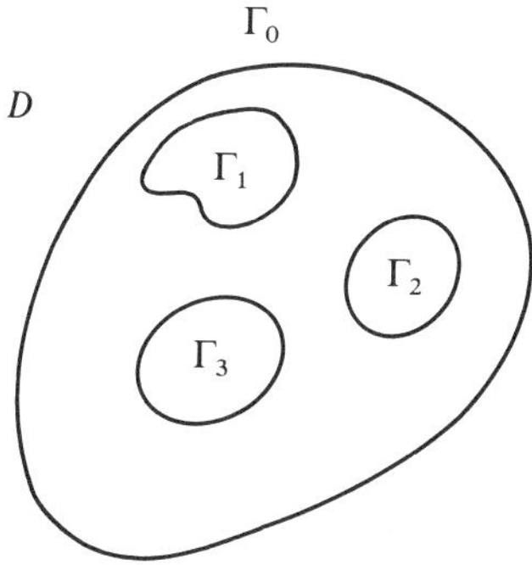

Figure 3.10

Remark 2. The existence theory developed in Chapter 2 requires that $\omega_0 \in L_\infty(D)$. To be more precise, the result of Theorem 2.2 should be formulated in the following way. For all $\varepsilon > 0$ there exists $\delta = \delta(\varepsilon)$ such that the conditions

$$\omega_0 \in L_\infty(D) \qquad \text{and} \qquad \|u_0 - u^*\|_2 + \|\omega_0 - \omega^*\|_2 < \delta, \tag{2.6}$$

(where we have posed $\omega_0 = \operatorname{curl} u_0$) imply

$$\|u_t - u^*\|_2 + \|\omega_t - \omega^*\|_2 < \varepsilon, \tag{2.7}$$

where u_t and ω_t are the velocity and vorticity field obtained by solving the Euler equation with initial datum given by u_0 and ω_0.

Before giving the proof of Theorem 2.2 we first discuss the underlying heuristic idea. Suppose we find a first integral $H(u)$ having a minimum (or also a maximum) at the stationary point u^*. Without loss of generality, we assume that $H(u^*) = 0$. A perturbed trajectory u_t will have the property that $H(u_t)$ will be small and constant during the motion. This gives us a control on the trajectory u_t in a norm which is the positive, quadratic form which arises in the development of H around u^*

$$H(u_t) = \tfrac{1}{2}Q(u_t - u^*)\ldots \tag{2.8}$$

(the first and second term in the expansion vanish because u^* is a stationary point in which H is vanishing and Q is positive because u^* is a minimum (Fig. 3.11)).

Therefore we seek such an H. The most general first integral is the following:

$$H(u) = \frac{1}{2}\int_D u^2 + \int_D \phi(\omega) + \sum_{i=0}^n a_i \oint_{\Gamma_i} u \cdot dl, \tag{2.9}$$

where $\omega = \operatorname{curl} u$, ϕ is a real measurable function, and $a_i \in \mathbb{R}$. ϕ and a_i are for the moment arbitrary, and will be determined on the basis of the condition that u^* is a minimum point for H. Proceeding formally, we compute the first

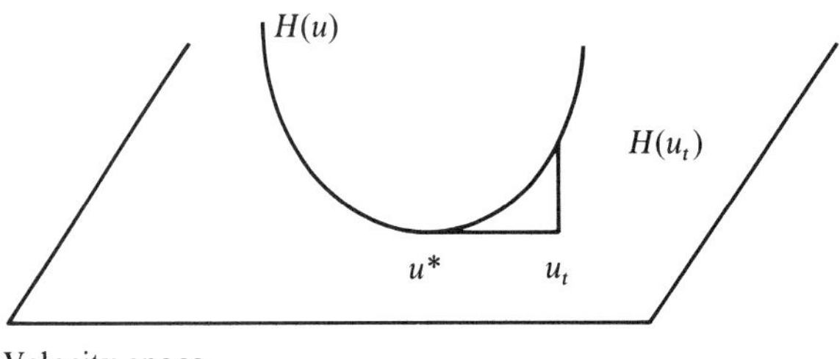

Figure 3.11

variation of H

$$\delta H(u^*) = \int_D u^*\delta u + \int_D \phi'(\omega^*)\delta\omega + \sum_{i=0}^{n} a_i \oint_{\Gamma_i} \delta u \cdot dl, \tag{2.10}$$

where δu is a divergence-free arbitrary variation and $\delta\omega = \text{curl } \delta u$.

From the relation

$$\phi'(\omega^*)\delta\omega = \text{curl}(\phi'(\omega^*)\delta u) + \phi''(\omega^*)\nabla^{\perp}\omega^* \cdot \delta u \tag{2.11}$$

we have that the condition

$$\delta H(u^*) = 0 \tag{2.12}$$

is verified, provided that

$$u^* = -\phi''(\omega^*)\nabla^{\perp}\omega^*, \tag{2.13}$$

$$a_i \oint_{\Gamma_i} \delta u \cdot dl = \oint_{\Gamma_i} \phi'(\omega^*)\delta u \cdot dl, \tag{2.14}$$

(here we have used the Stokes theorem)

Since u^* is stationary, ω^* must be constant on the boundary of D (in fact, $\nabla\omega^*$ is orthogonal to the boundary) so that the condition (2.14) is satisfied by putting

$$a_i = -\phi'(\omega^*)|_{\Gamma_i} \tag{2.15}$$

so that our major job is to find a function ϕ satisfying the condition (2.13).

Introducing the stream function ψ^* associated to the velocity field u^* ($u^* = \nabla^{\perp}\psi^*$), we see that $\nabla\psi^*$ and $\nabla\omega^*$ are collinear vector fields. Hence there exists a functional relation between ω^* and ψ^* because ω^* and ψ^* have the same equipotential lines

$$\psi^* = F(\omega^*). \tag{2.16}$$

We notice that, in general, F is multivalued. By (2.16)

$$u^* = \nabla^{\perp}\psi^* = F'(\omega^*)\nabla^{\perp}\omega^* \tag{2.17}$$

from which

$$\phi''(\omega^*) = -F'(\omega^*) \tag{2.18}$$

and hence ϕ is an indefinite integral of F. In conclusion, we can find ϕ and a_i

in such a way that (2.13) and (2.14) are satisfied. The second variation of the functional H follows immediately from (2.10):

$$\delta^2 H(u^*) = \int_D (\delta^2 u) + \int_D \phi''(\omega^*)(\delta^2 \omega). \tag{2.19}$$

This expression is positive whenever the condition (2.4) of the theorem is satisfied, as follows by (2.17) and (2.18). Thus H, with the above choice of ϕ and a_i, has a minimum in u^* which is expected to be stable in the norm (2.5). This norm arises from the form of the second variation (2.19). We now make rigorous the above considerations.

PROOF OF THEOREM 2.2. Let H be definite in (2.9) with a choice of ϕ and a_i such that conditions (2.13) and (2.14) are satisfied. From our hypotheses

$$0 < c_1 \le \phi''(\omega^*) \le c_2 < +\infty. \tag{2.20}$$

We extend ϕ in all R^1, outside the range of ω^* in which it is initially defined, in such a way as to be a smooth function satisfying the condition (2.20). Hence

$$H(u) - H(u^*) = \frac{1}{2} \int_D (u^2 - u^{*2}) + \int_D (\phi(\omega) - \phi(\omega^*)) + \sum_{i=0}^{n} a_i \int_{\Gamma_i} (u - u^*) \cdot dl$$

$$= \frac{1}{2} \int_D (u - u^*)^2 + \int_D u^* \cdot (u - u^*) + \int_D \phi'(\omega^*)(\omega - \omega^*)$$

$$+ \frac{1}{2} \int_D \phi''(\xi)(\omega - \omega^*)^2 + \sum_{i=0}^{n} a_i \int_{\Gamma_i} (u - u^*) \cdot dl, \tag{2.21}$$

where $\xi : D \to R^1$ is a suitable function arising from the expansion of ϕ up to the second order. By virtue of (2.13) and (2.14) we have

$$H(u) - H(u^*) = \frac{1}{2} \int_D (u - u^*)^2 + \frac{1}{2} \int_D \phi''(\xi)(\omega - \omega^*)^2 \tag{2.22}$$

so that, by (2.20), we can find two constants α and β for which

$$\alpha(\|u - u^*\|_2^2 + \|\omega - \omega^*\|_2^2) \le H(u) - H(u^*)$$

$$\le \beta(\|u - u^*\|_2^2 + \|\omega - \omega^*\|_2^2). \tag{2.23}$$

Making use of the invariance of H and of the inequalities (2.23), the stability result is easily achieved

$$(\|u_t - u^*\|_2^2 + \|\omega_t - \omega^*\|_2^2) \le \alpha^{-1}(H(u_t) - H(u^*))$$

$$\le \alpha^{-1}(H(u_0) - H(u^*)) \tag{2.24}$$

$$\le \alpha^{-1}\beta(\|u_0 - u^*\|_2^2 + \|\omega_0 - \omega^*\|_2^2). \qquad \square$$

The above theorem can be applied to the case of the periodic channel we have discussed so far. For stationary solutions of the type (2.3), condition

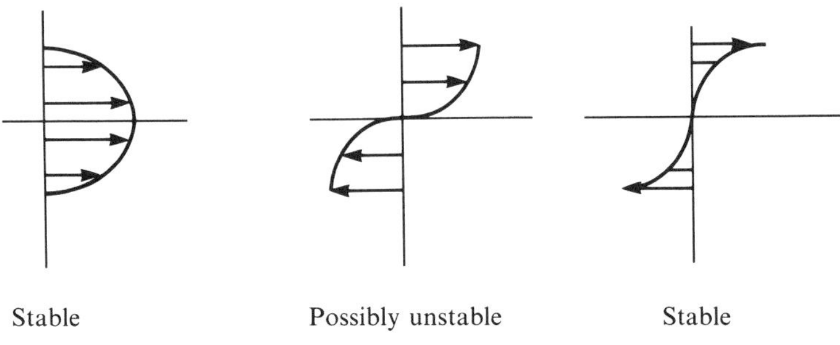

Figure 3.12

(2.4) becomes

$$0 \le c_1 \le \frac{u_1}{u_1''} \le c_2. \tag{2.25}$$

Since u_1 may always be chosen positive (if not, we can change the reference frame by a translation in such a way as to add a suitable constant factor to u_1), condition (2.25) is satisfied if u_1'' is strictly positive (or negative). Thus we have the Rayleigh condition requiring the absence of inflexion points. However condition (2.25) is more general. There are cases in which there is a unique inflexion point and the zero of u_1'' is suitably compensated by a zero of the velocity field u_1 (Fig. 3.12). As we will see in Section 3.4, this feature may also be found in the analysis of the linear stability.

The next theorem, also due to Arnold, deals with the case in which the ratio $-u^*/\nabla^\perp \omega^*$ is negative.

Theorem 2.3. *Under the same hypotheses of Theorem 2.2, let c_1 and c_2 be two constants for which*

$$0 < c_1 \le \frac{u^*}{\nabla^\perp \omega^*} \le c_2 < +\infty. \tag{2.26}$$

Then u^* is stable in the norm (2.5), provided that the inequality

$$\|\nabla \varphi\|_2^2 \le \alpha \|\Delta \varphi\|_2^2 \tag{2.27}$$

is verified for all $\varphi \in C^2(D)$ with $c_1 > \alpha > 0$.

PROOF. Proceeding as in the proof of Theorem 2.2 (see (2.22)), we arrive at the inequalities

$$-\tfrac{1}{2}\|u - u^*\|_2^2 + \frac{c_2}{2}\|\omega - \omega^*\|_2^2 \ge H(u^*) - H(u)$$

$$\ge -\tfrac{1}{2}\|u - u^*\|_2^2 + \frac{c_1}{c}\|\omega - \omega^*\|_2^2. \tag{2.28}$$

Denoting by φ_t the stream function relative to the perturbation $u_t - u^*$, i.e.,

$\nabla^{\perp}\varphi_t^* = u_t - u^*$, we have, for all t

$$\frac{(c_1 - \alpha)}{2}\|\Delta\varphi_t\|_2^2 \leq H(u^*) - H(u_t) \leq \frac{c_2}{2}\|\Delta\varphi_t\|_2^2 \tag{2.29}$$

from which

$$\frac{(c_1 - \alpha)}{4}\|\Delta\varphi_t\|_2^2 + \frac{(c_1 - \alpha)}{4\alpha}\|\nabla\varphi_t\|_2^2 \leq H(u^*) - H(u_t) \leq \frac{c_2}{2}(\|\Delta\varphi_t\|_2^2 + \|\nabla\varphi_t\|_2^2). \tag{2.30}$$

By using the invariance of H and the inequalities (2.29), we easily achieve stability in the norm (2.5). $\qquad\square$

Remark 1. In geometrical terms, in Theorem 2.3, H has a maximum in u^*, whereas in Theorem 2.2, H has a minimum.

Remark 2. Condition (2.27) is relative to the lowest eigenvalue of the Laplace operator. The value of α in (2.27) depends on the form and size of the domain D.

We give an example. Consider the two-dimensional flat torus $D = [0, L] \times [0, 2\pi]$, and the stationary velocity field in D

$$u^* = (\sin x_2, 0). \tag{2.31}$$

The ratio $u^*/u^{*''} = -1$ suggests the application of Theorem 2.3. Therefore we have to see whether the condition

$$\|\nabla\varphi\|_2^2 < \|\Delta\varphi\|_2^2 \tag{2.32}$$

is verified. We assume, without loss of generality, that $\int_D \varphi = 0$. In terms of the Fourier transform the above inequality becomes

$$\sum_{(k_1, k_2) \in Z}\left(k_2^2 + \left[\frac{2\pi}{L}k_1\right]^2\right)|\varphi^{\wedge}(k_1, k_2)|^2$$

$$< \sum_{(k_1, k_2) \in Z}\left(k_2^2 + \left[\frac{2\pi}{L}k_1\right]^2\right)^2|\varphi^{\wedge}(k_1, k_2)|^2. \tag{2.33}$$

For $L > 2\pi$, condition (2.33) is not verified. Also, for $L \leq 2\pi$, such a condition is not verified for all φ. Actually, some terms of the Fourier series give equality instead of the inequality we would need for stability. Such terms are

For $L = 2\pi$,

$$\varphi(x_1, x_2) = \sin x_1, \cos x_1, \sin x_2, \cos x_2. \tag{2.34}$$

For $L < 2\pi$, we have at least

$$\varphi(x_1, x_2) = \sin x_2, \cos x_2. \tag{2.35}$$

It follows that the stability condition (2.33) holds only in the subspaces orthogonal to those generated by (2.34) and (2.35). This implies the following result. If the solution is close at time zero to that subspace, then it will remain close, uniformly in time. Hence the subspaces (2.34) and (2.35) are stable. Nevertheless, the problem of stability of single stationary solutions belonging to these subspaces remains open. However, if we limit ourselves to perturbations in this subspace, then conditional stability is ensured.

We now discuss the range of applicability of Theorem 2.2. We will show that condition (2.4) cannot be verified in domains without boundary. In fact, from the collinearity of u^* and $\nabla^\perp \omega^*$, follows the existence of a function G such that

$$\Delta \psi^* = G(\psi^*). \tag{2.36}$$

Let us introduce the functional H of Theorem 2.2. The condition $\delta H = 0$ imposes

$$\phi''(\omega^*) = \frac{1}{G'(\psi^*)} \tag{2.37}$$

and hence it must be

$$0 < c_1' \le G'(\psi^*) \le c_2' < +\infty. \tag{2.38}$$

On the other hand, from (2.36),

$$\partial_{x_1} \Delta \psi^* = G'(\psi^*) \partial_{x_1} \psi^* \tag{2.39}$$

from which

$$\int_D \partial_{x_1} \psi^* \Delta \partial_{x_1} \psi^* = \int_D G'(\psi^*)(\partial_{x_1} \psi^*)^2. \tag{2.40}$$

By integrating by parts

$$\int_{\partial D} \partial_{x_1} \psi^* \nabla \partial_{x_1} \psi^* \cdot n - \int_D (\nabla \partial_{x_1} \psi^*)^2 = \int_D G'(\psi^*)(\partial_{x_1} \psi^*)^2. \tag{2.41}$$

We now analyze formula (2.41), and exclude the case in which $\partial_{x_1} \psi^*$ is identically zero (otherwise, we could repeat the argument replacing $\partial_{x_1} \psi^*$ by $\partial_{x_2} \psi^*$; the trivial case $\psi^* = $ const. is not considered here). We suddenly realize that the absence of the boundary term leads us to a contradiction: the second and third term in expression (2.41) have opposite signs, so that such an identity cannot be realized.

This remark excludes Theorem 2.2 from being applied to the domain $D = R^2$, which is of some interest in oceanographic and geophysic applications. On the other hand, we also notice that Theorem 2.3 cannot be applied to unbounded domains. In fact, in this case, $-\Delta$ does not have a minimum positive eigenvalue, strictly larger than zero, so that condition (2.27) cannot be realized. As a conclusion, the important domain $D = R^2$ is excluded by the considerations in this section. However, we will see in the next section how the difficulty arising from the application of Theorems 2.2 and 2.3 can be overcome by exploiting the symmetry properties of the domain D whenever they are present.

We conclude this section by extending the range of the validity of Theorem 2.2.

Theorem 2.4. *In the hypotheses of Theorem 2.2, the stability result is achieved if condition (2.4) holds with $c_1 \geq 0$.*

PROOF. The proof is based on the use of a family of Liapunov functions. We sketch the proof outlining only the differences from the proof of Theorem 2.2.

Let us define

$$H_\gamma(u) = \frac{1}{2}\int_D u^2 + \int_D \phi_\gamma(\omega) + \sum_{i=0}^n a_i \oint_{\Gamma_i} u \cdot dl \tag{2.42}$$

and choose ϕ_γ in such a way that $\delta H_\gamma(u^*)$ is not zero but proportional to γ, $\gamma \in [0, \frac{1}{2}]$,

$$\delta H_\gamma(u^*) = \gamma \int_D \nabla^\perp \omega^* \delta u. \tag{2.43}$$

Proceeding as in Theorem 2.2, we arrive at the condition

$$u^* = -\phi''(\omega^*)\nabla^\perp \omega^* + \gamma \nabla^\perp \omega^* \tag{2.44}$$

and hence

$$\phi''(\omega^*) = -\frac{u^*}{\nabla^\perp \omega^*} + \gamma. \tag{2.45}$$

On the other hand

$$Q(\delta u) := \delta^2 H(u^*) = \int_D (\delta^2 u) + \int_D \phi''(\omega^*)(\delta^2 \omega) \geq 0 \tag{2.46}$$

from which, proceeding as before, we obtain

$$H_\gamma(u) - H_\gamma(u^*) = \int_D \nabla^\perp \omega^* \delta u + \tfrac{1}{2}Q(u - u^*) + \cdots \tag{2.47}$$

so that the following inequalities hold:

$$|H_\gamma(u) - H_\gamma(u^*)| \leq \|u - u^*\|_2^2 + (c_2/2)\|\omega - \omega^*\|_2^2 + \gamma\|\nabla^\perp \omega^*\|_2 \|u - u^*\|_2 \tag{2.48}$$

$$|H_\gamma(u) - H_\gamma(u^*)| \geq \|u - u^*\|_2^2 + (\gamma/2)\|\omega - \omega^*\|_2^2 - \gamma\|\nabla^\perp \omega^*\|_2 \|u - u^*\|_2$$

$$= \{1 - (\gamma/2)\} \|u - u^*\|_2^2 + (\gamma/2)\{\|u - u^*\|_2^2 + \|\omega - \omega^*\|_2^2\}$$

$$- \gamma\|\nabla^\perp \omega^*\|_2 \|u - u^*\|_2. \tag{2.49}$$

Here we have used the Cauchy–Shwarz inequality and the bound $\phi'' \geq \gamma$ which follows from (2.45). Minimizing the sum of the first and third term in the right-hand side of the lower bound (2.49), we obtain

$$\{1 - (\gamma/2)\} \|u - u^*\|_2^2 - \gamma\|\nabla^\perp \omega^*\|_2 \|u - u^*\|_2 \geq -\frac{\gamma^2 \|\nabla^\perp \omega^*\|_2^2}{2(1 - \gamma)} \tag{2.50}$$

so that

$$|H_\gamma(u) - H_\gamma(u^*)| \geq -\frac{\gamma^2 \|\nabla^\perp \omega^*\|_2^2}{2(1-\gamma)} + (\gamma/2)\{\|u - u^*\|_2^2 + \|\omega - \omega^*\|_2^2\}$$

$$\geq -\gamma^2 \|\nabla^\perp \omega^*\|_2^2 + (\gamma/2)\{\|u - u^*\|_2^2 + \|\omega - \omega^*\|_2^2\}. \qquad (2.51)$$

Finally, choosing

$$\gamma = \frac{\|u - u^*\|_2^2 + \|\omega - \omega^*\|_2^2}{4\|\nabla^\perp \omega^*\|_2^2} \qquad (2.52)$$

(which is assumed to be smaller than $\frac{1}{2}$), we obtain

$$|H_\gamma(u) - H_\gamma(u^*)| \geq \frac{1}{8\|\nabla^\perp \omega^*\|_2^2}\{\|u - u^*\|_2^2 + \|\omega - \omega^*\|_2^2\}^2. \qquad (2.53)$$

In conclusion, from (2.53) and (2.48) (using also the time invariance of the energy), we have

$$\{\|u_t - u^*\|_2^2 + \|\omega_t - \omega^*\|_2^2\}^2$$

$$\leq (8\|\nabla^\perp \omega^*\|_2^2)\{\|u_0 - u^*\|_2^2 + (c_2/2)\|\omega_0 - \omega^*\|_2^2 + \gamma\|\nabla^\perp \omega^*\|_2\|u_0 - u^*\|_2\}, \qquad (2.54)$$

provided that

$$\frac{\|u_t - u^*\|_2^2 + \|\omega_t - \omega^*\|_2^2}{4\|\nabla^\perp \omega^*\|_2^2} \leq \frac{1}{2}. \qquad (2.55)$$

We realize that the left-hand side of (2.54), in the maximal interval in which (2.55) is verified, is arbitrarily small having chosen, at time zero, a perturbation u_0 sufficiently close to u^*. Hence, the inequality (2.55) is verified for all times and the proof is thus complete. $\qquad \square$

We conclude the section by showing a simple application of Theorem 2.4. We consider the periodic channel, $D = [0, L] \times [-A, A]$, periodic in the x_1 direction with period L, and with the usual impermeability conditions in the x_2 direction. We consider the stationary with the usual impermeability conditions in the x_2 direction. We consider the stationary solution:

$$u^* = (u_1(x_2), 0),$$
$$u_1(x_2) = x_2^3. \qquad (2.56)$$

The condition expressed by Theorem 2.2 requires

$$0 < c_1 \leq \frac{x_2^3 + \lambda}{6x_2} \leq c_2 < +\infty, \qquad (2.57)$$

where λ is an arbitrary constant that we can always add to the velocity field in order to maximize the possibilities of verifying the condition. However, since $c_2 < +\infty$, it must be $\lambda = 0$. As a consequence, $c_1 = 0$. Thus the stability is ensured by Theorem 2.4, while Theorem 2.2 does not give us a conclusive answer.

3.3. Stability in the Presence of Symmetries

We have seen in Chapter 1 how the presence of symmetries in the system (groups of transformations, leaving invariant the domain in which the fluid is confined), give rise to new first integrals beyond the kinetic energy of the system. This allows the construction of new Liapunov functions which could improve the stability results that we have established in the previous section. We will start by considering a domain which is translationaly invariant: the periodic channel, $D = [0, L] \times [-A, A]$. In this case, the ordinate of the center of vorticity

$$M_2 = \int_D x_2 \omega(x_1, x_2) \, dx_1 \, dx_2 \tag{3.1}$$

is a conserved quantity. As a Liapunov function we can choose

$$H(u) = \frac{1}{2} \int_D u^2 + \int_D \phi(\omega) + a_1 \int_0^L dx_1 \, u(x_1, A)$$

$$+ a_2 \int_0^L dx_1 \, u(x_1, -A) + \lambda \int_D x_2 \omega(x_1, x_2) \, dx_1 \, dx_2, \tag{3.2}$$

with a_1, a_2, λ real parameters. By repeating the arguments developed in the proof of Theorem 2.2 we find the following to be a sufficient condition for stability:

$$0 < c_1 \leq -\frac{u^* + \lambda}{\nabla^\perp \omega^*} \leq c_2 < +\infty. \tag{3.3}$$

Condition (3.3) does not say anything more than Theorem 2.2: actually condition (2.4) is modified by making use of the Galilean invariance.

For the rotationally invariant domain the moment of inertia of the system

$$I = \int_D x^2 \omega(x) \, dx \tag{3.4}$$

is conserved. Consider, for example, the domain enclosed between two circles (Fig. 3.13)

$$D = \{x \in R^2 | 0 \leq r_1 < |x| < r_2 < +\infty\}. \tag{3.5}$$

Suppose u^* to be of the form

$$u^* = f(|x|)e, \qquad e = \frac{x^\perp}{|x|}. \tag{3.6}$$

Let us take, as a Liapunov function,

$$H(u) = \frac{1}{2} \int_D u^2 + \int_D \phi(\omega) + \lambda \int_D x^2 \omega(x) \, dx + \sum_{i=1}^2 a_i \int_{C_i} u \cdot dl, \tag{3.7}$$

then the stability condition becomes

$$0 < c_1 \leq -\frac{u^* + \lambda |x| e}{\nabla^\perp \omega^*} \leq c_2 < +\infty. \tag{3.8}$$

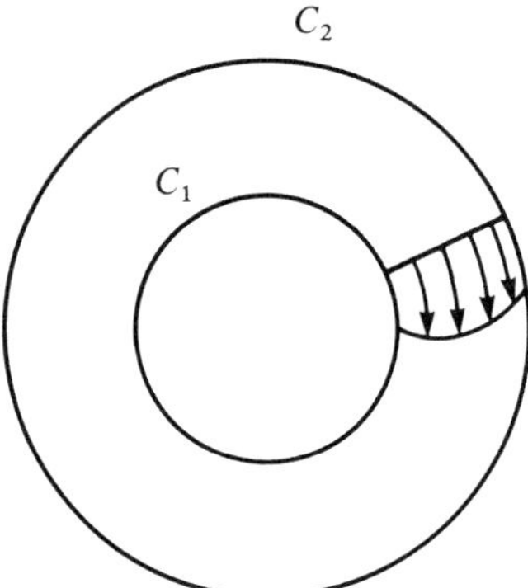

Figure 3.13

This result is not at all surprising: the rotational invariance allows us to add to the velocity field u^* a rigid rotation with angular velocity λ. However, this is an effective improvement of condition (2.4). In fact, for the stationary profile (3.6), condition (3.8) becomes

$$0 < c_1 \leq \frac{-f + \lambda r}{-\partial_r(f/r + f')} \leq c_2 < +\infty, \tag{3.9}$$

where $r = |x|$ and we have used the formula $(\text{curl})_3 = (1/r)(\partial_r(ru_\theta) - \partial_\theta u_r)$. Denoting by $F = F(r)$ the vorticity field, from (3.9) we have

$$0 < c_1 \leq \frac{-(1/r)\displaystyle\int_{r_1}^{r} Fr\, dr + \lambda r}{-F'} \leq c_2 < +\infty. \tag{3.10}$$

Suppose $F \geq 0$ and $F'(r_0) = 0$ for some r_0. The zero of F' cannot be compensated for by the numerator for $\lambda = 0$ (condition (2.4)) whereas, with a suitable choice of λ, condition (3.10) could also be satisfied in this case. Moreover, for $F \geq 0$ and $F' < 0$, condition (2.4) (which is (3.10) with $\lambda = 0$) can never be applied, so that we must use Theorem 2.3 with all problems arising from the condition on the minimum eigenvalue of the Laplace operator. Condition (3.10) can, however, be satisfied because the term λr can make the ratio positive.

A more handable condition than (3.10) is

$$0 < c_1 \leq \frac{\lambda r}{-F'} \leq c_2 < +\infty \tag{3.11}$$

which is obtained by neglecting the kinetic energy in the construction of the Liapunov function.

The Arnold method, summarized in Theorems 2.2, 2.3, and 2.4, depends, in an essential way, on the regularity of the stationary solution u^* and on the boundedness of the domain D. The regularity of u^* plays an explicit role in the proof of the above theorems and, for the moment, we do not discuss this aspect further. Concerning the boundedness of the domain, the situation is more subtle. As we remarked in the previous section, Theorem 2.3 cannot work in unbounded domains because there is not a lowest eigenvalue of the operator $-\Delta$. We want to investigate the applicability of Theorems 2.2 and 2.4: we will see how, except for very particular unbounded domains, these theorems also cannot work. We limit ourselves in discussing the spherically symmetric situation, namely, a domain of the form

$$D = \{x \in R^2 \mid 0 \le r_1 < |x| < +\infty\} \tag{3.12}$$

and a solution of the type (3.6). Let us investigate the ratio (3.10). The first term of the numerator may be bounded by the Cauchy–Shwarz inequality

$$\left| \frac{1}{r} \int_{r_1}^{r} Fr' \, dr' \right| \le \frac{\sqrt{(r^2 - r_1^2)/2}}{r} \left| \int_{r_1}^{r} F^2 r' \, dr' \right|^{1/2} \le \frac{1}{\sqrt{\pi}} \sqrt{\|\omega^*\|_2}. \tag{3.13}$$

The above expression is uniformly bounded in r, provided that the enstrophy of the stationary solution under consideration is bounded too. Therefore, the ratio (3.10) cannot be uniformly bounded if $\lambda \ne 0$, unless $F' \to +\infty$ as $r \to \infty$. On the other hand, F' must have, asymptotically, a definite sign, so that $F \to \pm\infty$. But this is not compatible with the hypothesis that $\|\omega^*\|_2$ is bounded. Hence $\lambda = 0$. Consider, first, the case in which F' never vanishes. Then, since F must vanish at infinity, F and F' have opposite signs, so that the ratio (3.10) cannot be positive. It remains to analyze the case in which there exists a point in which F' vanishes. Let r_M be the largest zero of F (an indefinite number of zeros cannot happen). It must be

$$\frac{1}{r_M} \int_{r_1}^{r_M} Fr' \, dr' = 0 \tag{3.14}$$

in order to compensate F'. However, for $r > r_M$, F is always positive or negative (it cannot oscillate), making the ratio (3.10) negative. In conclusion, inequality (3.10) can never be verified in this context.

A natural question arises. Are both the possible nonsmoothnesses of u^* and the unboundedness of the domain the features leading to instability, or is it the fact that the method we have developed so far does not work in these cases for technical reasons only? We will answer this question by showing stable stationary solutions which are nonsmooth and eventually confined in unbounded domains. A particularly meaningful case is that described by the following vorticity profile

$$\omega^* = \chi_\Lambda^*, \tag{3.15}$$

where χ_Λ is the characteristic function of the set Λ, and

$$\Lambda^* = \{x \mid |x| < 1\}. \tag{3.16}$$

ω^* is a stationary solution of the Euler equation in circularly symmetric domains as $D = R^2$ or in a circle. We want to study the stability ω^*. For a solution of this type the Arnold method does not apply since ω^* is not C^1. We will show that such a solution is stable in a suitable norm by considering, for the moment, only perturbations in the family P

$$P_R = \{\omega \in L_\infty | \, \omega = \chi_\Lambda, \, \Lambda \text{ measurable, meas } \Lambda$$

$$= \text{meas } \Lambda^*, \, \text{meas}(\Lambda \cap \{|x| > R\}) = 0\}, \tag{3.17}$$

$$P = \bigcup_R P_R. \tag{3.18}$$

From the above definition it follows that we deform the initial circle of radius one in an essentially bounded set of the same area. The analysis of the previous chapter tells us that the Euler dynamics is well defined for the elements $\omega \in P$, and leaves such a set stable, i.e., $\omega_t \in P$, if $\omega_0 \in P$. We equip P with the metric induced by the L_1 norm

$$d(\omega_1, \omega_2) = \|\omega_1 - \omega_2\|_1 = \text{meas}(\Lambda_1 \Delta \Lambda_2) = \text{meas}(\Lambda_1 \cup \Lambda_2 / \Lambda_1 \cap \Lambda_2), \tag{3.19}$$

where we have posed $\omega_i = \chi_{\Lambda_i}$.

Theorem 3.1. *For all $\varepsilon > 0$, there exists $\delta = \delta(\varepsilon, R)$ such that if*

$$\omega_0 \in P \qquad \text{and} \qquad d(\omega_0, \omega^*) < \delta, \tag{3.20}$$

then

$$d(\omega_t, \omega^*) < \varepsilon. \tag{3.21}$$

PROOF. The proof is based on the conservation of the following quantities

$$m_\Lambda = \int \chi_\Lambda, \qquad J = \int x^2 \chi_\Lambda \tag{3.22}$$

from which it follows immediately

$$J_{\Lambda_t} - J_\Lambda^* = J_{\Lambda_0} - J_\Lambda^* \le c_1 d(\omega_0, \omega^*) < c_1 \delta, \tag{3.23}$$

where we have posed $\omega_t = \chi_{\Lambda_t}$ and

$$c_1 = \text{ess sup}\{|x|^2 | x \in \Lambda_0\} \le R^2. \tag{3.24}$$

Moreover (this is the key point of the proof), we can see that J_Λ, restricted in P, has a minimum in ω^*

$$J_{\Lambda_t} - J_\Lambda^* = \int_{A_1} x^2 \, dx - \int_{A_2} x^2 \, dx, \tag{3.25}$$

where $A_1 = \Lambda_t / \Lambda^*$ and $A_2 = \Lambda^* / \Lambda_t$ (Fig. 3.14). Fixed, the area of A_1 and A_2 (it is the same!), the right-hand side of (3.25) is minimum whenever A_1 and A_2 are two circular annuli (Fig. 3.15). Since Λ^* has radius 1 and

$$\text{meas } A_1 = \text{meas } A_2 = \tfrac{1}{2} d(\omega^*, \omega_t), \tag{3.26}$$

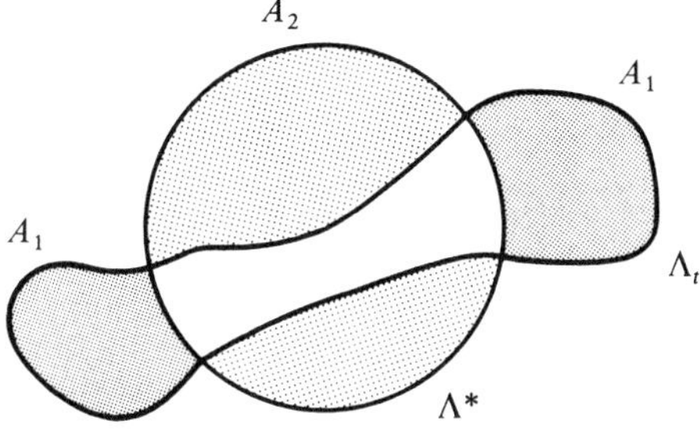

Figure 3.14

it follows that the radii of the circles enclosing the two anuli are

$$r_1 = \left(1 + \frac{\pi^{-1}}{2}d(\omega^*, \omega_t)\right)^{1/2}, \qquad r_2 = \left(1 - \frac{\pi^{-1}}{2}d(\omega^*, \omega_t)\right)^{1/2}. \quad (3.27)$$

Performing explicitly the integrals (3.26) in this extreme situation, we obtain

$$J_{\Lambda_t} - J_\Lambda^* \geq 2\pi \left\{ \int_1^{r_1} r^3\, dr - \int_{r_2}^1 r^3\, dr \right\}$$

$$= \frac{\pi}{2}\left\{ [(1 + (2\pi)^{-1}\, d(\omega^*, \omega_t))^2 - 1] - [1 - (1 - (2\pi)^{-1}\, d(\omega^*, \omega_t))^2] \right\}$$

$$= \frac{1}{4\pi} d(\omega^*, \omega_t)^2. \quad (3.28)$$

Combining (3.28) and (3.23) we achieve the proof of the theorem.   □

We have seen how the circular vortex patch χ_{Λ^*} is stable in the norm L_1 with respect to perturbation in the class P. This implies that the velocity fields $K\chi_{\Lambda^*}$ and $K\chi_{\Lambda_t}$ are close in the L_∞ norm, uniformly in time. Therefore, we have quite good control of the time evolution of the deviation of the perturbation from the stationary solution. However, we could introduce

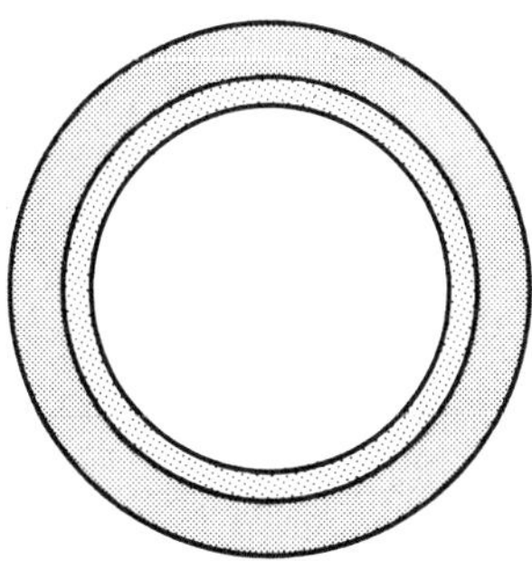

Figure 3.15

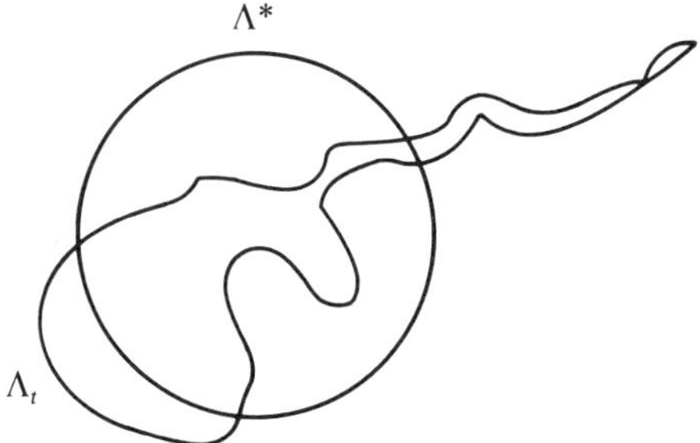

Figure 3.16

other norms such as

$$\inf_{x \in \partial \Lambda^*} \sup_{y \in \partial \Lambda_t} |x - y|. \tag{3.29}$$

Numerical experiments give a clear indication of the fact that X_{Λ^*} is unstable with respect to norms like (3.29): a small deformation of the boundary of Λ^*, at time zero, is magnified in time and, actually, long, thin vortex filaments (of very small measure, according to Theorem 3.1) are observed (Fig. 3.16).

We remark that the class of initial perturbations in Theorem 3.1 is restricted to suitable patches of compact support. However, the result can be extended to include L_1 perturbations which are essentially bounded and of compact support. The last hypothesis can also be removed by introducing suitable weighted norms (see Comments in Section 3.5).

We conclude here the discussion on the nonlinear stability of stationary solutions of the Euler equation. Recently, new interesting results on the argument have been obtained. However, the detailed discussion of these results is beyond the scope of this book: we will simply mention and quote them in Section 3.5.

3.4. Instability

As we have seen in Section 3.1, a natural way to investigate the instability of a stationary solution for an evolution problem in $\mathbb{R}^N$ is to look at the linearized equation. Such an approach is also natural for systems with infinitely many degrees of freedom, although in this case the instability for the full problem does not follow automatically from the instability of the linearized problem. In fact, in order to give a rigorous proof of the instability for the full problem, we must first find an eigenvalue of the linearized operator whose real part is positive. This is not enough because it is also necessary to find a norm, denoted by $\|\cdot\|$, for which the eigenfunctions associated to that eigenvalue have a finite $\|\cdot\|$-norm and, simultaneously, the nonlinear term behaves as $O(\|x\|^2)$. Thus, to obtain an instability result we must deal with two kind of difficulties. The first is related to the formal manipulations necessary to investigate the spectrum of the linearized operator. The second is to show that the linear instability is enough to prove the full instability as already mentioned. For these reasons, very few rigorous results are known, although many efforts have been devoted to the problem.

Let us begin by discussing some classical formal results in a particular setting. Let us consider as domain an infinite channel

$$D = \{x, y \,|\, -\infty < x < +\infty, \, -A < y < +A\} \tag{4.1}$$

(in this section we will use the notation x, y instead of x_1, x_2). Let us consider a stationary state given by the stream function $\psi^* = \psi^*(y)$ to which corresponds a velocity field, depending only on the y-coordinate,

$$u^* = \nabla^\perp \psi^* = U(y)e, \tag{4.2}$$

where e is the horizontal versor. We perturb this profile and express the solution of the Euler equation in terms of the stream function ψ as a sum of ψ^* and a perturbation ϕ

$$\psi(x, y, t) = \psi^*(y) + \phi(x, y, t). \tag{4.3}$$

Inserting the solution (4.3) into the Euler equation for the stream function (which is $\partial_t \Delta \psi + \nabla^\perp \psi \nabla \Delta \psi = 0$), and neglecting the second-order terms in ϕ, we immediately find

$$\partial_t \Delta \phi + U \Delta \partial_x \phi - (\partial_x \phi) U'' = 0, \tag{4.4}$$

which is the linearized Euler equation (around ψ^*).

We look for solutions of (4.4) of the form

$$\phi(x, y, t) = \theta(y)e^{ik(x-ct)}, \tag{4.5}$$

where $k \in \mathbb{R}$ and we use the convention of taking the real part of ϕ. $c = c_R + ic_I$, and its real part, has the dimension of a velocity. If $c_I > 0$ the solution is unstable: the perturbation grows exponentially in time for some k. We assume this hypothesis with the target of finding conditions on U that can be interpreted as necessary conditions for instability. Of course, this is not completely true. Strictly speaking, we will find necessary conditions on U for which the perturbations behave like (4.5) with $c_I > 0$. We will call this kind of instability *exponential instability* to stress that other kinds of instabilities (for which the perturbations grow, for instance, polynomially in time) are also possible. As a consequence, a necessary condition for exponential instability (see the Rayleigh and Fjortoft criteria below) *is not* a sufficient condition for stability as sometimes improperly claimed.

Inserting the expression (4.5) in (4.4) we find the Orr–Sommerfeld equation:

$$-ick(\theta'' - k^2\theta) + ikU(\theta'' - k^2\theta) - ik\theta U'' = 0 \tag{4.6}$$

from which, provided that k and c are different from zero, we obtain

$$\theta'' - k^2\theta - \frac{U''\theta}{U - c} = 0. \tag{4.7}$$

The boundary conditions for ψ are

$$\psi(x, -A, t) = \psi(x, A, t). \tag{4.8}$$

In fact, we have the identity

$$\psi(x, A, t) = \psi(x, -A, t) + \int_{-A}^{+A} u_1(x, y, t)\, dy, \tag{4.9}$$

where u_1 is the horizontal component of the velocity generated by ψ. Since ψ is constant on the boundary, the integral in the right-hand side of (4.9) is constant too, and we can always choose a Galilean frame making it vanish. Finally, from (4.8), we add a suitable constant to ψ in order to get

$$\theta(-A) = \theta(A) = 0. \tag{4.10}$$

We now integrate the identity (4.7) after multiplying by the complex conjugate of θ. Integrating by part we obtain

$$\int_{-A}^{+A} \{|\theta'|^2 + k^2|\theta|^2\}\, dy + \int_{-A}^{+A} \frac{U''}{U - c}|\theta|^2\, dy = 0. \tag{4.11}$$

Since both the real and imaginary part of the above expression must vanish, we have, for the imaginary part (the first integral is surely real)

$$c_I \int_{-A}^{+A} \frac{U''|\theta|^2}{|U - c|^2}\, dy = 0. \tag{4.12}$$

By hypothesis $c_I \neq 0$ so that the integral in the left-hand side of (4.12) must vanish. Thus:

Rayleigh Condition
In order to have exponential instability for the stationary solution ψ^*, U must exhibit, at the least, an inflexion point.

We now analyze the real part of the expression (4.11). We have

$$\int_{-A}^{+A} \{|\theta'|^2 + k^2|\theta|^2\}\, dy + \int_{-A}^{+A} \frac{U''(U - c_R)|\theta|^2}{|U - c|^2}\, dy = 0. \tag{4.13}$$

Moreover by (4.12)

$$(c_R - U_s) \int_{-A}^{+A} \frac{U''|\theta|^2}{|U - c|^2}\, dy = 0, \tag{4.14}$$

where U_s is the velocity in a point in which $U'' = 0$. By virtue of (4.13) we have

$$\int_{-A}^{+A} \frac{U''(U - U_s)|\theta|^2}{|U - c|^2}\, dy = -\int_{-A}^{+A} \{|\theta'|^2 + k^2|\theta|^2\}\, dy < 0. \tag{4.15}$$

This implies the

Fjortoft condition
In order to have exponential instability it is necessary that, in some point,

$$U''(U - U_s) < 0, \tag{4.16}$$

where U_s is the velocity in the point in which $U'' = 0$.

The converse of the Rayleigh and Fjortoft conditions, as a necessary condition for instability, would give (if rigorous) sufficient conditions for the stability. Actually, we have found the conditions given by the Arnold theorems. We see immediately that the result presented in Theorem 2.4 gives exactly the converse of condition (4.16). However, the nonlinear methods described in the previous two sections are more general, efficient, and elegant.

We have, up to now, discussed the necessary conditions for instability. To investigate sufficient conditions, we can again make use of the linearization methods to prove at least the instability of the linear problem. We know of only one case in which linear instability has been proved rigorously. The proof has been given by Tollmien, who obtained an eigenvalue with a positive real part by perturbing an eigenvalue with vanishing real part. We discuss this example in some detail.

Consider a fluid in an unbounded two-dimensional channel: $D = (-\infty, \infty) \times [-A, A]$. As we have seen, linear stability or instability can be studied by investigating the Rayleigh equation (4.7)

$$\varphi'' - k^2\varphi - \frac{U''}{U - c}\varphi = 0 \tag{4.17}$$

with the boundary conditions

$$\varphi(-A) = \varphi(A) = 0. \tag{4.18}$$

The linear instability will be proved by showing that:

(i) When $c_I = 0$ and $c = U_s$, problems (4.17) and (4.18) admit one eigenfunction φ_s (which is a neutral mode for the linearized Euler equation). We denote by k_s the corresponding value of k. We recall that U_s is the velocity at the inflexion point.
(ii) For k close to k_s and c close to U_s there exist unstable modes which are obtained by perturbing φ_s.

We start by proving point (i).
We suppose that U has an inflexion point at $\bar{y}$ and denote $U(\bar{y}) = U_s$. Moreover, let us assume $c_I = 0$ and $c_R = U_s$. We denote

$$\lambda = -k^2, \qquad F(y) = -\frac{U''(y)}{U(y) - c_r}, \tag{4.19}$$

(4.17) can be rewritten as

$$\varphi'' + (\lambda + F)\varphi = 0. \tag{4.20}$$

We suppose $U'(\bar{y}) \neq 0$. As a consequence, F is not singular in $\bar{y}$. We also suppose that $F(y) > 0$ for any $y \in [-A, A]$ (i.e., the Fjortoft condition is satisfied).

Equation (4.20) with boundary conditions (4.18) gives us a Sturm–Liouville problem. It admits an infinite set of eigenvalues which is bounded from below. We show that the minimun eigenvalue is negative. As is well known, this eigenvalue λ_m is given by a variational principle

$$\lambda_m = \min_{f \in S} \frac{\displaystyle\int_{-A}^{A} [f'^2(y) - F(y)f^2(y)] \, dy}{\displaystyle\int_{-A}^{A} f^2(y) \, dy}, \tag{4.21}$$

where $S = H_0^1$ (the space of square summable functions vanishing on the boundary, with a square summable derivative).

By using the well-known inequality (prove it by using the Fourier series)

$$(2A)^2 \int_{-A}^{A} f'^2(y) \, dy \geq \pi^2 \int_{-A}^{A} f^2(y) \, dy \tag{4.22}$$

we obtain that λ_m is negative if $F(y) > (\pi/2A)^2$.

Remark. Since F is positive, it follows that the Arnold criterion $U''/U > 0$ is violated. Therefore the profile U could, in principle, be unstable.

Under the assumption (4.22), equation (4.20) with boundary condition (4.10) has a solution denoted by φ_s. $k_s = \sqrt{-\lambda_m}$ is the square root of the associated eigenvalue. We want to prove the instability by a perturbative procedure. To find a solution with $k \neq k_s$, we follow the strategy of reviewing (4.17) as a perturbation of (4.20). Therefore the solution can be expressed by means of the variation of the constants method. For this we need to find another independent solution to (4.17) (in general, not satisfying the boundary conditions). We define

$$\varphi_z = \varphi_s \eta, \tag{4.23}$$

where

$$\eta = \int_0^y \frac{dt}{\varphi_s^2(t)}. \tag{4.24}$$

φ_z is a solution of (4.20) with $k = k_s$, as follows by direct inspection. Moreover, it is easy to verify that the functions φ_s and φ_z are independent by observing that their Wronskian is equal to 1. Thus (φ_s, φ_z) constitute a fundamental system of solutions and any further solution can be written in the form

$$\varphi(y) = c_s \varphi_s(y) + c_z \varphi_z(y). \tag{4.25}$$

It is easy to find the boundary condition on φ_z. By using the Wronskian we have

$$\varphi_z(-A) = -\frac{1}{\varphi_s'(-A)}, \qquad \varphi_z(A) = -\frac{1}{\varphi_s'(A)}. \tag{4.26}$$

We introduce the parameters $\varepsilon = k^2 - k_s^2$ and $\sigma = c - c_s$. Equation (4.17) becomes

$$\varphi''(y) - k_s^2 \varphi(y) + F(y)\varphi(y) = \left[\varepsilon - \frac{F(y)\sigma}{U(y) - c}\right]\varphi(y). \tag{4.27}$$

Equation (4.27) can be transformed in the integral equation

$$\varphi(y) = c_s\varphi_s(y) + c_z\varphi_z(y) + \int_{-A}^{y} N(t, y)\left[\varepsilon - \frac{F(t)\sigma}{U(t) - c}\right]\varphi(t)\, dt, \tag{4.28}$$

where c_s and c_z are two constants, and

$$N(t, y) = \varphi_s(t)\varphi_z(y) - \varphi_s(y)\varphi_z(t). \tag{4.29}$$

The boundary conditions give

$$\varphi(-A) = 0 \quad \Rightarrow \quad c_z = 0, \tag{4.30}$$

$$\varphi(A) = 0 \quad \Rightarrow \quad \int_{-A}^{A} N(t, A)\left[\varepsilon - \frac{F(t)\sigma}{U(t) - c}\right]\varphi(t)\, dt = 0. \tag{4.31}$$

We note that when $c_I \neq 0$, (4.28) is not singular and the problem has one and only one solution. Now our problem is to find a suitable $c_I > 0$, for a fixed $k \neq k_s$, in such a way that (4.31) is satisfied. We write the solution in the form

$$\varphi(y) = \varphi(y; \varepsilon, \sigma) = \varphi_s + f(y; \varepsilon, \sigma) \tag{4.32}$$

with $f(y, 0, 0) = 0$. Obviously, φ must satisfy the boundary conditions (4.30) and (4.31).

Inserting (4.32) in (4.31), we have

$$0 = I(\varepsilon, \sigma) \equiv \varepsilon I_1(\varepsilon, \sigma) + \sigma I_2(\varepsilon, \sigma), \tag{4.33}$$

where

$$I_1(\varepsilon, \sigma) = \int_{-A}^{A} N(t, A)[\varphi_s(t) + f(t; \varepsilon, \sigma)]\, dt, \tag{4.34}$$

$$I_2(\varepsilon, \sigma) = -\int_{-A}^{A} \frac{N(t, A)F(t)}{U(t) - c}[\varphi_s(t) + f(t; \varepsilon, \sigma)]\, dt. \tag{4.35}$$

We have

$$I_1(0, 0) = \int_{-A}^{A} N(t, A)\varphi_s(t)\, dt = \varphi_z(A)\int_{-A}^{A} [\varphi_s(t)]^2\, dt \neq 0. \tag{4.36}$$

The study of I_2 is more complicated

$$I_2(\varepsilon, \sigma) = -\int_{-A}^{A} \frac{N(t, A)F(t)}{U(t) - (c_R + ic_I)}[\varphi_s(t) + f(t; \varepsilon, \sigma)]\, dt. \tag{4.37}$$

Hence

$$I_2(0, \sigma) = -\int_{-A}^{A} \frac{N(t, A)F(t)}{[U(t) - c_R] - ic_I}[\varphi_s(t) + f(t; 0, \sigma)]\, dt. \tag{4.38}$$

We must find

$$\lim_{\sigma \to 0} I_2(0, \sigma) = I_2(0, 0). \tag{4.39}$$

(We remember that $\sigma \to 0 \Rightarrow c_R \to c_s$ and $c_I \to 0$.)

Now we introduce a new hypothesis on U. We suppose $U(y)$ to be monotonic, for instance, $U'(y) > 0$, $\forall y \in [-A, A]$. We denote by $U^{-1}(U)(y) = y$ the inverse function of U. We define

$$\alpha_1 \equiv U(-A), \qquad \alpha_2 \equiv U(A). \tag{4.40}$$

We suppose that $\alpha_1 < 0$ and $\alpha_2 > 0$.

Equation (4.38) becomes

$$I_2(0, \sigma) = \int_{\alpha_1}^{\alpha_2} \frac{N(U^{-1}(u), A)F(U^{-1}(u))}{(u - c_R) - ic_I}[\varphi_s(U^{-1}(u)) + f(U^{-1}(u); 0, \sigma)] \tag{4.41}$$

$$\frac{du}{U'(U^{-1}(u))} \equiv -[\mathscr{I}_1 + \mathscr{I}_2],$$

where

$$\mathscr{I}_1 = \int_{a_1}^{a_2} \frac{N(U^{-1}(v + c_R), A)F(U^{-1}(v + c_R))}{u'(U^{-1}(v + c_R))} \frac{\varphi_s(U^{-1}(v + c_R))\, dv}{v - ic_I}, \tag{4.42}$$

$$\mathscr{I}_2 = \int_{a_1}^{a_2} \frac{N(U^{-1}(v + c_R), A)F(U^{-1}(v + c_R))}{u'(U^{-1}(v + c_R))} \frac{f(U^{-1}(v + c_R), 0, \sigma)\, dv}{v - ic_I}, \tag{4.43}$$

where

$$v \equiv u - c_R, \qquad a_1 \equiv \alpha_1 - c_R, \qquad a_2 \equiv \alpha_2 - c_R.$$

(We consider systems such that $a_1 < 0$ and $a_2 > 0$.) We study $\mathscr{I}_1$ and we write it as

$$\mathscr{I}_1(0) = \int_{a_1}^{a_2} \frac{L(v, A)}{v - ic_I}\, dv$$

$$= \int_{a_1}^{a_2} \frac{L(0, A)}{v - ic_I}\, dv + \int_{a_1}^{a_2} \frac{L(v, A) - L(0, A)}{v - ic_I}\, dv. \tag{4.44}$$

Define

$$n(v, A) = \frac{L(v, A) - L(0, A)}{v}. \tag{4.45}$$

Then

$$\mathscr{I}_1 = L(0, A)\left\{\int_{a_1}^{a_2} \frac{v}{v^2 + c_I^2}\, dv + ic_I \int_{a_1}^{a_2} \frac{dv}{v^2 + c_I^2}\right\}$$

$$+ \int_{a_1}^{a_2} \frac{n(v, A)v(v + ic_I)}{v^2 + c_I^2}\, dv$$

$$= L(0, A)\left\{\frac{1}{2} \ln\left|\frac{a_2^2 + c_I^2}{a_1^2 + c_I^2}\right| + i\left[\arctan\frac{a_2}{c_I} - \arctan\frac{a_1}{c_I}\right]\right\}$$

$$+ \int_{a_1}^{a_2} \frac{n(v, A)v^2}{v^2 + c_I^2}\, dv + ic_I \int_{a_1}^{a_2} \frac{n(v, A)v}{v^2 + c_I^2}\, dv. \tag{4.46}$$

Then we study the limit $\sigma \to 0$. First, we show that the last term vanishes. In fact,

$$\left| c_I \int_{a_1}^{a_2} \frac{n(v, A)v}{v^2 + c_I^2} \, dv \right| \leq \frac{|c_I| M}{2} \left\{ \ln \frac{a_1^2 + c_I^2}{c_I^2} + \ln \frac{a_2^2 + c_I^2}{c_I^2} \right\}, \qquad (4.47)$$

where $M = \max_{v \in [a_1; a_2]} |n(v, A)|$, which vanishes as $c_I \to 0$.

The second term in (4.46) can be written in the limit $\sigma \to 0$ as

$$P \int_{a_1}^{a_2} \frac{L(v, A)}{v} \, dv - L(0, A) \ln \left| \frac{a_2}{a_1} \right|, \qquad (4.48)$$

where P means the Cauchy principal part.

In conclusion we have

$$\lim_{\sigma \to 0} \mathscr{I}_1(\sigma) = i\pi L(0, A) + P \int_{a_1}^{a_2} \frac{L(v, A)}{v} \, dv. \qquad (4.49)$$

In the limit $\sigma \to 0$ the term $\mathscr{I}_2(\sigma)$ vanishes, as we can see by similar calculations (remember that $f(y, 0, 0) = 0$). Then

$$I_2(0, 0) = -i\pi \frac{F(\bar{y})\varphi_s^2(\bar{y})\varphi_z(A)}{U'(\bar{y})} - P \int_{-A}^{A} \frac{N(t, A)F(t)\varphi_s(t)}{U(t) - c_R} \, dt. \qquad (4.50)$$

Hence (4.31) can be written as

$$I(\varepsilon, \sigma) = \varepsilon I_1(0, 0) + \sigma I_2(0, 0) + o(\varepsilon, \sigma). \qquad (4.51)$$

We have $I(0, 0) = 0$ and $\partial I(0, 0)/\partial \varepsilon = I_1(0, 0) \neq 0$ and so the implicit function theorem assures the existence of a neighborhood of $(0, 0)$ in which we can define a function $\sigma = \sigma(\varepsilon)$ such that $I(\varepsilon, \sigma(\varepsilon)) = 0$. In particular, (4.51) gives the first-order term

$$\sigma = -\varepsilon \frac{I_1(0, 0)}{I_2(0, 0)}$$

$$= -(k^2 - k_s^2) \frac{\displaystyle\int_{-A}^{A} \varphi_s^2(t) \, dt}{-i\pi U'(\bar{y})^{-1}[F(\bar{y})\varphi_s^2(\bar{y})] - P \displaystyle\int_{-A}^{A} (F(t)\varphi_s^2(t))/(U(t) - c_s) \, dt}. \qquad (4.52)$$

Hence

$$c_I = \pi(k_s^2 - k^2)F(\bar{y})\varphi_s^2(\bar{y})U'(\bar{y})^{-1}$$

$$\times \frac{\displaystyle\int_{-A}^{A} \varphi_s^2(t) \, dt}{\pi^2[F(\bar{y})U'(\bar{y})^{-1}\varphi_s^2(\bar{y})] + \left[P \displaystyle\int_{-A}^{A} (F(t)\varphi_s^2(t))/(U(t) - c_s) \, dt \right]^2} + o(\varepsilon). \qquad (4.53)$$

It is well known that the eigenfunction φ_s, corresponding to the lowest eigenvalue, can vanish only at the extremities $(-A, A)$. Then $\varphi_s(\bar{y}) \neq 0$, $U'(\bar{y}) > 0$, so that the conditions $F(\bar{y}) > 0$, $k_s^2 > k^2$, imply $c_I > 0$.

Thus we have obtained the example of linear instability.

3.5. Comments

The precise notion of stability with respect to perturbations of the initial condition is due to A.A. Liapunov, who also proved the basic Theorems 1.1, 1.2, and 1.4. For a review of the stability concept in the framework of ordinary differential systems, see [HiS 74].

The analysis of nonlinear stability presented in Theorems 2.2 and 2.3 is due to Arnold. In [Arn 65] there are, essentially, the general ideas of the method. A complete proof is presented in [Arn 69]. The use of the conservation laws to improve the results is discussed in $[\text{Arn } 66]_1$, while the geometrical aspects of the theory are analyzed in $[\text{Arn } 66]_2$. See also [Arn 78]. An excellent review paper on the Arnold method, with geometrical insights and applications to fluids and plasmas, is [HMR 85].

The use of the family of Liapunov functions has been introduced in [MaM 87].

The stability of a circular vortex patch has been proved in [WaP 85]. In this paper, the stability of a vortex patch in a circle is proved in L_1, along the same lines as Theorem 3.1, for all initial conditions in L_∞. Moreover, the stability of a circular vortex patch is also proved by means of a different Liapunov function: the energy. This latter proof is more complex, although it has the advantage of avoiding the use of rotational invariance, so that the same ideas can be applied to different contexts in the absence of symmetries. Along these lines Wan proved the stability of other stationary solutions (see, for instance, [Wan 86]).

The stability in the L_1 norm for the vorticity of a circular vortex patch does not imply a C^k control on the boundary (however, as we will see later, one has a L_2 control. For further information on the C^k regularity of the boundary, see the Section 3 in Chapter 6).

Numerical simulations show very interesting features such as a stretching of the vortex patch and growth of the curvature of its boundary, which suggest that a vortex patch is not stable if we measure the distance of the perturbation from the stationary solution by means of a C^k norm of the boundary ([MMZ 87], [Dri 88a]). However, we must remark that these features do not affect the velocity field in a sensitive way.

Dritschel [Dri 88b] remarked that an L_2 norm of the deviation of the boundary of a perturbed vortex patch (from a circular one) can be expressed by a combination of conserved quantities (see Exercise 10), so that we have stability in this norm. From this remark, we can prove the stability of a circularly symmetric monotonic distribution of vorticity. The same stability result was previously proved (in the L_1 norm of the vorticity) in [MaP 85] by generalizing the ideas of Theorem 3.1. Given a vorticity profile of the type $\omega(x) = f(|x|)$ with f monotonic, Theorem 3.1 can be extended rather easily to the case in which f is a step function with a finite number of jumps. The general case requires more care. The following theorem holds:

Theorem ([MaP 85]). *Suppose D to be a rotationally symmetric domain (possibly unbounded) and $\omega(x) = f(|x|)$ with f bounded monotonic and vanishing at infinity if the domain is unbounded. Then ω is L_1 stable with respect to perburbations in $L_1 \cap L_\infty$.*

All these results can be extended to the case of a periodic channel making use of the translational invariance and the maximum (or minimum) property of the first integral $\int \omega x_2 \, dx_1 \, dx_2$.

Theorem ([MaP 85]). *Let $D = [0, L] \times [-A, A]$ be periodic in the x direction and $\omega^*(x_1, x_2) = \xi(x_2)$, ξ monotone and essentially bounded. Then ω^* is stable in L_1 with respect to perturbations in L_∞.*

This result can be extended to the case of an infinite channel. The extension is not trivial because the stability relation (i.e., $\delta = \delta(\varepsilon)$) depends on L. However, the exponential decay of the Green function of the infinite channel allows us to formulate the following result.

Theorem ([CaM 86]). *Let $\omega^*(x_1, x_2) = \xi(x_2)$ be a stationary solution of the Euler equation in $D = [-\infty, +\infty] \times [-A, A]$. Suppose ξ essentially bounded and monotone. Then, for all $\varepsilon > 0$, there exists $\delta = \delta(\varepsilon) > 0$ such that for all $\omega_0 \in L_\infty$ satisfying*

$$\|\omega^* - \omega_0\|_1 < \delta \tag{5.1}$$

it follows:

$$|u(x_1, x_2, t) - u^*(x_2)| < \varepsilon, \tag{5.2}$$

where $u(x_1, x_2, t)$ is the solution of the Euler equation with initial datum given by the velocity field generated by ω_0, and u^* is the velocity profile of the stationary solution whose vorticity field is ω^*.

We notice that in this case the final norm (5.2) (i.e., the unifoml norm in the velocity field) differs from the initial one (the L_1 norm in the vorticity field). We also remark that the existence and uniqueness of the solutions of the Euler equation in the infinite channel are not obvious and do not follow directly from the analysis presented in Chapter 2. The solutions for this problem have been constructed in [CaM 86], again making use of the exponential decay of the Green function in the channel. Such solutions are constructed as a limit of corresponding solutions living in the periodic channel, in the limit $L \to \infty$.

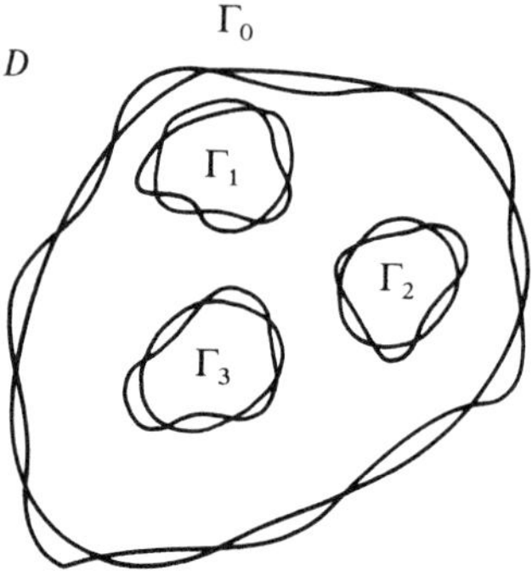

Figure 3.17

The Arnold method can be extended to include perturbation of the boundary as well as perturbation of the initial condition (Fig. 3.17). More precisely, consider a domain D enclosed in N regular boundaries C_0, C_1, ..., C_{N-1}. We perturb D arriving at a new domain D' with boundaries C_0', C_1', ..., C_{N-1}' and with the same topological structure as D. We define a distance between the two domains in the following way:

$$d(D, D') = \max_i \max_{x \in C_i} \min_{x' \in C_i'} |x - x'|. \tag{5.3}$$

The following result can be proved.

Theorem ([MaP 85]). *Under the hypotheses of Theorem 2.2, suppose that u_t is a solution of the Euler equation in D'. Then, for all $\varepsilon > 0$, there exists $\delta = \delta(\varepsilon)$ such that the condition*

$$\max\left\{d(D, D'), \int (u_0\chi_D - u^*\chi_{D'})^2 + (\omega_0\chi_D - \omega^*\chi_{D'})^2 < \delta\right\} \tag{5.4}$$

implies

$$\int (u_t\chi_D - u^*\chi_{D'})^2 + (\omega_t\chi_D - \omega^*\chi_{D'})^2 < \varepsilon. \tag{5.5}$$

The proof of this theorem is based on a combination of the techniques of Section 3.2 with the continuity properties of the Green function with respect to deformations of the boundary.

To extend some of the results we have discussed in this chapter to a fluid in a rotating sphere, with constant angular velocity, or in the so-called β-plane, is of particular interest for geophysic, atmospheric, and oceanographic applications. Let us consider the case of a rotating sphere. This system may be considered a good model for the atmospherical flows (see [Bat 67] and [Ped 79] for details). We choose a frame solidal to the rotating sphere, taking into account the centrifugal and Coriolis forces. The first has an irrelevant effect in the Euler equation: it derives from a potential so that it is absorbed by the pressure. The second introduces a linear term in the equation which becomes

$$\partial_t u + (u \cdot \nabla)u + 2\Omega \times u = -\nabla p. \tag{5.6}$$

Equation (5.6), written for the vorticity, gives rise to a conservation law

$$D_t(\omega + f) = 0, \tag{5.7}$$

where ω is, as usual, the vorticity, and

$$f = 2\Omega \cos \theta. \tag{5.8}$$

Here θ denotes the azimutal coordinate of a system of spherical coordinates with the z-axis coinciding with the rotation axis.

Let us take a point of the sphere and consider the tangent plane in this point. Consider a system of orthogonal coordinates in this plane with the ordinate axis coinciding with the projection of the rotation axis. Developing in the Taylor formula f, we obtain

$$f = f_0 + \beta y, \qquad f_0 = 2\Omega \cos \theta_0. \tag{5.9}$$

The Euler equation (5.7) in the plane with f given by (5.9) is a realistic approximation of an ideal flow in a rotating sphere. This approximation is usually called the β-plane. We can extend the Arnold method to this case ([BPV 82]) (see also [CaM 88]) by substituting the ratio $u^*/\nabla^\perp\omega^*$ with $u^*/\nabla^\perp(\omega^* + f)$. Also, the analysis based on the invariance principles can be extended to these cases by either using the translational invariance along the x-axis for the β-plane, or the rotational invariance for the rotating sphere ([CaM 88]). Moreover, we can also add a Rossby deformation term (see [Ped 79] for its physical meaning) and generalize in a straightforward way the stability results discussed in the present chapter.

We have underlined quite often how linear stability (i.e., the stability for the linearized system) does not imply full stability because nonlinear terms may play a decisive role. However, linear stability may be relevant in the applications, even if the steady state is unstable. In many such cases, the solution deviates from the steady state in a very slow scale of time so that, performing experiments in finite time, the instability is not recognizable.

We have discussed so far the stability of the velocity field in appropriate functional spaces. In other words, we have adopted the Eulerian point of view. A natural question is whether the stability of the velocity field implies the stability of the individual trajectories. The answer is in general negative. The study of the qualitative behavior of the trajectories is of interest for the applications. Consider, for example, situations in which the dispersion of tagged particles must be investigated for pollution problems.

Very few results concerning Lagrangian stability are known. We mention only that Lagrangian stability is more difficult to achieve than Eulerian stability. Arnold showed that an innocent periodic solution of the Euler equation can produce chaotic trajectories [Arn 78]. This kind of chaoticity is called Lagrangian turbulence. The Arnold's analysis goes further. It is well known that the free motion of a particle constrained on a surface with negative curvature is chaotic, namely each trajectory is exponentially unstable. On the other hand, the Euler flow can be thought of as a free motion on the manifold of the diffeomorphisms of the domain containing the fluid, which preserve the Lebesgue measure (recall Section 1 of Chapter 1). Thus it is natural to investigate the existence of pieces of such a manifold exhibiting negative curvature to outline the Lagrangian instability of suitable motions. It is possible to prove (see [Arn 78] for details) that such pieces do exist.

The hydrodynamical stability theory has many applications that go beyond the purposes of this book. We address the reader to the book [DrR 81] for a rather comprehensive study of the argument.

EXERCISES

1. Let $x(t)$ and $y(t)$ be two solutions of (1.1) with initial data given by x_0 and y_0, respectively. Suppose that $|\nabla F(x)| \le M < +\infty$. Prove that $|x(t) - y(t)|$ grows at most exponentially in time.

2. Let A be a matrix with real entries. Prove that A has either real eigenvalues or a pair of complex conjugate eigenvalues.

3. Draw a vector field in $\mathbb{R}^2$ for which the origin is unstable but attractive.

4. Prove the formula (1.12).

5. Consider the Hamiltonian system given by the Hamilton function

$$H = \tfrac{1}{2}k(x_1^2 + x_2^2) + R(x_1, x_2), \qquad x = (x_1, x_2) \in \mathbb{R}^2,$$

where $R(x) = O(|x|^3)$ and is a C^∞ function. Prove that the origin is stable.

6. Consider a Hamiltonian function as in the previous exercise. Assume that it can be multivalued and, for the sake of concreteness, has the form

$$H = \tfrac{1}{2}kr^2 + r^3\theta,$$

where r and θ are the polar coordinates in the plane. Prove that 0 is unstable. (*Hint:* Prove that H is a constant of motion and, for the initial condition, $r(0) = r_0$, $\theta(0) = 0$, we have $\tfrac{1}{2}kr^2 + r^3\theta = \tfrac{1}{2}kr_0^2$. From this it follows that the origin is attractive. Changing the versus of the time (i.e., reversing the velocity of the initial condition) we find that the origin is repelling.)

7. Consider the time-dependent Hamiltonian

$$H = \tfrac{1}{2}k(x_1^2 + x_2^2) + R(x_1, x_2; t), \qquad x = (x_1, x_2) \in \mathbb{R}^2,$$

where $R(x; t) = O(|x|^3)$ and is a C^∞ function. Discuss whether the origin can be unstable.

8. Let D be a bounded domain and $u(t)$ a solution of the Euler initial value problem with initial datum $u(0)$. Put $\omega(t) = \operatorname{curl} u(t)$, $t \ge 0$. Prove that, for all $\varepsilon > 0$, there

exists $\delta = \delta(\varepsilon)$ such that the condition

$$\|\omega(0)\|_\infty < \delta$$

implies

$$\|u(t)\|_\infty < \varepsilon.$$

9. Let x^* be an equilibrium point for a Hamiltonian system in $\mathbb{R}^{2N}$. Let L be a Liapunov function (see the definition in Theorem 1.4). Prove that L is a first integral. (*Hint*: Use the Liouville theorem for which the Lebesgue measure is preserved during motion and the continuity of L).

10. Prove the stability of a circular vortex patch with respect to perturbations in P with a metric given by the L_2 distance of the boundaries. (*Hint*: Express the L_2 distance of the boundaries in terms of first integrals.)

11. Discuss the stability for the Euler equation in $\mathbb{R}^3$ for axisymmetric stationary solutions.

CHAPTER 4

The Vortex Model

In this chapter we introduce and study the vortex model. Its relation with the Euler equation is investigated in some detail. Finally, we also discuss three-dimensional models.

4.1. Heuristic Introduction

In this chapter we analyze in some detail a rather natural situation: a vorticity profile is sharply concentrated around some points. As we will see, the two-dimensional situation is somehow more natural and richer than the three-dimensional one. We will concentrate most of our efforts in the two-dimensional case, and the whole chapter, except Section 4.5, is devoted to it. We begin with some heuristic considerations.

Suppose we have, in a domain D, an initial profile of vorticity given by a measure

$$\omega(x)\,dx = \sum_{i=1}^{N} a_i \delta_{x_i}(dx), \qquad (1.1)$$

where δ_{x_i} is the Dirac measure concentrated on the point x_i. The single component $a_i \delta_{x_i}(dx)$ of the measure (1.1) is called a *vortex* (sometimes, a *point vortex*). a_i (this is a real number) is the *intensity* of the vortex localized in x_i. The reason for this name (see Chapter 1, (2.22)) is the nature of the velocity field generated by a single vortex.

We want to study the time evolution for the initial data (1.1) according to the Euler equation. Since the initial data is not smooth, it is natural to make use of the weak form of the Euler equation

$$\frac{d}{dt}\omega_t(f) = \omega_t(u \cdot \nabla f), \tag{1.2}$$

$$u(x, t) = \int_D \nabla_x^\perp G_D(x, y)\omega_t\,(dy), \tag{1.3}$$

$$\omega_0\,(dx) = \sum_{i=1}^{N} a_i\delta_{x_i}\,(dx), \tag{1.4}$$

where

$$\omega_t(f) = \int_D \omega_t\,(dx)f(x) \tag{1.5}$$

and f is a smooth bounded function $f: D \to \mathbb{R}$. As usual, G_D denotes the Green function associated to the domain D.

As we will see in a moment, even this weak formulation of the Euler equation does not make sense. Assume, for simplicity, $D = \mathbb{R}^2$. Then the velocity field at time zero is

$$u(x, 0) = \sum_{i=1}^{N} \nabla_x^\perp G_D(x, x_i)a_i \tag{1.6}$$

$$= -\frac{1}{2\pi}\left(\sum_{i=1}^{N} \frac{(x - x_i)_2}{(x - x_i)^2}, -\sum_{i=1}^{N} \frac{(x - x_i)_1}{(x - x_i)^2}\right). \tag{1.7}$$

It is clear that u becomes singular whenever x tends to x_i. Since all the mass of the measure is concentrated just on the points x_i, it follows that the right-hand side of (1.2) does not make sense even at time zero. The term we are not able to deal with is the velocity generated by a single point vortex localized in x_i, computed exactly in x_i. However, physical intuition says that a single (approximately) point vortex, very far apart from the boundary, stays almost at rest. This follows also by the following consideration. Consider a sequence of smooth functions $\omega_n \in C^\infty$ approximating a vortex localized at the origin, i.e., such that $\omega_n \to \delta$ in the weak sense. Assume ω_n spherically symmetric and denote by $u_n = K_D * \omega_n$ the velocity field generated by ω_n. An easy computation shows that, due to the symmetry of the distribution, $u_n(0) = 0$ so that it is reasonable that a vortex does not move under the action of its own field. If we assume this statement, we are led to introduce the following ordinary differential system for which each vortex simply moves according to the velocity field generated by all other vortices:

$$\frac{d}{dt}x_i(t) = \nabla_i^\perp \sum_{j=1; j \neq i}^{N} a_j G(x_i(t), x_j(t))$$

$$= \sum_{j=1; j \neq i}^{N} a_j K(x_i(t), x_j(t)),$$

$$x_i(t = 0) = x_i. \tag{1.8}$$

Here $\nabla_i^\perp$ stands for $\nabla_{x_i}^\perp$, $\nabla^\perp = (\partial_2, -\partial_1)$, $G(x, y) = -(1/2\pi) \log |x - y|$ is the Green function on the plane and $K = \nabla^\perp G$.

Having solved the initial value problem (1.8) let us define

$$\omega_t(x) \, dx = \sum_{i=1}^{N} a_i \delta_{x_i(t)} \, (dx). \tag{1.9}$$

After an explicit computation, we obtain

$$\frac{d}{dt} \omega_t(f) = \omega_t(u_r \cdot \nabla f), \tag{1.10}$$

where u_r is a regularized velocity field defined by the following expression:

$$u_r(x, t) = \int_D (\nabla_x^\perp G(x, y) \chi(\{x \neq y\}) \omega_t \, (dy), \tag{1.11}$$

where $\chi(\{ \ \})$ denotes the characteristic function of the set $\{ \ \}$. We notice that (1.10) is equivalent to the Euler equation for absolutely continuous profiles of vorticity, and so we have established a sort of equivalence between the vortex model, as expressed by (1.8), and the Euler equation.

Let us see now what happens in the presence of a boundary. The Green function has the form

$$G_D(x, y) = G(x, y) + \gamma(x, y), \tag{1.12}$$

where γ is a symmetric C^∞ function in $D \times D$. Therefore the logarithmic divergence, yielding the self-interaction problem we have discussed in the case of the whole plane, is still present. Repeating the arguments given for the plane, by introducing a sequence of spherically symmetric, smooth vorticity profiles ω_n approximating a δ-function localized at the origin (we are assuming $0 \in D$), we arrive easily at the conclusion that

$$\lim_{n \to \infty} u_n(0) = \nabla_x^\perp \gamma_D(x, 0)|_{x=0} = \tfrac{1}{2} \nabla^\perp \gamma_D(0), \tag{1.13}$$

where $\gamma_D(x) = \gamma_D(x, x)$. Therefore a single vortex moves because of the presence of the walls. This is not at all surprising. For example, we have seen in Chapter 1 how, in the presence of cylindrical symmetry, the Euler equation reduces to a two-dimensional motion in the half-plane (with a modified continuity equation). Therefore an annulus of vorticity in the space is equivalent to a two-dimensional distribution of vorticity sharply concentrated around a point. It is a common fact that a smoking ring moves along its axis, so that a two-dimensional blob in the half-plane is expected to move in a direction approximately parallel to the boundary, as it actually does according to (1.13). The analogy is not so strict, because the two-dimensional motion arising from a cylindrical symmetry gives rise to an equation which is different from the two-dimensional Euler equation in the half-plane. In particular, the velocity of an annulus of vorticity is inversely proportional to the section of the annulus, and diverges when the section goes to zero (the total charge is concentrated in a circle), as we will discuss in Section 4.5. On the contrary, (1.13) shows that the velocity of an infinitely extended tube of vorticity, with its axis parallel to the plane constituting the boundary of the half-space, remains bounded when the section of the tube goes to zero, and the vorticity concentrates uniformly on a line.

As a consequence of (1.13) the evolution equations for a system of vortices in a domain D are

$$\frac{d}{dt} x_i(t) = \nabla_i^{\perp} \sum_{j=1; j \neq i}^{N} a_j G_D(x_i(t), x_j(t)) + \tfrac{1}{2} a_i \nabla_i^{\perp} \gamma_D(x_i)$$

$$= \sum_{j=1; j \neq i}^{N} a_j K_D(x_i(t), x_j(t)) + \tfrac{1}{2} a_i \nabla_i^{\perp} \gamma_D(x_i),$$

$$x_i(t = 0) = x_i. \tag{1.14}$$

However, the justification of the vortex model we have given is unsatisfactory both in the whole plane and in the domain D. The crucial point is that we simply skipped the self-interaction term on the basis of the authority principle. The symmetry argument which we have used is not at all convincing. A small perturbation can completely destroy the symmetry, and even if we could construct an initial datum formed by perfectly symmetric "blobs" of vorticity around the points x_i, the dynamics would immediately break such a symmetry. In conclusion, the argument we have given is misleading. In Section 4.4 we will present a deeper discussion of the problem and will give a rigorous derivation of the vortex model. For the present we accept the vortex model and discuss its properties.

4.2. Motion of Vortices in the Plane

We want to study, in some detail, the motion of the vortex system (1.8). First of all, we remark that we are dealing with a Hamiltonian system. Actually, (1.8) can be written in the following way:

$$a_i \frac{d}{dt} x_{i_1} = \frac{\partial}{\partial x_{i_2}} H,$$

$$a_i \frac{d}{dt} x_{i_2} = -\frac{\partial}{\partial x_{i_1}} H, \tag{2.1}$$

where $x_{i\alpha}$, $\alpha = 1, 2$, are the components of x_i and

$$H = -\frac{1}{4\pi} \sum_{ij=1; i \neq j}^{N} a_i a_j \log |x_i - x_j|. \tag{2.2}$$

Choosing, as canonical conjugate variables, $\sqrt{|a_i|}\, x_{i1}$ and sign $a_i \sqrt{|a_i|}\, x_{i2}$ (but another possible choice is also x_{i1} and $a_i x_{i2}$), equations (2.1) are the

Hamilton equations associated with the Hamiltonian (2.2). Important features are a consequence of this fact. In particular, by the Liouville theorem, the Lebesgue measure

$$\prod_{i=1}^{N} dx_i \qquad (2.3)$$

is preserved during the vortex motion. Moreover, the Hamiltonian H, being time invariant, is a constant of the motion. Furthermore, due to the invariance of the Hamiltonian with respect to the translations and rotations groups, there are other first integrals of the motion as a consequence of the Noether theorem. They are

$$M = \sum_{i=1}^{N} a_i x_i \qquad \text{(translation invariance)}, \qquad (2.4)$$

$$I = \sum_{i=1}^{N} a_i x_i^2 \qquad \text{(rotational invariance)}. \qquad (2.5)$$

M is proportional to the center of vorticity B of the system: $M = (\sum a_i)B$. The quantity I is called the moment of inertia. The reader unfamiliar with analytical mechanics can verify the invariance of the quantities (2.2), (2.4), and (2.5) by directly using the equations of motion (2.1). The first integrals H, M, I allow us to determine explicitly the motion of two or three vortices. For $N = 2$, the solution of equations (2.1) is rather simple. The Hamiltonian is a function of the distance of the two vortices which remains constant. Moreover, the position of the center of vorticity must also remain constant. It lies, obviously, on the line joining the positions of the two vortices. If $a_1 \neq -a_2$ the motion is a rotation around the center of vorticity (see Figs. 4.1 and 4.2). If $a_1 = -a_2$, the two vortices move along two parallel lines (Fig. 4.3). In all cases, the speed of each vortex is inversely proportional to the distance between the two vortices and directly proportional to the vorticity intensity of the other.

The three-body problem is also integrable. The motion can be completely determined in all details (see the references in the last section): the motion is not simple, but may be analytically described by using the first integrals (2.2), (2.4), and (2.5).

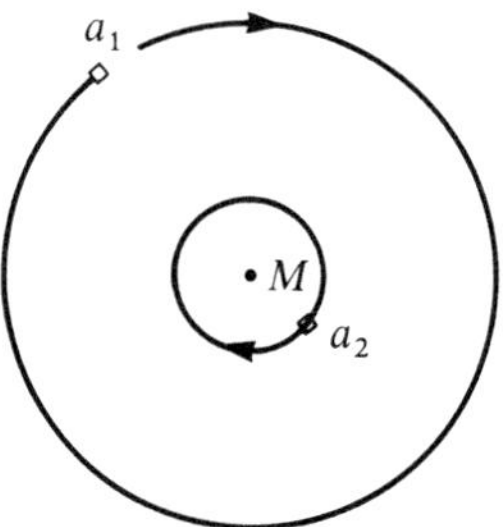

Figure 4.1

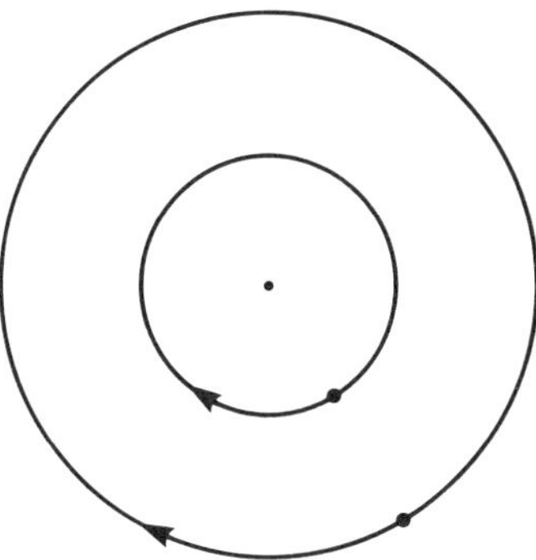

Figure 4.2

For a higher number of vortices, the problem of understanding the main features of the motion is a very complicated task. However, before studying the qualitative behavior of the many vortices system we have to show that the solution to the Cauchy problem exists and is unique. Although (2.1) is an ordinary differential system, the existence and uniqueness problem is not trivial because of the logarithmic divergence of the Hamiltonian, which can cause catastrophes whenever two particles stay at the same point. In this case, the second member of the equation of motion no longer makes sense. Another possibility is when a particle reaches an infinite velocity in a finite time. Are those pathologies possible? When all the vortices have intensities of the same sign the answer to this question is negative. The existence of the first integrals (2.2) and (2.5) allow us to exclude the presence of any kind of collapse: if two particles are very close, the logarithm of their distance diverges negatively. The conservation of the energy implies that, in order to compensate such a negative divergence, two particles must be very far apart from each other. However, this is prevented by the conservation of the momentum of inertia; thus, we conclude that the motion develops in a bounded region and that there is a minimum distance between any pair of vortices. The diameter of the maximal region and the minimal distance can be estimated by the initial condition only. We leave it as an exercise for the reader.

Very different is the case in which we deal with vortices of different signs. We show an example in which three vortices collapse to the same point in such a way that all divergences in the energy are compensated. Consider the

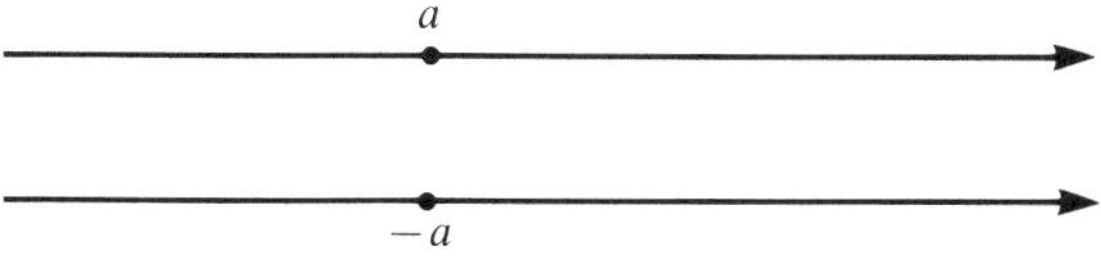

Figure 4.3

three-vortices system in which $a_1 = 2$, $a_2 = 2$, $a_3 = -1$. The initial conditions are $x_1 = (-1, 0)$, $x_2 = (1, 0)$, $x_3 = (1, \sqrt{2})$.

From the evolution equations (2.1) we obtain

$$\frac{d}{dt} l_{ij}^2 = \frac{2}{\pi} A a_k \left[\frac{1}{l_{jk}^2} - \frac{1}{l_{ki}^2} \right], \tag{2.6}$$

where $l_{ik} = |x_i - x_k|$ and i, j, k are the names of the three vortices appearing in counterclockwise order, and A is the area of the triangle whose vertices are the positions of the vortices. It can be verified that in our case the ratios between the distances of a pair of vortices are preserved in time (see Section 4.6 for more details), and that

$$\frac{d}{dt} l_{ij}^2(t) = -\frac{1}{3\sqrt{2\pi}} l_{ij}(0) \tag{2.7}$$

from which

$$l_{ij}^2(t) = l_{ij}^2(0) \sqrt{1 - \frac{t}{3\sqrt{2\pi}}}. \tag{2.8}$$

Therefore $l_{ij}(t) = 0$ at time $t = 3\sqrt{2\pi}$. In Fig. 4.4 we show the trajectories leading to the collapse. We notice that the solution we have constructed is self-similar: the initial triangle formed by the three vortices maintains the same form, but is rotating and contracting in time.

The possibility of having collapses for an initial value problem for the vortex system excludes the possibility to obtain a general theorem, ensuring existence and uniqueness of the solutions for all times and initial conditions.

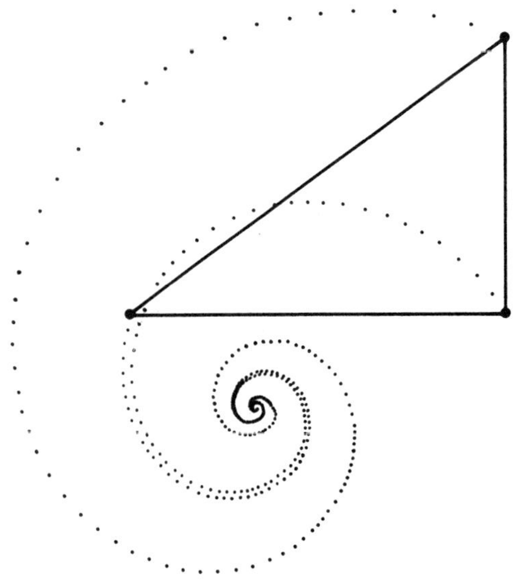

Figure 4.4

However, the example we have shown is very particular and we can hope that, even if collapses are present, they may be, in some sense, exceptional. As a measure of the exceptionality of an event, we take the (normalized) Lebesgue measure on the initial conditions. To be more precise, consider a bounded, measurable set A in R^{2N}. Consider also the set B (contained in A) of all initial conditions in A leading to a collapse in a finite time. We say that the Lebesgue measure of B, divided by the Lebesgue measure of A (in formulas: $|B|/|A|$), is the probability of having a collapse starting from A. It is clear that if such a probability is zero, a collapse is a very exceptional event which can hardly be observed. We will prove this statement (i.e., $|B|/|A| = 0$) by studying a related problem which is the exceptionality (in the same sense explained above) of an ε-collapse. An ε-collapse is an event in which two vortices arrive at a distance less than ε. We want to estimate the probability of an ε-collapse. We notice that such a problem is even more interesting from a physical point of view. Actually, it is not possible to distinguish between a mathematical collapse and a real ε-collapse for ε sufficiently small. We will find that the probability of an ε-collapse vanishes when $\varepsilon \to 0$ and, in particular, this proves that $|B|/|A| = 0$.

In what follows it is useful to introduce a regularization of the dynamics defined by (1.8). For $\varepsilon < 1$, let $\ln_\varepsilon$ be a $C^\infty(R)$ function of $|x|$ satisfying:

(i) $$\ln_\varepsilon x = \ln |x| \qquad \text{for} \quad |x| > \varepsilon,$$

(ii) $$|\ln_\varepsilon x| \le |\ln |x|| \qquad \text{for} \quad x \in R,$$

(iii) $$\left| \frac{d}{dx} \ln_\varepsilon x \right| \le |x|^{-1}.$$

$$(2.9)$$

We define a regularized Green function

$$G_\varepsilon(x - y) = -\frac{1}{2\pi} \ln_\varepsilon(x - y) \tag{2.10}$$

and the following initial value problem:

$$\frac{d}{dt} x_i^\varepsilon(t) = \nabla_i^\perp \sum_{j=1; j \ne i}^N a_j G_\varepsilon(x_i^\varepsilon(t), x_j^\varepsilon(t))$$

$$= \sum_{j=1; j \ne i}^N a_j K_\varepsilon(x_i^\varepsilon(t) - x_j^\varepsilon(t)),$$

$$x_i^\varepsilon(t = 0) = x_i, \tag{2.11}$$

where $K_\varepsilon = \nabla^\perp G_\varepsilon$.

It is immediately verified that the dynamics defined by (2.11) (it makes sense for all times because the vector field in the right-hand side of (2.11) is globally Lipschitz) is Hamiltonian. Moreover, up to the first ε-collapse (before the two particles arrive at a distance smaller than ε), the dynamics defined by (2.11) coincide with the true vortex dynamics (1.8).

As we have claimed above, our target is to prove that the measure of the initial data leading to an ε-collapse, according to the dynamics (2.11) (and hence also according to the dynamics (1.8)), is infinitesimal in ε. To prove this, we first need to prove a preliminary property of the vortex dynamics which is interesting in itself. Let us suppose that, initially, the N vortices are contained in the circle of radius R, around the origin: $\Sigma_R = \{x \in R^2 \, | \, |x| < R\}$. We want to prove that, for a fixed time T, there exists a larger circle, of radius $R(T)$, containing the N vortex systems for all times $t \leq T$ independently of ε. Such boundedness properties cannot be true for all choices of the a_i's. Indeed, for two vortices of opposite intensities this property is false because they go more and more quickly when the distance of their initial points goes to zero. We give a condition ensuring the result. We will assume

$$\sum_{i \in P(N)} a_i \neq 0, \tag{2.12}$$

where $P(N)$ denotes the family of all subsets of the first N integers. We notice that this condition is generic in the space R^N of the intensities a_i's.

Theorem 2.1. *Under the hypothesis (2.12) there exists a constant C, depending on N, T, $a_1, \ldots, a_N$ but independent of ε and the initial conditions, for which*

$$\max_{i=1,\ldots,N} \sup_{0 \leq t \leq T} |x_i^\varepsilon(t) - x_i| \leq C. \tag{2.13}$$

Here $\{x_i^\varepsilon(t)\}_{i=1}^N$ denotes the solution of the regularized initial value problem (2.11).

PROOF. The proof is based on the conservation law (2.4) and proceeds by induction. Let us denote by $y_i^{\varepsilon,k}(t)$, $i = 1, \ldots, k$, the solution of the initial value problem of the k-vortex system under the action of a given smooth external field F

$$\frac{d}{dt} y_i^{\varepsilon,\kappa}(t) = \nabla_i^\perp \sum_{j=1; j \neq i}^k a_j G_\varepsilon(y_i^{\varepsilon,\kappa}(t) - y_j^{\varepsilon,\kappa}(t)) + F(y_i^{\varepsilon,\kappa}(t), t), \qquad t \in [0, T], \tag{2.14}$$

$$y_i^{\varepsilon,\kappa}(t = 0) = x_i.$$

We suppose

$$|F(y_i^{\varepsilon,\kappa}(t), t)| \leq 1. \tag{2.15}$$

By the induction hypothesis, we assume that there exists a constant C_k, independent of ε and the initial conditions, such that

$$\max_{i=1,\ldots,k} \sup_{t \leq T} |y_i^{\varepsilon,h}(t) - x_i| \leq C_k, \qquad h \leq k. \tag{2.16}$$

We want to prove that there exists

$$C_{k+1} \geq C_k, \qquad C_{k+1} < +\infty, \tag{2.17}$$

such that

$$\sup_{\varepsilon; X} \max_{i=1,\ldots,k} \sup_{t \leq T} |y_i^{\varepsilon,k+1}(t) - x_i| \leq C_{k+1}. \tag{2.18}$$

Suppose "ab absurdo" that

$$C_{k+1} = +\infty, \tag{2.19}$$

then there exists an arbitrarly large S (to be fixed later) and, consequently, $\varepsilon \in (0, 1)$, $t^* \in [0, T]$, and an index i_1, for which

$$|y_{i_1}^{\varepsilon,k+1}(t^*) - x_{i_1}| = S. \tag{2.20}$$

We will find a contradiction for a sufficiently large S, and hence the inequality (2.18) will be achieved.

We set

$$a = \max_{i=1,\ldots,N} |a_i|, \tag{2.21}$$

$$A = \min_{P(N)} \left| \sum_{i \in P(N)} a_i \right|, \tag{2.22}$$

$$M_k^{\varepsilon}(t) = \sum_{j=1}^{k} a_j y_j^{\varepsilon,k}(t). \tag{2.23}$$

The center of vorticity is almost conserved by the regularized dynamics with the external field. Actually we have, for all $h \leq N$,

$$|M_h^{\varepsilon}(t) - M_h^{\varepsilon}(0)| = \left| \sum_{j=1}^{h} a_j [y_j^{\varepsilon,h}(t) - x_j] \right|$$

$$= \left| \sum_{j=1}^{h} a_j \int_0^t F(y_j^{\varepsilon,h}(\tau), \tau) \, d\tau \right| \leq haT \leq NaT \equiv b. \tag{2.24}$$

Notice that the mutual interaction among the vortices disappears because of the conservation of the center of vorticity. On the other hand,

$$|M_{k+1}^{\varepsilon}(t^*) - M_{k+1}^{\varepsilon}(0)| = \left| \sum_{j=1}^{k+1} a_j [y_{i_1}^{\varepsilon,k+1}(t^*) - x_{i_1}] + \sum_{j=1}^{k+1} a_j [z_j(t^*) - z_j] \right|, \tag{2.25}$$

where

$$z_j(t) = y_j^{\varepsilon,k+1}(t) - y_{i_1}^{\varepsilon,k+1}(t), \qquad t \in [0, T], \qquad z_j = z_j(0). \tag{2.26}$$

From (2.25), (2.24), (2.20) and by the triangle inequality we have

$$\left| S \sum_{j=1}^{k+1} a_j \right| - \left| \sum_{j=1}^{k+1} a_j [z_j(t^*) - z_j] \right| \leq b \tag{2.27}$$

from which there exists an index i_2 such that

$$|z_{i_2}(t^*) - z_{i_2}| \geq \frac{SA - b}{aN}. \tag{2.28}$$

Therefore the distance between the two particles with indices i_1 and i_2 is larger than $(SA - b)/2aN$ either at time zero, or at time t^*. Thus we have found an instant $t^{\wedge} \in [0, T]$ for which the particles, at such an instant, are divided into two clusters at a distance d larger than $(SA - b)/2aN^2$. Actually, in the worst case we could have a chain of $N - 2$ particles, equally spaced, connecting the two with indices i_1 and i_2. The (regularized) velocity field generated by each cluster on the other is smaller than

$$\frac{Na}{2\pi d} \leq \frac{N^3 a^2}{\pi(SA - b)}. \tag{2.29}$$

By choosing S large enough, we can make the above expression smaller than one. Thus we can use the inductive hypothesis. Each cluster (formed by a number of vortices smaller than $k + 1$) at time $t^\wedge$ is moving under the action of its own vortices, and under the action of a small external field (which is the field generated by the vortices of the other cluster). Consider now the maximal interval of time around $t^\wedge$ for which the two clusters remain sufficiently far apart so that they generate (each on the other) an external field smaller than one. According to the inductive hypothesis, each particle is displaced, in such an interval of time, by a distance at most C_k. By choosing S so large, we find that the gap between the two clusters cannot be filled for the whole time interval $[0, T]$. In conclusion, we find a contradiction because the maximal displacement of each particle is bounded by $C_k < S$. From (2.17) the proof of the theorem easily follows by putting $C = C_N$ since (2.16) is trivially verified for $k = 1$ (by putting $C_1 = T$). $\qquad\square$

We remark that in the above theorem an hypothesis like (2.12) is essential, as we realize by studying the simple system composed by two opposite vortices. As a corollary we have the boundedness property which is essential in proving the exceptionality of the collapses.

Corollary 2.1. *Let the condition* (2.12) *be verified. Then, for all R and T, there exists $R^*(T)$ for which N vortices, initially in Σ_{R^*}, cannot leave the circle $\Sigma_{R^*(T)}$ for all initial data and $\varepsilon \in (0, 1)$.*

PROOF. The proof follows from the previous theorem by choosing $R^*(T) = R + C$. $\qquad\square$

We are now in a position to prove the fundamental theorem on the ε collapses. Let us put

$$d_T^\varepsilon(X) = \min_{i \neq j} \inf_{0 \leq t \leq T} |x_i^\varepsilon(t) - x_j^\varepsilon(t)|, \tag{2.30}$$

where $X = \{x_1, \ldots, x_N\}$ denotes the initial condition.

Theorem 2.2. *Let condition* (2.12) *be verified, and let*

$$\lambda(dX) = dx_1, \ldots, dx_N/(\pi R^2)^N \tag{2.31}$$

be the normalized Lebesgue measure on Σ_R^N. Then

$$\lim_{\varepsilon \to 0} \lambda(\{X \,|\, d_T^\varepsilon(X) < \varepsilon\}) = 0. \tag{2.32}$$

PROOF. The proof is based on the invariance of the Lebesgue measure under the flow generated by (2.11). We choose a function ϕ which is singular in the presence of collapses, but which is integrable with respect to the measure $d\lambda$. We prove that its time evolution is still integrable, by the conservation of the measure (2.31) so that, at a fixed time t, collapsing configurations form a set of negligible λ-measures. In order to prove that the set of initial data yielding collapses in a fixed time interval are of λ-measure zero, we evaluate the derivative along the trajectories of ϕ and prove its integrability with respect to $d\lambda$.

Define

$$\phi_\varepsilon(X) = \frac{1}{2} \sum_{i;j=1, i \neq j}^{N} \ln_\varepsilon |x_i - x_j|, \qquad X = \{x_1, \ldots, x_N\}. \tag{2.33}$$

Denoting by $S_t^\varepsilon X = \{x_1^\varepsilon(t), \ldots, x_N^\varepsilon(t)\}$ the configuration solution of the initial value problem (2.11) with initial datum X, we have

$$\frac{d}{dt} \phi_\varepsilon(S_t^\varepsilon X) = \frac{1}{2} \sum_{i;j=1, i \neq j}^{N} \nabla_i \ln_\varepsilon |x_i^\varepsilon(t) - x_j^\varepsilon(t)| \cdot \frac{d}{dt}(x_i^\varepsilon(t) - x_j^\varepsilon(t)). \tag{2.34}$$

By using the equations of motion we obtain

$$\left| \frac{d}{dt} \phi_\varepsilon(S_t^\varepsilon X) \right| \leq h(X), \tag{2.35}$$

where

$$h(S_t^\varepsilon X) = a \sum_{i=1}^{N} \sum_{j=1; i \neq j}^{N} \sum_{k=1; k \neq i; k \neq j}^{N} |\nabla_i \ln_\varepsilon(x_i^\varepsilon(t) - x_j^\varepsilon(t))| \, |\nabla_i^\perp \ln_\varepsilon(x_i^\varepsilon(t) - x_k^\varepsilon(t))|. \tag{2.36}$$

Notice that thanks to the obvious identity

$$\nabla f(X) \nabla^\perp f(X) = 0, \qquad f \in C^1(\mathbb{R}^2), \tag{2.37}$$

we eliminate the most singular term in (2.36).

Therefore

$$h(S_t^\varepsilon X) = \text{const.} \sum_{i=1}^{N} \sum_{j=1; i \neq j}^{N} \sum_{k=1; k \neq i; k \neq j}^{N} |x_i^\varepsilon(t) - x_j^\varepsilon(t)|^{-1} |x_i^\varepsilon(t) - x_k^\varepsilon(t)|^{-1}. \tag{2.38}$$

We observe that for a bounded measurable set $A \subset \Sigma_R^N \subset \mathbb{R}^{2N}$ we can find an increasing function of R, denoted by $F(R)$, not depending on ε, such that

$$\int_A |\phi_\varepsilon(X)| \lambda(dX) + \int_A h(X) \lambda(dX) < F(R). \tag{2.39}$$

Therefore

$$\int_{\Sigma_R^N} \lambda(dX) \sup_{0 \le t \le T} |\phi_\varepsilon(S_t^\varepsilon(X)| \le \int_{\Sigma_R^N} \lambda(dX)|\phi_\varepsilon(X)|$$

$$+ \int_{\Sigma_R^N} \lambda(dX) \int_0^T dt \left| \frac{d}{dt} \phi_\varepsilon(S_t^\varepsilon X) \right|$$

$$\le \int_{\Sigma_R^N} \lambda(dX)|\phi_\varepsilon(X)| + \int_{\Sigma_R^N} \lambda(dX) \int_0^T dt \, |h(S_t^\varepsilon X)|$$

$$\le \int_{\Sigma_R^N} \lambda(dX)|\phi_\varepsilon(X)| + \int_0^T dt \int_{S_t^\varepsilon \Sigma_R^N} \lambda(dX)|h(X)|, \tag{2.40}$$

where $S_t^\varepsilon \Sigma_R^N = \{ S_t^\varepsilon X \,|\, X \in \Sigma_R^N \}$. In the last step of (2.40) we used the Fubini theorem and the invariance of the measure λ with respect to the flow S_t^ε.

By Theorem 2.1 we know that

$$S_t^\varepsilon \Sigma_R^N \subset \Sigma_{R(T)}^N \tag{2.41}$$

so that, by (2.39),

$$\int_{\Sigma_R^N} \lambda(dX) \sup_{0 \le t \le T} |\phi_\varepsilon(S_t^\varepsilon(X)| \le \int_{\Sigma_R^N} \lambda(dX)|\phi_\varepsilon(X)| + \int_0^T dt \int_{\Sigma_{R(T)}^N} \lambda(dX)|h(X)|$$

$$\le F(R) + TF(R(T)) \le H, \tag{2.42}$$

where H is a positive constant depending on T, N, R, but not on ε.

We observe now that, for a sufficiently small $\varepsilon > 0$,

$$\{ X \,|\, d_T^\varepsilon(X) < \varepsilon \} \subseteq \left\{ X \,\middle|\, \sup_{0 < t < T} |\phi_\varepsilon(S_t^\varepsilon X)| > -\tfrac{1}{2} \ln \varepsilon \right\}. \tag{2.43}$$

In fact, if two particles (say i and j) arrive in the time interval $[0, T]$ at a distance smaller than ε, the two terms in the sum involving the particles i and j give a contribution $-\ln \varepsilon$. This cannot be compensated for by positive contributions because the distance between the pairs of particles is bounded by Theorem 2.1. By the Chebychev inequality

$$\lambda(\{ X \,|\, d_T^\varepsilon(X) < \varepsilon \}) \le \lambda \left(\left\{ X \,\middle|\, \sup_{0 \le t \le T} |\phi_\varepsilon(S_t^\varepsilon X)| > -\tfrac{1}{2} \ln \varepsilon \right\} \right)$$

$$\le H(-\tfrac{1}{2} \ln \varepsilon)^{-1}. \tag{2.44}$$

$$\square$$

The above theorem allows us to construct a global flow S_t, almost everywhere defined, by putting

$$S_t X = S_t^\varepsilon X \qquad \text{for all } X \text{ for which } \quad d_T^\varepsilon(X) \in \varepsilon. \tag{2.45}$$

Since, by Theorem 2.2, λ-almost all $X \in \Sigma_R$ has the property that $d_T^\varepsilon(X) > \varepsilon$ for some sufficiently small ε, the flow S_t is well defined. Moreover, since the initial value problem (2.11) coincides with the original vortex flow whenever any pair of particles does not get closer than ε in the interval of time $[0, T]$, we can conclude:

Corollary 2.2. *Outside a set of initial conditions of Lebesgue measure zero, the initial value problem associated to the vortex equations (2.1) has a global smooth solution, provided that condition (2.12) is verified.*

PROOF. We first fix a time T and then take a sequence $R_n \to \infty$. For almost all initial conditions in $\Sigma_{R_n}^N$ we are able to construct a smooth global flow. Therefore the set of all collapsing configurations is a countable union of sets of negligible measure, and hence it is a set of Lebesgue measure zero. $\qquad\square$

We remark that the probability estimate (2.44) controlling the ε-collapses can be considerably improved. Actually, we can prove (see Exercise 1) that the probability of an ε-collapse goes to zero as $C(1 + T)\varepsilon^{1-\delta}$ where $\delta \in (0, 1)$ is arbitrary and C is increasing with δ. This latter estimate gives a better idea of how exceptional the ε-collapses are.

After the construction of the vortical flow S_t it is natural to investigate the qualitative properties of the motion. As we mentioned above, the motion of three vortices is an integrable system. This means that the orbits of the system can be (at worst, by means of implicit formulas) analytically determined. The integrability of the system was first established by Poincaré. The explicit expression of the orbits was determined recently. However, if we consider the motion of a fourth vortex of zero vorticity intensity, in the velocity field generated by the three vortices system (called the reduced four-vortex problem in analogy with the analogous problem in celestial mechanics which is called the reduced three-body problem), it is possible to outline situations in which the motion of the fourth vortex is chaotic. This strongly resembles the Lagrangian turbulence which we have discussed in Chapter 3. By means of this observation, making use of perturbative techniques, it has been proved that the motion of four vortices is, in general, chaotic. Nevertheless, it has also been shown that, for some initial conditions, the motion of four vortices is quasi-periodic, that is, it is possible to find a suitable system of coordinates moving independently as an harmonic oscillator (for the above statements see the references in Section 4.6). We give here the main idea of the proof. The reader can find in the literature the rigorous details (for references see Section 4.6). Consider four vortices x_1, x_2, x_3, x_4 of equal intensity a. We put, initially, the pair x_1, x_2 very far from the pair x_3, x_4. Then the total Hamiltonian H may be written as the sum $H_0 + V$, where H_0 is the interaction of the two pairs of vortices, and the interaction between the center of vorticity of the first pair considered as a point of intensity $2a$ with the center of vorticity of the second pair considered as a point of intensity $2a$,

$$H = -\frac{a^2}{4\pi} \sum_{i,j=1; i \neq j}^{4} \ln |x_i - x_j|, \tag{2.46}$$

$$H_0 = -\frac{a^2}{2\pi} \ln |x_1 - x_2| - \frac{a^2}{2\pi} \ln |x_3 - x_4| - \frac{(2a)^2}{2\pi} \ln \left| \frac{x_1 + x_2}{2} - \frac{x_3 + x_4}{2} \right|. \tag{2.47}$$

$V = H - H_0$ is the remaining interaction. If we neglect the effect of V, the only term H_0 in the Hamiltonian gives rise to a quasi-periodic motion. Obviously, the whole Hamiltonian produces a more complicated motion, in general, nonintegrable. However, if the perturbation V is very small, we can apply the KAM theorem that ensures the existence (for V small enough) of a positive measure set of initial data for which the whole motion is quasi-periodic. V can be made small enough by choosing initially the two pairs of vortices very far apart.

A similar result can be obtained for vortices of different intensity, the only requirement being that H_0 gives rise to a quasi-periodic motion. Keeping in mind the explicit solution of the case in which only two vortices are present, this requirement implies that $a_1 \neq -a_2, a_3 \neq -a_4, a_1 + a_2 \neq -(a_3 + a_4)$.

By induction, we can extend a similar result for any number of vortices. Thus we have obtained a region of initial data for which the motion is quasi-periodic. We can prove that its measure is positive. A natural question arises: how large is it? The rigorous estimates require that the perturbation V must be extremely small to apply the KAM theorem, while numerical experiment suggests that the threshold of integrability is higher. (This gap between the rigorous estimates and the reality is common in almost all applications of the KAM theorem.)

For the vortex system, we can extend many considerations of classical mechanics. Here we analyze in some detail one of these, interesting in itself, and containing some ideas which will be useful in the sequel. We want to prove the analogue of the classical theorem of the center of mass and the only difficulty is related to the singularity of the interaction.

We consider a family of N vortices localized in the points $x_1^\delta, \ldots, x_N^\delta$. $\delta \in (0, 1)$ is a parameter. Suppose all the intensities $a_1, \ldots, a_N$ are positive. We divide the set of vortices in n clusters according to the partition $J_1, \ldots, J_n$ of the first N integers $1, \ldots, N$. Let us denote

$$z_k^\delta = \frac{1}{A_k} \sum_{i \in J_k} a_i x_i^\delta, \qquad k = 1, \ldots, n, \tag{2.48}$$

$$A_k = \sum_{i \in J_k} a_i, \tag{2.49}$$

the center of vorticity and the total charge of the kth cluster. We assume that, initially, the clusters have a small size

$$\{x_i^\delta | i \in J_k\} \subset \Sigma_\delta(z_k^\delta), \tag{2.50}$$

where, according to the previous notation, $\Sigma_\delta(z_k^\delta)$ denotes the circle of radius δ around z_k^δ. We suppose also that all the $\Sigma_\delta(z_k^\delta)$ are disjoint and that the limit

$$\lim_{\delta \to 0} z_k^\delta \equiv z_k \tag{2.51}$$

exists for all $k = 1, \ldots, n$.

We denote by $\{x_i^\delta(t)\}_{i=1,\ldots,N}$ and by $\{z_i^\delta(t)\}_{i=1,\ldots,n}$ the time evolution of the vortex system with initial data given by $\{x_i^\delta(0)\}_{i=1,\ldots,N} \equiv \{x_i^\delta\}_{i=1,\ldots,N}$ and the time evolution of the centers of vorticity of the clusters, respectively. Moreover, we introduce the following reduced dynamics:

$$\frac{d}{dt} z_k(t) = \nabla_k^\perp \sum_{i=1; i \neq k}^{n} A_k G(z_k(t) - z_i(t)),$$

$$z_i(t = 0) = z_i^\delta, \tag{2.52}$$

which is the vortex dynamics in which all the clusters are identified with their centers of vorticity. The following theorem shows that the dynamics expressed by (2.52) is close to the real dynamics. In other words, in situations in which the vortices cluster in groups, the true dynamics can be described by the evolutions of a system with a smaller number of degrees of freedom, in which each group is replaced by a single point vortex with a charge which is the sum of the charges of the vortices constituting the group. This result has important practical consequences: we are authorized (up to some accuracy) to consider a more natural, simpler system, in all situations in which the vortex system clusters into groups sharply concentrated around some points.

Theorem 2.3. *Under the above hypothesis, we have*

$$\lim_{\delta \to 0} z_\kappa^\delta(t) = z_\kappa(t) \qquad \text{for all} \quad k = 1, \ldots, n. \tag{2.53}$$

Moreover, the position of each vortex converges, at time $t > 0$, to the position of the center of vorticity of its cluster.

PROOF. The center of vorticity of each cluster evolves according to the equation

$$\frac{d}{dt} z_\kappa^\delta(t) = \frac{1}{A_k} \sum_{i \in J_k} a_i \sum_{h=1; h \neq k}^{n} \sum_{j \in J_h} \nabla_i^\perp a_j G(x_i^\delta(t) - x_j^\delta(t)). \tag{2.54}$$

Notice that if G could be replaced by G_ε (see definition (2.10)) the proof of Theorem 2.3 would be trivial. Actually, it would follow by the continuity of the solutions with respect to the initial data (which is true by the smoothness of G_ε), and the fact that the two particles sitting in the same point at time zero perform exactly the same trajectory. In our case, the difficulty in proving the theorem lies in the fact that the interaction is unbounded, so that, it is not at all obvious that the vortices of each group collapse to the center of vorticity of the group itself. We have to take into account that, because of the particular structure of the interaction, each vortex has the tendency to run around the center of vorticity of its cluster. This important feature is quantitatively expressed by the approximate conservation of the momentum of inertia of each cluster.

The strategy of the proof is the following. We first prove that a single cluster which initially concentrates in a point, if it evolves according to the vortex dynamics with an additional smooth external field, would remain concentrated around a time-dependent point up to some arbitrary time. In doing this, we make use of the conservation of the moment of inertia. We then apply this result to our case because the action on each cluster of all the others is a smooth field, it being impossible for any pair of clusters to merge.

Let $F = F(x, t)$ be a divergenceless vector field satisfying, globally, the Lipschitz condition

$$|F(x, t) - F(y, s)| \le L\{|t - s| + |x - y|\}. \tag{2.55}$$

Consider the cluster of vortices $\{x_1, \ldots, x_m\}$ and the equation of motion

$$\frac{d}{dt} x_k(t) = \nabla_i^\perp \sum_{i=1; i \ne k}^m a_k G(x_k(t), x_i(t)) + F(x_k(t), t), \tag{2.56}$$

$$x_k(t = 0) = x_k.$$

According to (2.56), the center of vorticity $z(t)$ of the above system evolves obeying the equation

$$\frac{d}{dt} z(t) = \frac{1}{A} \sum_{i=1}^m a_k F(x_i(t), t), \tag{2.57}$$

where

$$A = \sum_{i=1}^m a_k. \tag{2.58}$$

We consider the momentum of inertia $I(t)$ with respect to the center of vorticity

$$I(t) = \sum_{k=1}^m a_k(x_k(t) - z(t))^2 \tag{2.59}$$

and easily compute its time derivative

$$\frac{d}{dt} I(t) = 2 \sum_{k=1}^m a_k(x_k(t) - z(t)) \cdot \left[F(x_k(t)) - \frac{d}{dt} z(t) \right]$$

$$= 2 \sum_{k=1}^m a_k(x_k(t) - z(t)) \cdot F(x_k(t), t) - F(z(t), t), \tag{2.60}$$

where the last step is due to the definition of the center of vorticity implying

$$\sum_{k=1}^m a_k(x_k(t) - z(t)) \cdot -\frac{d}{dt} z(t) = 0, \tag{2.61}$$

$$\sum_{k=1}^m a_k(x_k(t) - z(t)) \cdot F(z(t), t) = 0. \tag{2.62}$$

By virtue of (2.60), (2.59), and (2.55), we conclude that

$$\left|\frac{d}{dt}I(t)\right| \le 2LI(t). \tag{2.63}$$

Finally, by the Gronwall lemma,

$$I(t) \le I(0)e^{2Lt}. \tag{2.64}$$

Suppose now that, at time zero, for a fixed point $z_0 \equiv z(0)$

$$\{x_i | i = 1, \ldots, m\} \subset \Sigma_\delta(z_0). \tag{2.65}$$

Then $I(0)$ is bounded by const. δ^2 and, by (2.64), the same bound also holds for $I(t)$. Since $I(t)$ is formed by positive contributions it follows that $|x_i(t) - z(t)|$ is of order δ. Thus the cluster does concentrate around $z(t)$.

Let us come back to our problem. It is convenient to introduce new dynamics $\{y_1(t), \ldots, y_N(t)\}$ defined as follows. The vortices of the same cluster interact, as usual, via the Green function G, while the vortices of different clusters interact via the regularized Green function G_ε with ε to be fixed later. This is described by the following equations:

$$\frac{d}{dt}y_i(t) = \sum_{k=1; k \ne h}^n \sum_{j \in J_k} a_j K_\varepsilon(y_i(t) - y_j(t)) + \sum_{j \in J_h} a_j K(y_i(t) - y_j(t)),$$

$$y_i(t = 0) = x_i, \qquad i \in J_h. \tag{2.66}$$

Fixed an arbitrary $T > 0$, we choose $\varepsilon > 0$ in such a way that

$$\min_{h \ne k} \inf_{0 \le t \le T} |z_h(t) - z_k(t)| > 10\varepsilon. \tag{2.67}$$

This is possible since the positivity of the charges A_k implies the existence of a minimal distance among the $z_k(t)$'s evolving according to (2.52), as discussed previously in this section.

Finally, we denote by $w_k(t)$ the center of vorticity of the cluster $\{y_i(t)\}_{i \in J_k}$. By using the equation of motion in integral form, we have

$$|w_h(t) - z_h(t)|$$

$$\le |w_h(0) - z_h| + \int_0^t ds \left\| \left[\frac{1}{A_h} \sum_{j \in J_h} a_j \sum_{k=1; k \ne h}^n \sum_{i \in J_k} a_i \nabla_j^\perp G_\varepsilon(y_j(s) - y_i(s)) \right.\right.$$

$$\left.\left. - \sum_{k=1; k \ne h}^n A_k \nabla_h^\perp G(z_h(s) - z_k(s)) \right] \right\|$$

$$\le |w_h(0) - z_h| + L_\varepsilon \int_0^t ds \sum_{k=1; k \ne h}^n A_k \{|w_h(s) - z_h(s)| + |w_k(s) - z_k(s)|\}$$

$$+ L_\varepsilon \int_0^t ds \left[\frac{1}{A_h} \sum_{j \in J_h} a_j \sum_{k=1; k \ne h}^n A_k |y_j(s) - w_h(s)| \right.$$

$$\left. + \sum_{k=1; k \ne h}^n \sum_{i \in J_k} a_i |y_i(s) - w_k(s)| \right]. \tag{2.68}$$

Here L_ε denotes a positive constant depending only on ε (related to the Lipschitz constant of G for the argument larger than 10ε). Using the Cauchy–Schwarz inequality, the last integral in the above estimate can be bounded by

$$\text{const. } L_\varepsilon \int_0^t ds \left\{ \frac{1}{A_h} \sum_{k=1;k \neq h}^n A_k \left[\sum_{j \in J_h} a_j |y_j(s) - w_h(s)|^2 \right]^{1/2} \sqrt{A_h} \right.$$

$$\left. + \sum_{k=1;k \neq h}^n \left[\sum_{i \in J_k} a_i |y_i(s) - w_k(s)|^2 \right]^{1/2} \sqrt{A_k} \right\}. \tag{2.69}$$

We sketch further steps. Equation (2.64) implies that the expression (2.69) vanishes as $\delta \to 0$. So, by the Gronwall lemma, by (2.68) we have the proof of the theorem for the approximated evolution. But when the clusters remain separated enough, the approximate dynamics coincide with the real one and this is the case if δ is small enough. $\square$

In this theorem we made the hypothesis that all vortices have the same sign. This is too restrictive. In fact, we can allow the clusters to have different signs (at least before the appearance of collapses). However, we must suppose that in each cluster there are vortices with the same sign. In fact, if in the same cluster vortices of different signs are present, the center of vorticity of the cluster can be very far from the geometrical center of the cluster itself, so the theorem is false.

4.3. The Vortex Motion in the Presence of Boundaries

In this section we want to study, in some more detail, the vortex motion in the presence of boundaries. We have established the equations of motion in Section 4.1 (see (1.14)). We notice primarily that a vortex system in a domain D is Hamiltonian. In fact, it can be written in the form

$$a_i \frac{d}{dt} x_{i_1} = \frac{\partial}{\partial x_{i_2}} H,$$

$$a_i \frac{d}{dt} x_{i_2} = -\frac{\partial}{\partial x_{i_1}} H, \tag{3.1}$$

where

$$H = \frac{1}{2} \sum_{i;j=1,i \neq j}^N a_i a_j G_D(x_i, x_j) + \frac{1}{2} \sum_{i=1}^N a_i^2 \gamma_D(x_i). \tag{3.2}$$

As in the full-plane case, a choice for the canonical conjugate variables is $\sqrt{|a_i|} x_{i_1}$ and $\sqrt{|a_i|} x_{i_2}$ sign a_i. Obviously, the Liouville theorem holds so that the Lebesgue measure

$$\prod_{i=1}^N dx_i \quad \text{restricted on } D^N \tag{3.3}$$

is preserved during the motion.

Figure 4.5

To give a physical intuition of the vortex motion in a general domain D, we consider some examples. We first observe that a single vortex is not always at rest as in the whole-plane case. For the half-plane case ($x_1 > 0$), the Green function $G(x, y)$ is the sum of the free Green function plus a contribution given by an image charge symmetrically located with respect to the x_2-axes

$$G(x, y) = -\frac{1}{2\pi} \ln |x - y| + \frac{1}{2\pi} \ln |x - y^*|, \qquad (3.4)$$

where $y^* = (-y_1, y_2)$. It follows that a single vortex obeys the equation

$$\frac{d}{dt} x = \frac{1}{4\pi} \nabla^\perp \ln |2x_1| = \left(0, -\frac{1}{2\pi x_1} \right). \qquad (3.5)$$

Therefore the motion of a single vortex in the half-plane $x_1 > 0$ is a uniform motion, parallel to the x_2-axes, with velocity proportional to the inverse abscissa.

Two vortices give rise to a four-vortex system in the plane: the two given vortices plus the two images. Consider, for example, two vortices of the same intensity. If they are quite far apart from the boundary its action is weak. The motion of the two vortices is almost circular. However, to this motion must to be added a uniform translation of the center of vorticity along an axis parallel to the boundary (Fig. 4.5). If one of the two vortices is sufficiently close to the boundary, the motion can be qualitatively very different: if they are rather far apart, the motion is, asymptotically, a uniform translation along two axes parallel to the boundary (Fig. 4.6). If they are sufficiently close to each other an intermediate case can occur (see Exercise 4) (Fig. 4.7).

Figure 4.6

Figure 4.7

For a higher number of vortices the motion is much more complicated. We notice also that in the half-plane case, we have, in addition to the energy H, only another first integral of the motion

$$B_2 = \frac{\sum_{i=1}^{N} a_i x_{i_2}}{\sum_{i=1}^{N} a_i}, \tag{3.6}$$

all other symmetries, except the translation along the x_2-axis, being broken by the presence of the boundary.

Another interesting exarnple is given by a circular domain of radius R. In this case, the Green function is

$$G_D(x, y) = -\frac{1}{2\pi} \ln \frac{|x - y| R}{|y| |x - y^*|}, \tag{3.7}$$

where

$$y^* = \left(\frac{R^2 y_1}{|y|^2}, \frac{R^2 y_2}{|y|^2} \right), \qquad y = (y_1, y_2), \tag{3.8}$$

is the point conjugate to y.

A single vortex of unit intensity satisfies the equation

$$\frac{d}{dt} x = \frac{1}{4\pi} \nabla^{\perp} \ln \{ |x| |x - y^*| \}$$

$$= \frac{1}{4\pi} \nabla^{\perp} \ln \{ R^2 - |x|^2 \}. \tag{3.9}$$

Therefore the motion is circular and the velocity diverges when the initial data of the vortex approaches the boundary. In this case, the only first integral surviving, beyond the energy, is

$$I = \sum_{i=1}^{N} a_i x_i^2. \tag{3.10}$$

By the existence of such first integrals, it follows that the two vortex motion is integrable (this follows by the Liouville theorem on integrable Hamiltonian systems). However, such motion is not geometrically simple (Fig. 4.8).

When the number of vortices increases, the motion is difficult to investigate and, in general, is presumably chaotic. For general domains, due to a

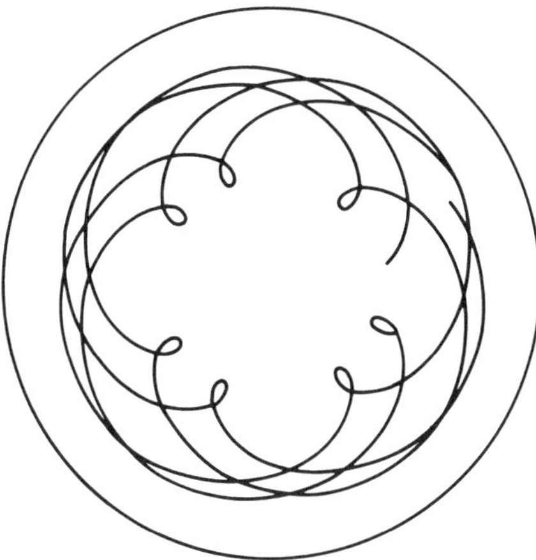

Figure 4.8

complete lack of the first integrals related to the symmetries of the domain D, the motion of two or more vortices is not expected to be integrable.

The motion of a single vortex, even if integrable, may be not completely trivial to understand. This is a Hamiltonian system whose equation is

$$\frac{d}{dt}x(t) = \nabla^{\perp}\gamma(x(t)), \qquad (3.11)$$

where $\gamma(x) = \frac{1}{2}\gamma(x, x)$. From (3.11) we know that the trajectories of the vortex are the curves $\gamma = $ const. (which, in general, are not explicitly known). However, the motion can be qualitatively understood in some cases. In convex relatively compact domains, $\nabla^{\perp}\gamma$ has a unique critical point and the domain D is spanned by the curves running around the critical point in a situation which is topologically similar to that of a circle (Fig. 4.9). The existence of a

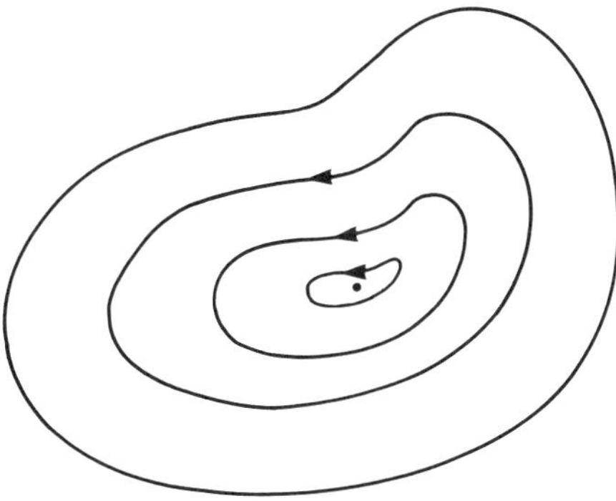

Figure 4.9

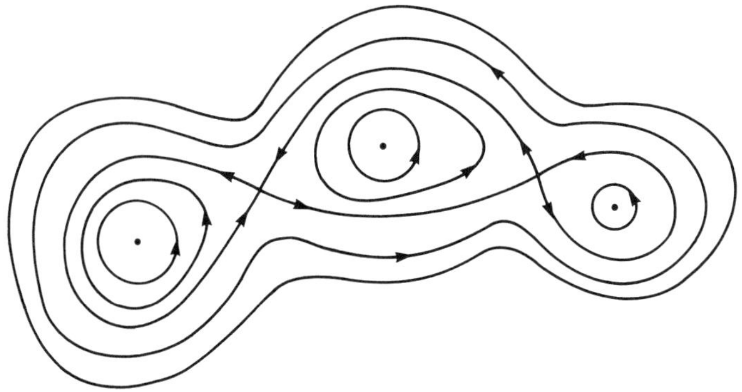

Figure 4.10

unique critical point for a convex domain (excluding the infinite strip) follows by arguments from two-dimensional potential theory (see Section 4.6 for references).

Nonconvex domains could give rise to several critical points for which the structure of the trajectories is more complicated (Fig. 4.10). An interesting domain which has already been considered many times, and for which we spend more words in connection with the vortex motion, is the flat torus $D = [-\pi, \pi]^2$ (with periodicity conditions). The Green function is explicitly known by means of a series expansion. We note that the series is not absolutely convergent. In this case, the motion of a single vortex does not make sense; in fact, by the circulation theorem and the periodicity of the velocity field, it follows that the total vorticity must vanish. Therefore, we can consider only vortex configurations for which the total vortex intensity is zero. By using electrostatic language, we will call such systems "neutral."

Another way to visualize the situation is to introduce the periodic images of the domain under consideration (Fig. 4.11). It is easy to realize that the motion of a vortex system in D is equivalent to the motion in the whole plane of the periodic system constituted by infinitely many vortices obtained by the original one by periodicity. In order to make sense of such a system, the velocity field generated by the whole system must be finite computed on a

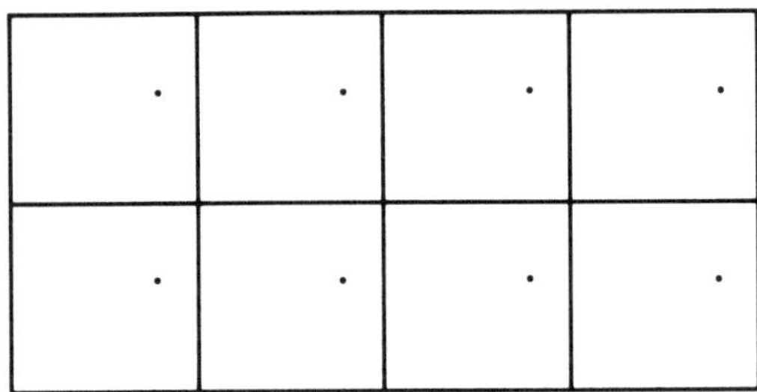

Figure 4.11

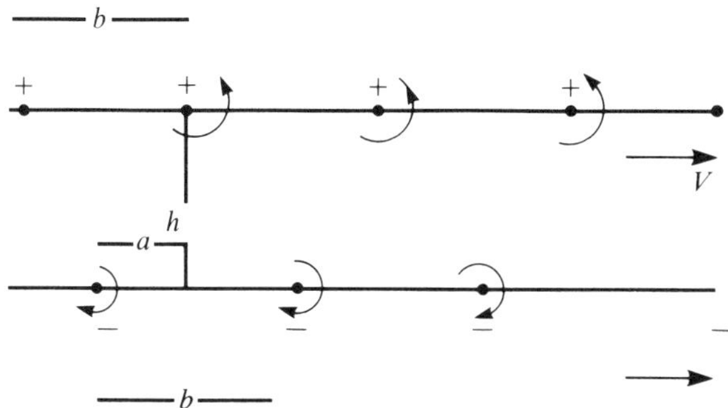

Figure 4.12

single vortex and is finite. Since the velocity decays at infinity like $1/|x|$, it is suddenly seen that the sum of all contributions is certainly diverging if the system is not neutral. In the case of neutrality, the dominant term decays like $1/|x|^2$ and the series is still absolutely (logarithmically) diverging. However, in this case, we can sum the contributions in such a way as to take advantage of the particular geometry.

We conclude this section, briefly describing a feature of real fluids, called the Von Karman street, which has been widely investigated. This is a special vortex configuration, consisting of an infinite sequence of positive and negative vortices, of the same intensity according Fig. 4.12. We denote by h and b the distances between the two straight lines on which the vortices are placed and the vortex distance, respectively. Moreover, a denotes the shift between the negative and positive configurations. Notice that the velocity field computed in each vortex is the Cauchy sum of the field produced by the vortices of opposite sign, and that the whole configuration is moving by a rigid translation of constant speed V.

Configurations of this type arise in real flows past obstacles. Notice that the viscosity is responsible for the formation of this array, however, its time evolution is essentially inviscid for suitable scales of times, after which the viscosity destroys everything. The linear stability of the Von Karman street has been established for suitable values of the parameters a, b, h. All the other values make the configuration unstable.

4.4. A Rigorous Derivation of the Vortex Model

In Section 4.1 we established the vortex model in a heuristic way: when the initial profile of vorticity is sharply concentrated around some points $x_1, \ldots, x_N$, its time evolution is expected to be approximately described by the solution of the vortex equations. However, the fundamental equation governing the evolution of an incompressible nonviscous flow is the Euler equation derived in Chapter 1. This equation, as we have seen in Chapter 2, makes sense for an essentially bounded vorticity profile, or it is conceivable at most for data which have locally bounded energy. It seems not easy to give a simple meaning to the Euler equation for data whose vorticity is a linear combination of δ functions, which is exactly the situation we are dealing with when we want to describe the vortex evolution. Thus the following fundamental question arises: Is the vortex model something different from the genuine Euler evolution, requiring new "ad hoc" physical hypotheses, or can the model itself be explained in terms of the Euler equation? Were the last conjecture true, we should be able to prove that, when the vorticity is concentrated on very small regions (where it is very large), the Euler dynamics is, in some way, similar to the vortex dynamics in which the point vortices are localized in these small regions, and the vortex intensities are the total vorticity associated to such small domains. To be more precise, a rigorous derivation of the vortex model (by means of the Euler dynamics) would be a proof of the following fact. Suppose that at time zero

$$\omega_0(x)\, dx \to \sum_{i=1}^{N} a_i \delta_{x_i}\, (dx), \tag{4.1}$$

then

$$\omega_t(x)\, dx \to \sum_{i=1}^{N} a_i \delta_{x_i(t)}\, (dx), \tag{4.2}$$

where we denote by ω_t the solution of the Euler equation with initial datum given by ω_0, and by $\{x_i(t)\}$ the solution of the initial value problem associated to the vortex equation with initial datum given by $\{x_i\}$. The convergences (4.1) and (4.2) have to be understood in the sense of weak convergence of measures. If the convergence (4.2) follows from (4.1) only, we could be sure that the vortex model is nothing other than the Euler dynamics seen in some asymptotic regime. Property (4.2) is not trivial to prove and presents subtle features which we will discuss throughout this section.

We will start our analysis by considering the case of a single vortex in the whole plane, under the action of a given smooth divergenceless external field, and we choose a special sequence of approximations for the δ function at time zero. The following theorem expresses a property of localization of the solution of the Euler equation which is the basis of the validity of the vortex model which will be discussed later on.

Theorem 4.1. *Let Λ_ε, $\varepsilon \in (0, \varepsilon_0)$, be a family of open sets satisfying the conditions*

$$\text{meas } \Lambda_\varepsilon = \varepsilon^2, \qquad \Lambda_\varepsilon \subset \Sigma(x^*, \alpha\varepsilon), \tag{4.3}$$

for some $\alpha > 0$, where $\Sigma(x^*, R)$ denotes the disk of radius R centered in x^*. Denote by

$$\omega_{\varepsilon,0}(x) = \varepsilon^{-2}\chi_{\Lambda_\varepsilon}(x) \tag{4.4}$$

(where $\chi_{\Lambda_\varepsilon}$ denotes the characteristic function of the set Λ_ε) an initial profile of vorticity, and by $\omega_{\varepsilon,t}(x) = \varepsilon^{-2}\chi_{\Lambda_\varepsilon(t)}(x)$ the solution of the Euler equation, in

weak form, associated to the initial condition (4.4)

$$\frac{d}{dt}\,\omega_{\varepsilon,t}(f) = \omega_{\varepsilon,t}([u_{\varepsilon,t} + F]\cdot\nabla f), \tag{4.5}$$

$$u_{\varepsilon,t} = K * \omega_{\varepsilon,t}. \tag{4.6}$$

(Here, as usual, we use the notation $\omega_{\varepsilon,t}(f) = \int \omega_{\varepsilon,t}(x)f(x)\,dx = \int_{\Lambda_\varepsilon(t)} f(x)\,dx$
and f *denotes a smooth function.)*

Let F *be a divergence-free, uniformly bounded, time-dependent vector field
satisfying the Lipschitz condition*

$$|F(x, t) - F(y, t)| \le L|x - y| \tag{4.7}$$

for some $L > 0$. *Then, for an arbitrary fixed* $T > 0$,

(i)
$$\lim_{\varepsilon \to 0} B_\varepsilon(t) = B(t), \tag{4.8}$$

where $B_\varepsilon(t) = \int dx\, \omega_{\varepsilon,t}(x)x$ *is the center of vorticity of the patch* $\Lambda_\varepsilon(t)$, *and* $B(t)$
is the solution of the initial value problem

$$\frac{d}{dt}\,B(t) = F(B(t), t),$$
$$\tag{4.9}$$
$$B(0) = x^*.$$

(ii)
$$\lim_{\varepsilon \to 0} \omega_{\varepsilon,t}(f) = f(B(t)), \qquad t \in [0, T]. \tag{4.10}$$

(iii) *For all* $d > 0$, *we can choose* $\varepsilon_0(d, T) > 0$ *such that, if* $\varepsilon < \varepsilon_0$, *then*

$$\Lambda_\varepsilon(t) \subset \Sigma(B_\varepsilon(t), d), \qquad t \in [0, T]. \tag{4.11}$$

Remark 1. The above theorem does not assert that the motion of the fluid
particles supporting the vorticity ω_ε converges, in the limit $\varepsilon \to 0$, to $B(t)$. This
is, in general, false. The motion of such fluid particles, due to the singularity
of the kernel $K(x - y)$, is very irregular and does not converge at all. However, the motion of the center of vorticity converges to the motion of a single
point vortex in the velocity field F (see statement (i)).

Remark 2. The above theorem is a preliminary step in proving the validity of
the vortex model: we are looking at the behavior of a single vortex, assuming
that the field generated by all the others is given and smooth.

Remark 3. Statements (i) and (ii) assert that the patch $\Lambda_\varepsilon(t)$ is mostly localized
around $B_\varepsilon(t)$. However, they do not exclude the existence of filaments of
vanishing measure going very far away from $B_\varepsilon(t)$. Statement (iii) says more:
$\Lambda_\varepsilon(t)$ is strictly localized around $B_\varepsilon(t)$ and this will be essential in proving the
validity of the vortex model.

PROOF. The difficulty of the proof arises from the singularity of the kernel K
which forces a fluid particle to rotate with a very large velocity around the

center of vorticity. To overcome this difficulty we study the motion of the center of vorticity which will turn out to be much more regular than the motion of a given fluid particle. Moreover, the angular momentum is almost conserved during the motion, so that we can also control the spreading of the vorticity distribution around the center of vorticity.

The proof is rather technical and will be divided into three steps. In the first, we will prove that the moment of inertia around $B_\varepsilon(t)$

$$I_\varepsilon(t) = \varepsilon^{-2} \int_{\Lambda_\varepsilon(t)} dx \, (x - B_\varepsilon(t))^2 \tag{4.12}$$

is vanishing in the limit $\varepsilon \to 0$ because it is almost conserved.

As a second step we prove (i) and (ii) which will be straightforward. The last step consists in proving the localization property (iii). In doing this, we control the flux of vorticity through any circle around $B_\varepsilon(t)$. This is not too large because the field is essentially tangent to the boundary.

Step 1. By taking the time derivative of B_ε, we have

$$\frac{d}{dt} B_\varepsilon(t) = \int dx \, F(x, t)\omega_{\varepsilon,t}(x) = \varepsilon^{-2} \int_{\Lambda_\varepsilon(t)} dx \, F(x, t). \tag{4.13}$$

To obtain (4.13) we have used the Euler equation and the identity

$$\int_{\Lambda_\varepsilon(t)} dx \int_{\Lambda_\varepsilon(t)} dy \, K(x - y) = 0. \tag{4.14}$$

If the external field F would vanish, then both I_ε and B_ε would be constant in time. In general

$$\frac{d}{dt} I_\varepsilon(t) = 2 \int dx \, (x - B_\varepsilon(t)) \cdot F(x, t)\omega_{\varepsilon,t}(x) \tag{4.15}$$

as follows by a direct computation, making use of the identity

$$\int_{\Lambda_\varepsilon(t)} dx \int_{\Lambda_\varepsilon(t)} dy \, x \cdot K(x - y) = \frac{1}{2} \int_{\Lambda_\varepsilon(t)} dx \int_{\Lambda_\varepsilon(t)} dy \, (x - y) \cdot K(x - y) = 0 \tag{4.16}$$

due to the antisymmetry of K.

Making use of the fact that $\int dx \, (x - B_\varepsilon(t)) \cdot F(B_\varepsilon(t), t)\omega_{\varepsilon,t}(x) = 0$ and the Lipschitz continuity of F, we have

$$\left| \frac{d}{dt} I_\varepsilon(t) \right| \leq 2L \int dx \, (x - B_\varepsilon(t))^2 \omega_{\varepsilon,t}(x) = 2LI_\varepsilon(t) \tag{4.17}$$

from which

$$I_\varepsilon(t) \leq I_\varepsilon(0) \exp 2Lt. \tag{4.18}$$

Therefore,

$$\lim_{\varepsilon \to 0} I_\varepsilon(t) = 0 \qquad \text{(uniformly in } t \in [0, T]), \tag{4.19}$$

since by the weak convergence at time zero it follows that $I_\varepsilon(0) \to 0$ as $\varepsilon \to 0$.

Step 2. We have

$$
\begin{aligned}
|B(t) - B_\varepsilon(t)| &\le |x^* - B_\varepsilon(0)| + \int_0^t ds \left| F(B(s), s) - \varepsilon^{-2} \int_{\Lambda_\varepsilon(s)} dx\, F(x, s) \right| \\
&\le |x^* - B_\varepsilon(0)| + \int_0^t ds\, |F(B(s), s) - F(B_\varepsilon(s), s)| \\
&\quad + \int_0^t ds \left| F(B_\varepsilon(s), s) - \varepsilon^{-2} \int_{\Lambda_\varepsilon(s)} dx\, F(x, s) \right| \\
&\le |x^* - B_\varepsilon(0)| + L \int_0^t ds\, |B(s) - B_\varepsilon(s)| \\
&\quad + L\varepsilon^{-2} \int_0^t ds \int_{\Lambda_\varepsilon(s)} dx\, |B_\varepsilon(s), s) - x| \\
&\le |x^* - B_\varepsilon(0)| + L \int_0^t ds\, |B(s) - B_\varepsilon(s)| + LT \sup_{0 \le t \le T} \sqrt{I_\varepsilon(t)}.
\end{aligned}
$$

(4.20)

By the Gronwall lemma, because the third and first terms in the right-hand side of (4.20) are vanishing in the limit $\varepsilon \to 0$, we finally achieve the proof of (i). The proof of (ii) follows from (4.19).

Step 3. We first prove that the amount of vorticity crossing the boundary of a small disk around B_ε is small. We show then that, as a consequence of this, the radial part of the velocity field is also small so that the particle paths cannot go far apart from B_ε.

To control the vorticity flux we find it convenient to introduce the following function $W_R \in C^\infty(\mathbb{R}^2)$, depending only on $|r|$, such that:

$$
W_R(r) = \begin{cases} 1 & \text{if } |r| \le R, \\ 0 & \text{if } |r| > 2R, \end{cases}
$$

(4.21)

such that, for some $C_1 > 0$,

$$
|\nabla W_R(r)| \le \frac{C_1}{R},
$$

(4.22)

$$
|\nabla W_R(r) - \nabla W_R(r')| \le \frac{C_1}{R^2} |r - r'|.
$$

(4.23)

Define the quantity

$$
\mu_R(t) = 1 - \varepsilon^{-2} \int_{\Lambda_\varepsilon(0)} dx\, W_R(B_\varepsilon(t) - x_\varepsilon(t)),
$$

(4.24)

where $x_\varepsilon(t)$ are the particle paths leaving x at time zero.

Notice that if $\Lambda_\varepsilon(t) \subset \Sigma(B_\varepsilon(t), R)$, then $\mu_R(t) = 0$. Thus we choose μ_R as a measure of the localization of Λ_ε around B_ε (for R sufficiently large compared with ε, but infinitesimal in ε). We then evaluate the time derivative

$$\frac{d\mu_R(t)}{dt} = -\varepsilon^{-2} \int_{\Lambda_\varepsilon(0)} dx\, \nabla W_R(B_\varepsilon(t) - x_\varepsilon(t)) \cdot \frac{d}{dt}[B_\varepsilon(t) - x_\varepsilon(t)]$$

$$= \varepsilon^{-4} \int_{\Lambda_\varepsilon(0)} dx\, \nabla W_R(B_\varepsilon(t) - x_\varepsilon(t)) \cdot \int_{\Lambda_\varepsilon(0)} dy\, K(x_\varepsilon(t) - y_\varepsilon(t))$$

$$+ \varepsilon^{-4} \int_{\Lambda_\varepsilon(0)} dx\, \nabla W_R(B_\varepsilon(t) - x_\varepsilon(t)) \cdot \int_{\Lambda_\varepsilon(0)} dy\, F(x_\varepsilon(t), t) - F(y_\varepsilon(t), t).$$

$$(4.25)$$

We now estimate the first term in the right-hand side of (4.25). By the Liouville theorem and the antisymmetry of K, we can write it as

$$\frac{\varepsilon^{-4}}{2} \int_{\Lambda_\varepsilon(t)} dx \int_{\Lambda_\varepsilon(t)} dy\, \{\nabla W_R(B_\varepsilon(t) - x) - \nabla W_R(B_\varepsilon(t) - y)\} \cdot K(x - y). \quad (4.26)$$

To estimate the term (4.26) we split the integration domain into four parts

$$T_1 = \{(x, y) | x \in \Sigma(B_\varepsilon(t), R)^c,\, y \in \Sigma(B_\varepsilon(t), \gamma)\},$$

$$T_2 = \{(x, y) | x \in \Sigma(B_\varepsilon(t), R)^c,\, y \notin \Sigma(B_\varepsilon(t), \gamma)\},$$

$$T_3 = \{(x, y) | y \in \Sigma(B_\varepsilon(t), R)^c,\, x \in \Sigma(B_\varepsilon(t), \gamma)\},$$

$$T_4 = \{(x, y) | y \in \Sigma(B_\varepsilon(t), R)^c,\, x \notin \Sigma(B_\varepsilon(t), \gamma)\},$$

where Σ^c denotes the complement of Σ and $\gamma = R^6$, and from now on we suppose $R < 1$. Moreover, we denote by $S_i = T_i \cap \Lambda_\varepsilon(t) \times \Lambda_\varepsilon(t)$. Therefore, since $\gamma < R$ and $\nabla W_R(x) \cdot K(x) = 0$, we have

$$\frac{\varepsilon^{-4}}{2} \left| \int_{S_1} dx\, dy\, \nabla W_R(B_\varepsilon(t) - x) \cdot \{K(x - y) - K(B_\varepsilon(t) - x)\} \right|$$

$$\leq \frac{\varepsilon^{-4}}{2} C_1 \left| \int_{S_1} dx\, dy\, \frac{|B_\varepsilon(t) - y|}{R(R - \gamma)^2} \right|$$

$$\leq \tfrac{1}{2} C_1 m_t(R) \frac{\gamma}{R(R - \gamma)^2} \leq C_2 m_t(R) R^3. \quad (4.27)$$

Here we set

$$m_t(R) = \varepsilon^{-2} \operatorname{meas}(\Lambda_\varepsilon(t) \cap \Sigma(B_\varepsilon(t), R)^c) \quad (4.28)$$

which is the amount of vorticity outside $\Sigma(B_\varepsilon(t), R)$. In the second step we used (4.22) and the fact that $|x - y| > R - \gamma$. Finally, in the last ones, we used that $\varepsilon^{-2} \operatorname{meas} \Lambda_\varepsilon(t) = 1$ and that $R - R^6 > \tfrac{1}{2}R$ if R is sufficiently small.

To estimate the contribution over S_2 we use, thanks to the obvious inequality $|K(x)| \leq C|x|^{-1}$, the bound

$$|\{\nabla W_R(B_\varepsilon(t) - x) - \nabla W_R(B_\varepsilon(t) - y)\} \cdot K(x - y)| \leq C_3 R^{-2} \quad (4.29)$$

from which we estimate the integral on S_2 by

$$C_4 \frac{m_t(\gamma)m_t(R)}{R^2}. \tag{4.30}$$

The integrals over S_3 and S_4 can be handled in exactly the same manner changing the role of x and y.

To achieve the estimate of the time derivative of μ_R we evaluate the second integral in the right-hand side of (4.25). It is

$$\varepsilon^{-4} \int_{\Lambda_\varepsilon(t)} dx \, \nabla W_R(B_\varepsilon(t) - x) \cdot \int_{\Lambda_\varepsilon(t)} dy \, \{F(x, t) - F(y, t)\}. \tag{4.31}$$

We split the domain of integration in y into two regions $|B_\varepsilon(t) - y| > R$ and its complement. The first contribution is bounded by

$$2C_1 \|F\|_\infty \frac{m_t(R)^2}{R} \tag{4.32}$$

while the second one is certainly bounded by

$$C_5 m_t(R) \tag{4.33}$$

since in this region

$$|\nabla W_R(B_\varepsilon(t) - x)| \, |x - y| \le \text{const.}$$

Before collecting all the above estimates we estimate $m_t(R)$ in terms of $I_\varepsilon(t)$

$$m_t(R) \le \frac{1}{(R\varepsilon)^2} \int_{\Lambda_\varepsilon(t)} x^2 \chi(|B_\varepsilon(t) - x| > R) \le \frac{I_\varepsilon(t)}{R^2} \le C_6 \frac{\varepsilon^2}{R^2} \tag{4.34}$$

(here we used (4.18)) so that

$$\left| \frac{d\mu_R}{dt} \right| \le C_5 m_t(R) + A(R, \varepsilon), \tag{4.35}$$

where

$$A(R, \varepsilon) = C_7 \varepsilon^2 (R + \varepsilon^2 R^{-16}). \tag{4.36}$$

On the other hand, we can bound $m_t(R)$ in terms of $\mu_{R/2}$

$$m_t(R) = \frac{1}{\varepsilon^2} \int_{\Lambda_\varepsilon(t)} \chi(|B_\varepsilon(t) - x| > R) \, dx$$

$$= 1 - \frac{1}{\varepsilon^2} \int_{\Lambda_\varepsilon(t)} \chi(|B_\varepsilon(t) - x| \le R)$$

$$\le 1 - \frac{1}{\varepsilon^2} \int_{\Lambda_\varepsilon(t)} W_{R/2}(B_\varepsilon(t) - x) \tag{4.37}$$

so that from (4.35) we obtain the integral inequality

$$\mu_R(t) \le TA(R, \varepsilon) + C_5 \int_0^t d\tau \, \mu_{R/2}(\tau). \tag{4.38}$$

Notice that the above inequality is valid for all R sufficiently small but large enough that $\mu_R(0) = 0$, i.e., $R \geq 2\alpha\varepsilon$. Therefore we can iterate the inequality (4.38) k times, if k satisfies the condition

$$2^{-k}R \geq 2\alpha\varepsilon. \tag{4.39}$$

We now choose $R = \varepsilon^{1/100}$ and $k = $ integer part of $D|\log \varepsilon|$ ($D = \frac{1}{3}$, for instance) so that (4.39) is certainly satisfied for ε sufficiently small. Hence, for ε sufficiently small,

$$\mu_R(t) \leq \frac{C_5^k t^k}{k!} + \sum_{s=0}^{k-1} T^{s+1} C_5^s A(R2^{-s}, \varepsilon) \tag{4.40}$$

$$\leq C_8 \varepsilon^{100} + C_9 \varepsilon^\beta \leq C_{10}\varepsilon^\beta, \qquad \beta > 2, \tag{4.41}$$

where we have chosen D small enough. Thus the amount of vorticity escaping the disk $\Sigma(B_\varepsilon(t), \varepsilon^{1/100})$ is vanishing at least as ε^β with $\beta > 2$. This information allows us to conclude the proof. Consider the disk $\Sigma_2 = \Sigma(B_\varepsilon(t), \varepsilon^{1/300})$. A particle localized in x, outside the boundary of such a disk, is moving under the action of three fields: one generated by the vorticity inside the disk $\Sigma(B_\varepsilon(t), \varepsilon^{1/100}) = \Sigma_1$, say u_1; another one, u_2, generated by the vorticity outside the disk Σ_1; and u_3 due to the external field. Let n be the versor in the direction $B_\varepsilon(t) - x$. Then

$$|u_1(x) \cdot n| = \left| n \cdot \varepsilon^{-2} \int_{\Sigma_1 \cap \Lambda_\varepsilon(t)} dy\, K(x - y) \right|$$

$$= \left| n \cdot \varepsilon^{-2} \int_{\Sigma_1 \cap \Lambda_\varepsilon(t)} dy\, \{K(x - y) - K(x - B_\varepsilon(t))\} \right|$$

$$\leq C_{11} \frac{\varepsilon^{1/100}}{(\varepsilon^{1/300} - \varepsilon^{1/100})^2} \to 0 \qquad \text{as} \quad \varepsilon \to 0. \tag{4.42}$$

Moreover

$$|u_2(x)| = \left| \varepsilon^{-2} \int_{\Sigma_1^c \cap \Lambda_\varepsilon(t)} dy\, K(x - y) \right| \leq \left| \varepsilon^{-2} \int_{|x-y|<r(\varepsilon)} \frac{1}{|x - y|} \right|. \tag{4.43}$$

Here we majorize the field by putting all the vorticity outside Σ_1 in a circular configuration of radius $r(\varepsilon)$. By a simple calculation it follows that $r(\varepsilon) = $ const. $\varepsilon^{(\beta+2)/2}$ from which

$$|u_2(x)| \leq C_{12}\varepsilon^{\beta/2-1} \to 0 \qquad \text{as} \quad \varepsilon \to 0. \tag{4.44}$$

Finally, the Lipschitz condition on the extemal field ensures that $u_3 \to 0$ as $\varepsilon \to 0$.

Therefore, the fields which could be responsible for carrying the particle paths far from $B_\varepsilon(t)$ are arbitrarily small and it follows that after a finite time T, $\Lambda_\varepsilon(t)$ must be contained in a fixed circle $\Sigma(B_\varepsilon(t), d)$, for an arbitrary d, provided that ε is sufficiently small. $\qquad\qquad\square$

Remark. Notice that estimate (4.18) on the moment of inertia is not enough to prove that u_2 is vanishing. Actually, it yields only that the amount of vorticity outside a disk of a fixed radius d around $B_\varepsilon(t)$ is of order ε^2. It generates a bounded field not vanishing a priori. Thus we need the more sophisticated analysis described in Step 3 to prove the complete localization (iii).

The above result can be used for the study of many vortices. Single out a blob, and consider the action of the others on it as an external perturbation field. If all the other blobs are rather far apart, they generate a Lipschitz vector field. On the other hand, we have proved that the blobs remain localized. Thus, it is not difficult to achieve our program.

Theorem 4.2. *Let us define*

$$\omega_\varepsilon(x) = \varepsilon^{-2} \sum_{i=1}^{N} a_i \chi_{\Lambda_\varepsilon^i}(x), \tag{4.45}$$

where Λ_ε^i are disjoint open regions such that

$$\text{meas } \Lambda_\varepsilon^i = \varepsilon^2, \qquad \Lambda_\varepsilon^i = \Sigma(x_i, \alpha\varepsilon), \qquad \alpha > 0 \tag{4.46}$$

and a_i are real numbers. Then, for all $T > 0$, the following holds:

$$\lim_{\varepsilon \to 0} \omega_{\varepsilon,t}(t) = \sum_{i=1}^{N} f(x_i(t)), \tag{4.47}$$

where $\omega_{\varepsilon,t}$ denotes the (weak) solution of the Euler equation with an initial condition given by ω_ε, and $\{x_i(t)\}_{i=1,\ldots,N}$ are the solutions of the vortex model with vorticity intensity $\{a_i\}_{i=1,\ldots,N}$ and initial condition given by $\{x_i\}_{i=1,\ldots,N}$, provided that the solution of the vortex model does not develop collapses within the time T.

PROOF. The proof follows readily by the following considerations. Let b be the minimal distance at which any couple of vortices can arrive in time T, according to the vortex dynamics, for the initial datum $\{x_i\}_{i=1,\ldots,N}$. We choose $\alpha\varepsilon \ll b$. Consider, for the initial condition ω_ε, the following regularized dynamics for the Euler equation: two disjoint blobs interact via a kernel K_η, $\eta \ll b$. K_η is defined in (2.11). Each blob interacts with itself via the singular K (see the analogy with the dynamics defined in (2.11)). By Theorem 4.1, it is not difficult to prove Theorem 4.2 with the Euler dynamics replaced by the regularized dynamics. In fact, the blobs remain localized around their center of vorticity, and hence around $x_i(t)$ (see (i) of Theorem 4.1). We can choose ε small that the minimal distance between two blobs is larger than $b/2$. On the other hand, if η is chosen sufficiently small, these regularized dynamics coincide with the real one. This concludes the proof. $\square$

Supplementary comments and references will be given in the Section 4.6.

4.5. Three-Dimensional Models

The vortex model constitutes an important tool of investigation for two-dimensional fluid dynamics. The reduction to an ordinary differential system, in principle simpler than the original Euler equation, gives an important insight into the nature of two-dimensional flows. So that a natural question arises: Is it possible to establish an analogous model for three-dimensional flows?

Let us suppose that, initially, the vorticity field is concentrated on regions of small measure, for example, a vorticity tube with a small section. For simplicity, we will work in $\mathbb{R}^3$. The conservation of the flux of vorticity implies that a vortex tube must be closed (for instance, a vorticity ring) or infinitely extended. Let us begin by considering a vortex ring with a small section in the limit when the vorticity concentrates on a circle. According to the Biot–Savart law (see Chapter 1) we can compute the velocity field generated by such a vorticity configuration. Suppose that the circular vortical filament is posed on the plane $z = 0$ with the center in the origin. Then

$$u(Q) = \frac{1}{4\pi} \int \frac{\omega \wedge QP}{|QP|^3}\, dP, \tag{5.1}$$

where P is the generic point on the circle, dP is the line measure, and ω is the vorticity intensity which is tangent to the circle.

Passing to cylindrical coordinates, putting Q on the circle $Q = (R, 0, 0)$, $P = (R, \theta, 0)$ (here R denotes the radius of the circle), we have

$$|QP| = 2R \sin \frac{\theta}{2}, \tag{5.2}$$

$$|\omega \wedge QP| = A2R \sin^2 \frac{\theta}{2}, \tag{5.3}$$

where $A = |\omega|$. Assuming $A = $ const., the modulus of the velocity in Q, which is directed along z, is

$$\frac{A}{16\pi R^2} \int_0^{2\pi} \frac{1}{\sin(\theta/2)}\, d\theta. \tag{5.4}$$

This expression is manifestly divergent.

We could try to subtract this infinity, due to the self-interaction, in analogy with the two-dimensional case, singling out a suitable regular part. However, this seems a difficult job. It is easy to realize that if the vortical ring has a finite cross section, the velocity of each element of the ring is finite but the velocity is increasing when the cross section goes to zero. In other words, the familiar smoke ring moves along its symmetry axis with a speed which increases when the section of the ring decreases, and in the limit situation when the ring becames a filament, the speed diverges. This is a matter of experience, so that a naive theory of vortex filaments, which is mathematically consistent and describes real physics, is impossible. From the expression (5.4) we remark that the only possibility to give sense to the speed of a filament is that $R = \infty$, which means that the filament is a line of constant vorticity intensity. Thus we are in the same two-dimensional situation discussed in the last four sections.

A more direct generalization of the vortex model is the so-called vorton model. Here we assume that the vorticity field is described by a vector valued measure which is a combination of δ functions

$$\omega_t^N(dx) = \sum_{k=1}^{N} \alpha_k(t)\delta(x - x_k(t)). \tag{5.5}$$

Here $\alpha_k(t)$ and $x_k(t)$ denote the vorticity intensity and the position, respectively, of the kth vorton. It must be understood as a moving spin in the velocity field generated by all the other vortons. Accordiing to the Euler equation for the vorticity

$$(\partial_t + u \cdot \nabla)\omega = (\omega \cdot \nabla)u \tag{5.6}$$

we postulate an evolution equation for the vorton system which is

$$\frac{d}{dt}x_k = u(x_k), \tag{5.7}$$

where

$$u(x_k) = -\frac{1}{4\pi} \sum_{i \neq k} \frac{\alpha_k \wedge (x_k - x_i)}{|x_k - x_i|^3} \tag{5.8}$$

and

$$\frac{d}{dt}\alpha_k = (\alpha_k \cdot \nabla) \cdot u(x_k). \tag{5.9}$$

The linear equation (5.9) takes into account the stretching term while (5.7) describes the transport of the vortons.

The system (5.7), (5.8), and (5.9) describes a finite-dimensional system. It violates most of the conservation laws which are valid for the Euler equation. Therefore, there is a major criticism which makes the vorton model a much less fundamental model than its analogue in two dimensions: there is no physical vorticity field which approaches a vorticity distribution like (5.5). Indeed, any physical ω must satisfy

$$\nabla \cdot \omega = 0 \tag{5.10}$$

so that for any smooth test function f we have $(\omega, \nabla f) = 0$. On the other hand, for ω of the type (5.5) we have

$$(\omega, \nabla f) = \sum_{k=1}^{N} \alpha_k(t) \cdot \nabla f(x_k(t)) \neq 0. \tag{5.11}$$

As a consequence of the above considerations, it follows that the distribution

(5.5) cannot be weakly approached, even at time zero, by a sequence of physically conceivable vorticity distributions, so that the vortons cannot be proposed as a model describing certain reasonable situations. For these reasons, we do not even pose the localization and desingularization problems discussed in the previous section.

From another, and in some sense opposite, point of view, the vorton model is useful. If N is very large in (5.5), we can choose the values x_k and α_k at time zero, to approach, weakly, a smooth vorticity profile ω. At later times t, $\omega^N(dx, t)$ given by (5.5), with $x_k(t)$ and $\alpha_k(t)$ obtained by the solution of the vorton model, is expected to approach the solution of the Euler equation with initial datum ω, at least for short times. If so, the vorton model can be used for numerical simulations of real flows. Indeed, it has been used for many simulations and seems very effective when the stretching plays a significant role. We will say something more in the next chapter when we present some approximation methods.

The vorton model we have introduced has many evident incongruities, related to the nonconservation of the total vorticity, the helicity, etc. To overcome (in part) these difficulties other models have been introduced. They start from the observation that the Euler equation can be written, besides the expression (5.6), as

$$(\partial_t + u \cdot \nabla)\omega = (\omega \cdot \nabla^{\mathrm{T}})u, \tag{5.12}$$

where

$$(\omega \cdot \nabla^{\mathrm{T}})u \quad \text{means} \quad \sum_{i=1}^{3} \omega_i \nabla u_i. \tag{5.13}$$

Indeed (5.12) follows from the identity

$$\sum_{i=1}^{3} \omega_i \partial_i u - \sum_{i=1}^{3} \omega_i \nabla u_i = 0 \tag{5.14}$$

which can be proved by a direct verification.

Then, starting from (5.12), we introduce a new model (called the adjoint vorton model) in which (5.9) is replaced by the equation

$$\frac{d}{dt}\alpha_k = (\alpha_k \cdot \nabla^{\mathrm{T}})u(x_k). \tag{5.15}$$

Another model is obtained by replacing (5.9) with

$$\frac{d}{dt}\alpha_k = \tfrac{1}{2}[\alpha_k \cdot (\nabla + \nabla^{\mathrm{T}})] \cdot u(x_k). \tag{5.16}$$

These models conserve some quantities, but not others, and, as in the genuine vorton model, they also are not weak solutions of the Euler equation. However, they could be interesting as approximation tools in the same spirit of the approach discussed in Section 3 of Chapter 5.

4.6. Comments

The point vortex system discussed in this chapter was introduced by Helmholtz in the nineteen century [Hel], and a first study of its properties and behavior was made by Kirchhoff [Kir], Poincaré [Poi] (who first proved the integrability of the three-vortex system), and Kelvin [Kel 10]. These results are contained in almost all books on fluid mechanics. For a classic monograph fully devoted to the vortex motion, see [Lin 43].

In the second half of this century, the problem has been reconsidered with the hope of having some insight into the problem of turbulence. With this problem in mind, the statistical mechanics of a system of point vortices has been studied, as will be seen in some detail in Chapter 7.

The system of the ordinary equations governing the point vortex evolution has been studied in detail for different N. For $N = 3$, the problem is completely integrable and the solutions have been explicitly computed in [Are 79]. For $N \geq 4$, the system is not integrable in general (see Ziglin in [Kha 82] and also [CMN 93]). However, there are integrability regions following the KAM theory as discussed in Sections 4.2 and 4.3 (see references later). A lot of numerical experiments in this field suggest the existence of a threshold for the advent of chaos for this system (see [Are 83]).

Many specific problems for the point vortex system have been studied. For instance, there are studies on the configurations of vortices which remain equal in form but rotate with a constant velocity. For three vortices with the same sign, it is enough to put the vortices on the vertices of an equilateral triangle to obtain an equilibrium configuration which is also stable ([Nov 75]). For $N \geq 4$ the problem is more complicated. We can put the vortices on the vertices of a regular polygon but this configuration is stable only for small N.

We have seen in Section 4.2 that two vortices with opposite intensity translate uniformly. We can ask if there exists, for the Euler equation, initial smooth configurations with the same behavior. The answer is affirmative and is given by the so-called "Batchelor's couple" that is discussed in [BAT 67], page 534.

The dynamics of the vortex system for a limited number of vortices is easily implementable on a personal computer. There is available a special simulation program, due to H. Aref, devoted to this, called MacVortex.

We have seen in Section 4.2 an example of the collapse of three vortices. It is natural to ask what happens when we gently perturb this catastrophic configuration. We might believe that the vortices arrive close to the point of collapse, and then remain indefinitely in a small region around this point to form a sort of single vortex whose intensity is the algebraic sum of the intensities of the whole system. In general, this is not so; in fact, using the Poincaré recurrence theorem we observe that for almost any initial condition the system will return in the future close to the initial configuration in which the vortices are far apart. We now give an explicit proof of the collapse. The initial positions are, $a_1 = 2$, $x_1 = (-1, 0)$, $a_2 = 2$, $x_2 = (1, 0)$, $a_3 = -1$, $x_3 = (1, \sqrt{2})$. Let L_{ij} be the distance between the i and j vortices, then the equation of motion gives

$$\frac{d}{dt} L_{ij}^2(t) = 2\{x_i(t) - x_j(t)\} \cdot \left\{\frac{d}{dt} x_i(t) - \frac{d}{dt} x_j(t)\right\}$$

$$= \frac{a_k}{\pi}\left[\frac{(x_i - x_j)\cdot(x_j - x_k)^\perp}{|x_j - x_k|^2} - \frac{(x_i - x_j)\cdot(x_i - x_k)^\perp}{|x_i - x_k|^2}\right]$$

$$= \frac{2Aa_k}{\pi}\left[\frac{1}{L_{jk}^2} - \frac{1}{L_{ki}^2}\right], \tag{6.1}$$

where A is the area of the triangle whose vertices are the position of the vortices.

We observe that the quantity

$$F(t) = \frac{1}{2}\sum_{i;j=1}^{N} a_i a_j L_{ij}(t)^2 \tag{6.2}$$

is a constant of motion. (We remark, incidentally, that this first integral is obviously dependent on the known ones.) We have $F = 0$.

Defining

$$b_1 = \frac{L_{23}^2}{a_1}, \qquad b_2 = \frac{L_{31}^2}{a_2}, \qquad b_3 = \frac{L_{12}^2}{a_3}, \tag{6.3}$$

we have

$$b_1 + b_2 + b_3 = \frac{F}{a_1 a_2 a_3} = 0. \tag{6.4}$$

Moreover

$$\frac{d}{dt}\frac{b_1}{b_2} = \frac{2A}{\pi b_2^2 b_3}[b_1 + b_2 + b_3] = 0 \tag{6.5}$$

and so

$$\frac{b_1}{b_2} = \text{const.} \tag{6.6}$$

From (6.4) and (6.6) we see that the triangle, whose vertices give the positions of the three vortices, remains equal in form and changes only in size by similitude. The right-hand side of (6.1) remains constant. We calculate it initially and obtain (2.7), hence the collapse.

The problem of evaluating the measure of the initial positions giving rise to a collapse is a classical problem in celestial mechanics. For the vortex system this study was first done in [DuP 82] in a flat torus and in [MaP 84] in the plane. When the vortices are confined in a bounded domain it is reasonable to try a similar proof. However, there is now an additional difficulty: a vortex can hit the boundary and in this situation the Green function becomes infinite. In fact, in the neighborhood of the boundary the Green function is similar to that of a half-plane, and diverges as a logarithm. This expresses a real fact: a vortex close to the border slides tangentially along the boundary with infinite speed. So we need to prove that not only a collapse of the vortices, but also a collision with the boundary, are exceptional. This proof is obtained explicitly when the domain containing the fluid is a circle ([MaP 84]), but we believe that this proof, combined with reasonable estimates on the Green function, also works in general.

The motion of a single vortex in the presence of a boundary has been studied in some detail. For instance, the equilibrium points of the system have been searched. In general, these critical points form a complicated set. However, it is possible to prove that a vortex, moving in a convex domain different from an infinite strip, has only one equilibrium point ([Hae 51], [Gus 79]).

The first application of the KAM theorem to the point vortex system in $\mathbb{R}^2$ is done in [Kha 82]. It is possible to extend the result of the existence of a region for which the motion is quasi-periodic for a system moving in a bounded region. Obviously, in this case, we can no longer use the idea to put the cluster very far apart because of the boundedness of the region. Nevertheless, we can use scaling properties of the interaction to construct weakly interacting clusters. The idea is to change the scale of space and time in such a way that the equations of motion remain almost unchanged. More precisely, we introduce the new variables z_i, τ, defined as

$$z_i = \alpha x_i, \qquad \tau = \alpha^2 t, \qquad \alpha \gg 1. \tag{6.7}$$

In these new variables the domain appears very large, so that we can form separated clusters in a similar way as in the case of the whole plane and then we can apply the KAM theorem. Some more details are given in [MaP 84].

The rigorous connection between the vortex system and the Euler equation was first proved in [MaP 83] for a short time. A result holding globally in time for a single vortex in a bounded domain was proved in [Tur 87]. This result was extended for two vortices of different sign in [MaP 86]. In [Mar 88] a global result for vortices of the same sign was established. In this paper was introduced the idea of the function W_R to control the vorticity flux across a disk aroud the center of vorticity. The version we gave in Section 4.4 of the present book is inspired by [Mar 88] and [MaP 93]. We notice that in [MaP 93] we proved the convergence for more general initial distributions than steps functions.

We now discuss what happens in the presence of a small viscosity and in the vanishing viscosity limit. In the presence of a viscosity v the fluid evolves according to the Navier–Stokes equation, and we have seen in Chapter 2, Section 7 how, even for smooth initial data, the vanishing viscosity limit is difficult in the presence of boundaries, so that we study the problem in $\mathbb{R}^2$. We suppose that the vorticity is initially concentrated in small regions and evolve the fluid according to the Navier–Stokes equation. Suppose also that the size of the initial regions vanish. A natural question arises: Do the limiting dynamics coincide with the point vortex evolution? The question is not trivial because the small perturbation given by the viscosity term acts on a singular situation and could, in principle, largely affect the motion of the system. We want to prove that this does not happen. We notice that there is no way to use the proof of the theorem of Section 4.4 because here the viscosity, no matter how small it is, destroys the spatial localization of the vorticity. On the contrary, the proof of the theorem given in [Mar 88] holds with minor modifications.

Theorem 6.1 ([Mar 90]). *Let the initial data be the same as Theorem 4.2 and suppose $a_i > 0$. Let $f: \mathbb{R}^2 \to \mathbb{R}$ be any continuous bounded function. For any fixed $t \geq 0$, we have*

$$\lim_{\varepsilon \to 0} \int f(x)\omega_{\varepsilon,v}(x, t)\, dx = \sum_{i=1}^{N} a_i f(z_i(t)), \tag{6.8}$$

where $v = v(\varepsilon)$ vanishes when $\varepsilon \to 0$ *with* any *rate.* $\omega_{\varepsilon,v}(x, t)$ *is the evolution via the Navier–Stokes equation that in the weak form reads*

$$\partial_t \omega[f] = \omega[u \cdot \nabla f] + v\omega[\Delta f], \qquad \omega[f] = \int dx\, \omega(x, t)f(x), \tag{6.9}$$

and $z_i(t)$ is given by the point vortex model.

This theorem means that in the vanishing viscosity limit the evolved measure essentially concentrates in N Dirac measures centered in the point $z_i(t)$ given by the point vortex model, as happens in the inviscid case.

We remark that v is allowed to vanish in an arbitrary way with ε. However, we do not consider the limit $v \to 0$ when the distribution is initially a linear combination of δ functions, because in this case the Navier–Stokes equations are not well investigated. In fact, for the Navier–Stokes equation with singular initial data only partial results are available. More precisely, in $\mathbb{R}^2$ the existence and uniqueness of the solution starting from a sum of Dirac measures is proved only for viscosity v large enough ([BEP 85]) and for arbitrary v only the existence is guaranteed by a compactness theorem ([GMO 86], [Cot 86], [CoS 88]). These last results hold for more general measures and in $\mathbb{R}^3$, locally in time.

We note that the proof of the previous theorem tells us something more: when v is small but finite an initially concentrated measure remains supported essentially in a circle of radius $\sqrt{v}$ whatever the evolution generated by the Navier–Stokes equation is (if the uniqueness does not hold). In conclusion, for large v the uniqueness is proved. For small v the problem is still open but even though the uniqueness fails all the possible solutions must converge for $v \to 0$ to the same limit.

The vorton model has been widely investigated. Its energy can be defined from the square of the velocity field subtracting the (infinite) self-energy term. We obtain a quantity which is not conserved during the motion.

The case of only two vortons is not trivial and the intensity of each vorton can increase. When many vortons are present we can exhibit configurations which give rise to a collapse. Our considerations on the vorton model originate from [Nov 83], [ANO 85], [SaM 86], [WiL 88], and [GrT 88]. The vorton model is useful for numerical computations as we will see in Chapter 5.

Another situation that gives rise to a vortex-like system is the three-dimensional case discussed in Chapter 1, Section 5, (5.38). There we have considered a special three-dimensional flow: a genuine three-dimensional potential flow, and a rotational part with a planar symmetry. In this case, when the initial conditions are sharply concentrated in small regions, using the arguments developed in the first four sections, we arrive at the following system:

$$\frac{d}{dt} z_i(t) = -\frac{1}{2\pi} \sum_{j=1, j \neq i}^{N} a_j \nabla_i^{\perp} \ln |z_j(t) - z_i(t)| + \Phi(z_i(t)),$$

(6.10)

$$z_i(0) = z_i, \qquad z_i = (x_i, y_i), \qquad \Phi = (\varphi_1, \varphi_2), \qquad a_j \in \mathbb{R}.$$

This system of equations cannot, in general, be put in a Hamiltonian form

$$\frac{d}{dt} x_i = \frac{\partial H}{\partial y_i},$$

$$\frac{d}{dt} y_i = -\frac{\partial H}{\partial x_i}.$$

(6.11)

In fact, the existence of a function H implies that

$$\partial_1 \varphi_1 = -\partial_2 \varphi_2$$

(6.12)

that is,

$$\nabla \cdot \Phi = 0.$$

(6.13)

In general, $\nabla \cdot \Phi$ is not zero and the measure $\prod_{i=1}^{N} dx_i \, dy_i$ is not conserved during the motion (we remark that its variation is related to $\nabla \cdot \Phi$ by the Liouville theorem). The lack of this conservation is related to the physical fact that the potential field contracts or dilates and distorts the two dimensional velocity field. We do not enter here into a study of the system (6.10), which is discussed in [Mar 89]. We only observe that to have a global solution we need to limit the growth at infinity of Φ. Even so, special initial conditions can produce collapses when the vortices have different signs. However, here too we could prove the exceptional nature of this situation. In fact, in Section 4.2, we used the constancy of the Liouville measure $\prod_{i=1}^{N} dx_i \, dy_i$ as an essential tool, here it is not so. However, we can control, by an explicit computation, its growth. Finally, we could prove rigorously the relation between this vortex-like system and the Euler equation.

In this chapter we have studied the case in which the vorticity is sharply concentrated in N regions and we have introduced the vortex model. In Chapter 2 we have studied the opposite case, when the initial vorticity is a bounded function and we have given an existence and uniqueness theorem global in time. It is possible to consider the mixed problem in which the singular part ("vortex") and the bounded part (that we call "wave") are both present. The problem is not trivial because the proof of the existence of the time evolution for the bounded part is based on a quasi-Lipschitz condition for the velocity field, while the field produced by the point vortices is infinite in some points. To overcome this difficulty we use an a priori estimate on the possible approach of a trajectory to the point vortices.

To be more precise, the model we want to study is the initial value problem for the two-dimensional Euler equation when the initial vorticity profile is constituted by a finite number of point vortices over an essentially bounded vorticity background. The absolutely continuous part of the vorticity distribution is convected by a velocity that is generated by the bounded vorticity part as well as by the point vortices. The velocity of the vortex is produced by the bounded part of the vorticity and by all the other vortices (excluding itself). In conclusion, for a single vortex (of vorticity intensity one) the model reads

$$\frac{d}{dt}\Phi_t(x) = u(\Phi_t(x), t) + K(\Phi_t(x) - x_v(t)), \qquad x \neq x_v, \qquad (6.14a)$$

$$\frac{d}{dt}x_v(t) = u(x_v(t), t), \qquad (6.14b)$$

$$u(\cdot, t) = (K * \omega)(\cdot, t), \qquad (6.14c)$$

$$\omega(\Phi_t(x), t) = \omega_0(x), \qquad (6.14d)$$

with $x_v(0)$ and ω_0 given.

For this system we can prove an existence result valid for any time:

Theorem 6.2. *Suppose $\omega_0 \in L_1 \cap L_\infty(\mathbb{R}^2)$ and $x_v(0) \in \mathbb{R}^2$. Then there exists $\omega \in L_\infty(\mathbb{R}^2 \times \mathbb{R})$ satisfying (6.14d) where Φ is a measure-preserving flow satisfying (6.14a). Moreover, $u(\cdot, t)$ is a divergence-free vector field given by (6.14c) and x_v satisfies (6.14b).*

PROOF. We only sketch the proof, outlining the main a priori estimate. The details can be found in [MaP 91].

We introduce the following approximating sequence:

$$(\Phi^n, x_v^n, u^n, \omega^n) \qquad (6.15)$$

defined by the following differential system:

$$\frac{d}{dt}\Phi_t^n(x) = u^n(\Phi_t^n(x), t) + K(\Phi_t^n(x) - x_v^n(t)), \qquad x \neq x_v(0), \quad \Phi_0^n(x) = x, \quad (6.16a)$$

$$\frac{d}{dt}x_v^n(t) = u^n(x_v^n(t), t), \qquad (6.16b)$$

$$u^n(\cdot, t) = (K * \omega^n)(\cdot, t), \qquad (6.16c)$$

$$\omega^n(x, t) = \omega_0(\Phi^{n-1}_{-t}(x)), \tag{6.16d}$$

starting with the identity flow

$$\Phi^0_t(x) = x. \tag{6.17}$$

The real dynamics (of Theorem 6.2) are obtained when $n \to \infty$.

We show that in any approximated dynamics a particle of fluid, initially separated from the point vortex x_v, will never fall on it. Having solved the problem at level $n - 1$, we notice that

$$\|\omega^n(\cdot, t)\|_p = \|\omega_0\|_p, \qquad p = 1, \ldots, \infty, \tag{6.18}$$

since $\nabla \cdot u^{n-1} = 0$ by (6.16c).

Moreover, as we have discussed in Chapter 2, we can find a constant c for which

$$\|u^n(\cdot, t)\|_\infty \le c(\|\omega_0\|_\infty + \|\omega_0\|_1), \tag{6.19}$$

$$|u^n(x, t) - u^n(y, t)| \le c(\|\omega_0\|_\infty + \|\omega_0\|_1)\varphi(|x - y|), \tag{6.20}$$

where

$$\varphi(|x - y|) = \begin{cases} |x - y|(1 - \ln|x - y|) & \text{if } |x - y| < 1, \\ 1 & \text{if } |x - y| \ge 1. \end{cases} \tag{6.21}$$

From (6.19) and (6.20) it can be proved that the initial problem (6.16) may be solved uniquely. To find the (global) solution of problem (6.16a) we must be sure that

$$|\Phi^n_t(x) - x^n_v(t)| > 0 \tag{6.22}$$

for all times. In fact,

$$\frac{1}{2}\frac{d}{dt}|\Phi^n_t(x) - x^n_v(t)|^2 = (\Phi^n_t(x) - x^n_v(t))$$

$$\cdot \{(u(\Phi^n_t(x), t) - u(x^n_v(t), t) + K(\Phi^n_t(x) - x^n_v(t))\}. \tag{6.23}$$

Using the following crucial remark:

$$x \cdot K(x) = 0 \qquad \text{for all} \quad x \in \mathbb{R}^2 \tag{6.24}$$

and inequality (6.20), from (6.23) we obtain

$$\frac{1}{2}\frac{d}{dt}|\Phi^n_t(x) - x^n_v(t)|^2 \le c|\Phi^n_t(x) - x^n_v(t)|\,\varphi(|\Phi^n_t(x) - x^n_v(t)|)$$

$$\le c\varphi(|\Phi^n_t(x) - x^n_v(t)|^2). \tag{6.25}$$

It is easily understood that a lower bound on $|\Phi^n_t(x) - x^n_v(t)|^2$ is obtained when the equality sign holds in (6.26). (A rigorous proof is similar to the proof of the Gronwal lemma given in Appendix 1 of Chapter 2.) So we consider the initial value problem

$$\frac{d}{dt}z = -2c\varphi(z),$$

$$z(0) = z_0 = |x - x_v(0)|^2, \tag{6.26}$$

which has the unique solution

$$z(t) = \begin{cases} z_0 \exp\{2ct\} \exp\{1 - \exp\{2ct\}\} & \text{if } z < 1, \\ z_0 - 2c|t - t_0| & \text{if } z \geq 1. \end{cases} \tag{6.27}$$

So that we conclude that

$$|\Phi_t^n(x) - x_v^n(t)|^2 \geq z(t) > 0. \tag{6.28}$$

The remaining steps are simple: we consider the fluid particle, initially distant from x_v more than $\eta > 0$. Then, by compactness, we can find a subsequence for $n \to \infty$ such that the limit satisfies Theorem 6.2. Finally, by the usual diagonal trick, the condition $\eta > 0$ can be removed. $\qquad\square$

The result can be extended without effort to many vortices of the same sign. Until now, we have used a compactness method and so the uniqueness of the solution is not guaranteed. However, if we add the condition that initially, the bounded part of the vorticity field does not overlap vortices, that is,

$$\text{support}\{\omega_0\} \cap x_v(0) = 0, \tag{6.29}$$

a uniqueness and regularity result can be easily obtained. For more details see [MaP 91].

It is natural now to study the relation between this new mathematical model and the Euler equation. It is easy to apply the techniques of Section 4.4 to give a validity proof of the model established in (6.14) whenever condition (6.29) is verified.

EXERCISES

1. Improve (2.42) showing that the probability of an ε-collapse vanishes faster than $C(1 + T)\varepsilon^{1-\delta}$, where $\delta \in (0, 1)$ and C increases with δ. (*Hint*: The proof is similar to that of Theorem 2.1. Now we assume a function more divergent than Φ_ε (given by (2.33)) when the vortices collapse, but are still integrable in λ for any time. A function with these properties is given by

$$\Phi_\varepsilon(X) = \frac{1}{2} \sum_{i;j=1,i\neq j}^{N} F(-\ln_\varepsilon |x_i - x_j|), \tag{E.1}$$

where

$$F(r) = \exp\{(1 - \delta)r\}, \qquad \delta \in (0, 1). \tag{E.2}$$

2. Prove the linear stability of a configuration of three equal vortices at the vertices of an equilateral triangle.

3. Prove that equal vortices at the vertices of a regular polygon give rise to a rotating stationary solution.

4. Study the case of two vortices in an half-plane. In particular, take two vortices x_1, x_2 of the same sign and prove that:
   (1) when the vortices are initially together (with respect to the boundary) their motion is the sum of a rotation and a translation; and

(2) when they are far apart each of them asymptotically uniformly translates. (*Hint*: To study the second case the following observation may be useful: if the distance of the vortice increases like t, the interaction between them decreases as t^{-1} and the component to the velocity orthogonal to the boundary (this component is responsible for the rotation) decreases as t^{-2}, that is, in an integrable way.)

5. Consider the system (6.10). Show Φ for which the solution blow up in finite time. (*Hint*: $\Phi = x^2 y + y^2 x - \frac{1}{3}(x^3 + y^3)$.)

6. Study the system (6.10) when $\varphi_1 = Ax$, $\varphi = By$, $A, B \in \mathbb{R}$. Show that for $A = -B$ the Hamiltonian is

$$H = -\frac{1}{4\pi} \sum_{i;j=1,i \neq j}^{N} a_i a_j \ln|z_i - z_i| + A \sum_{i=1}^{N} a_i x_i y_i.$$

Moreover, prove that for $a > 0$ the differential system (6.10) has a solution global in time. Finally, show a situation in which a collapse happens in finite time even if the potential field dilates ($A, B > 0$). (*Hint*: To prove that the solution is global, study the time evolution of the quantities

$$H = -\frac{1}{4\pi} \sum_{i;j=1,i \neq j}^{N} a_i a_j \ln|z_i - z_i|,$$

$$I = \sum_{i=1}^{N} a_i |x_i|^2.$$

For the collapse, put $A = B$, take the initial condition discussed in Section 4.2 and use the explicit relation between the present problem and the two-dimensional Euler equation given in Chapter 1 (5.45), (5.46).)

CHAPTER 5

Approximation Methods

In this chapter we introduce the spectral and the vortex methods as finite-dimensional approximation schemes for the Euler equation in two dimensions. We discuss the convergence of these approximations to the solutions.

5.1. Introduction

The explicit evaluation of the solutions of the Euler equation is an important problem of practical interest. There are very few cases in which the Euler equation is explicitly solvable, and these explicit solutions are quite far from describing the typical behavior of real fluids in many physically relevant situations such as, for instance, in the case of turbulent motion. On the other hand, recent developments of modern computational tools give us the possibility to simulate the time evolution of real flows by means of numerical integration of the equation of motion. Thus the development of efficient algorithms devoted to the simulation of the solutions of the Euler equation is a subject of great theoretical and practical interest.

By an algorithm for the numerical integration of the equation of the motion we mean a N-dimensional dynamical system (e.g., an ordinary differential system in $\mathbb{R}^N$) whose solutions approximate well the Euler flow and are practically implementable from a numerical point of view. In addition, the scheme should exhibit at least a reasonable rate of convergence: the a priori error measuring the deviation between the true flow, and the approximate solution, should vanish as $N \to \infty$ in a sufficiently fast manner, in order to give satisfactory answers to practical simulation problems, with a limited amount of computational work.

Due to the relevance of the argument and the large variety of situations for which numerical schemes are needed, there is a wide literature on the subject, a detailed analysis of which is beyond the scope of this book. Here we do not intend to review all the computational approaches to the Euler equation. We will, however, explain the underlying ideas of two methods which seem to be more specific and appropriate to simulate ideal flows: the spectral and the vortex methods. They are finite-dimensional approximations to the Euler flow which take into account the particular structure of the fluid motion. For this reason, the interest of these methods is somehow intrinsic and is going beyond numerical purposes. These methods are based on the Faedo–Galerkin projection introduced in Chapter 2, Section 5, and on the vortex system discussed in Chapter 4.

We limit ourselves to two-dimensional analysis. Results concerning three-dimensional results will be discussed in the last section.

5.2. Spectral Methods

As we have seen in Chapter 2, Section 5, the Euler equation in a flat torus can be projected into a finite-dimensional subspace, by considering the motion of a finite subset of the Fourier coefficients of the velocity field. In this way, it is possible to construct a finite-dimensional dynamical system approximating the Euler flow. This procedure can be generalized to more general domains by substituting the Fourier basis with the basis of the eigenfunctions of the Laplace operator. This is the basic idea underlying the spectral methods. In this section we want analyze, in some detail, the convergence of the spectral scheme in the simplest case: the motion of a two-dimensional flow in a flat torus.

Consider $D = [-\pi, \pi]^2$, the two-dimensional flat torus. The velocity and the vorticity field can be expanded by means of the Fourier transform

$$\hat{u}_t(k) = \left(\frac{1}{2\pi}\right) \int dx \, \exp\{-ik \cdot x\} u_t(x), \tag{2.1}$$

$$\omega_t^{\wedge}(k) = \left(\frac{1}{2\pi}\right) \int dx \, \exp\{-ik \cdot x\} \omega_t(x), \qquad k \in \mathbb{Z}^2. \tag{2.2}$$

Let us first notice that $\hat{u}_t(0) = 0$, if we assume at time zero that

$$\hat{u}_0(0) = \left(\frac{1}{2\pi}\right) \int dx \, u_0(x) = 0. \tag{2.3}$$

In fact, the average flow $\hat{u}_t(0)$ is a constant of motion as follows by a direct inspection of the Euler equation. Condition (2.3) simply means that we choose a reference framework in motion with the average flow.

Moreover, by periodicity, the circulation theorem gives

$$\omega_0^{\wedge}(k) = \left(\frac{1}{2\pi}\right) \int dx \, \omega_0(x) = 0. \tag{2.4}$$

Thus the following relation expresses the velocity in terms of the vorticity

$$\hat{u}_t(k) = \frac{k^\perp}{k^2}\omega_t^\wedge(k), \qquad k^\perp = (k_2, -k_1). \tag{2.5}$$

For the Fourier coefficients the Euler equation for the vorticity reads

$$\frac{d}{dt}\omega_t^\wedge(k) = -\frac{i}{2\pi}\sum_{h\in Z^2}\hat{u}_t(k-h)\cdot h\omega_t^\wedge(h). \tag{2.6}$$

The above equation constitutes a system of infinitely many coupled ordinary differential equations which can be truncated by simply neglecting the large frequencies.

Let Λ_N be the subset of Z^2 consisting of all k's satisfying the relation $-N \le k_i \le N$ for $i = 1, 2$. Denote by $\omega_{N,t}^\wedge(k)$ the solution of the ordinary initial value problem in C^{Λ_N}

$$\frac{d}{dt}\omega_{N,t}^\wedge(k) = -\frac{i}{2\pi}\sum_{h\in\Lambda_N}\hat{u}_{N,t}(k-h)\cdot h\omega_{N,t}^\wedge(h)), \qquad k \in \Lambda_N,$$

$$\omega_0^\wedge(k) = \left(\frac{1}{2\pi}\right)\int dx\,\exp\{-ik\cdot x\}\omega_0(x), \qquad k \in \Lambda_N. \tag{2.7}$$

Here $\hat{u}_{N,t}(k)$ is related to $\omega_{N,t}^\wedge(k)$ by the relation

$$\hat{u}_{N,t}(k) = \frac{k^\perp}{k^2}\omega_{N,t}^\wedge(k). \tag{2.8}$$

In other words, $\hat{u}_{N,t}(k)$ are the Fourier coefficients of a velocity field, denoted by $u_{N,t}$, whose vorticity $\omega_{N,t}$ has a Fourier transform given by $\omega_{N,t}^\wedge(k)$.

As already remarked in Section 5, Chapter 2, if ω_0 is real, then both $u_{N,t}$ and $\omega_{N,t}$ are real. System (2.7) is equivalent to the following equation:

$$\partial_t\omega_{t,N} + P_N[(u_{t,N}\cdot\nabla)\omega_{t,N}] = 0,$$

$$\omega_{N,0} = P_N\omega_0, \tag{2.9}$$

where P_N is the orthogonal projector in the subspace of $L_2(D)$ generated by the functions $\{\exp[ik\cdot x]\}$ with $k \in \Lambda_N$. This remark, which follows easily by taking the Fourier transform of (2.9), allows us to prove that (2.7) has a global solution. In fact, by the identity

$$(\omega_{N,t}, P_N(u_{N,t}\cdot\nabla)\omega_{N,t}) = 0, \tag{2.10}$$

which is an easy consequence of the fact that $u_{N,t}$ and $\omega_{N,t}$ are real valued and that $u_{N,t}$ is a divergence-free vector field, it follows (conservation of the truncated enstrophy)

$$\frac{d}{dt}|\omega_{N,t}|^2 = 0, \tag{2.11}$$

where we have denoted by $|\cdot|$ and by $(\cdot,\cdot)$ the norm and the scalar product, respectively, in $L_2(D)$.

The conservation law (2.11) follows, as usual, by taking the scalar product of (2.9) by $\omega_{N,t}$ and applying identity (2.10). By (2.11) we know that

$$|\omega_{t,N}^{\wedge}(k)|^2 \leq \sum_{k \in \Lambda_N} |\omega_{t,N}^{\wedge}(k)|^2 = \sum_{k \in \Lambda_N} |\omega_{0,N}^{\wedge}(k)|^2, \tag{2.12}$$

so that we have an a priori control on the growth in time of $|\omega_{N,t}^{\wedge}(k)|^2$, which allows us to extend, to arbitrary times, the local existence theorem for the solutions of the system (2.7) which holds for general arguments.

We now wish to investigate the convergence of the solutions of the system (2.7) to the solutions of the Euler equation. In doing this, we will use, basically, the same technique exploited in Section 5, Chapter 2, the only difference being that here we are in a two-dimensional framework, so that we have an a priori estimate on the vorticity allowing us to reach arbitrary times. Moreover, we will use the regularity properties of the flow to get good estimates on the rate of convergence of the approximations.

We fix an initial vorticity profile $\omega_0 \in C^p(D)$. We know, by the analysis developed in Chapter 2, Section 4, that the solution of the Euler equation ω_t belongs to $C^p(D)$. Let $\omega_{N,t}$ be the solution of the initial value problem (2.10) (with initial datum $P_N \omega_0$). Then

$$\frac{1}{2}\frac{d}{dt}|\omega_t - \omega_{N,t}|^2 = \frac{d}{dt}(\omega_{N,t}, \omega)$$

$$\text{(by the conservation of } |\omega_t| \text{ and } |\omega_{N,t}|)$$

$$= -(P_N(u_{N,t} \cdot \nabla)\omega_{N,t}, \omega_t) - ((u_t \cdot \nabla)\omega_t, \omega_{N,t})$$

$$+ ((u_{N,t} \cdot \nabla)\omega_{N,t}, \omega_t) - ((u_{N,t} \cdot \nabla)\omega_{N,t}, \omega_t)$$

$$= A(N) + ((u_{N,t} - u_t) \cdot \nabla)\omega_t, \omega_{N,t})$$

$$= A(N) + ((u_{N,t} - u_t) \cdot \nabla)\omega_t, \omega_{N,t} - \omega_t), \tag{2.13}$$

where

$$A(N) = ([1 - P_N](u_{N,t} \cdot \nabla)\omega_{N,t}, \omega_t) \tag{2.14}$$

and the last step in (2.13) is a consequence of the identity

$$((u_{N,t} - u_t) \cdot \nabla)\omega_t, \omega_t) = 0. \tag{2.15}$$

We now estimate the two terms of the right-hand side of (2.13). We have

$$|A(N)| \leq |(1 - P)\omega_t| \, |(u_{N,t} \cdot \nabla)\omega_{N,t}|, \tag{2.16}$$

$$|(1 - P)\omega_t|^2 = \sum_{k \notin \Lambda_N} |\omega_t^{\wedge}(k)|^2 = \sum_{k \notin \Lambda_N} |\omega_t^{\wedge}(k)|^2 \frac{|k|^{2p}}{|k|^{2p}}$$

$$\leq N^{-2p} \sum_{k \notin \Lambda_N} |\omega_t^{\wedge}(k)|^2 |k|^{2p} \leq |\omega_t|_p N^{-2p}. \tag{2.17}$$

On the other hand

$$|(u_{N,t} \cdot \nabla)\omega_{N,t}|^2 \leq \sum_{k,h \in \Lambda_N} \left|\omega_{N,t}^{\wedge}(k)\omega_{N,t}^{\wedge}(h)\frac{(k-h)^{\perp} \cdot h}{|k-h|^2}\right|^2 \leq C|\omega_0|^4 N^2. \tag{2.18}$$

The last inequality follows by the conservation of the enstrophy for the truncated dynamics. Finally

$$|((u_{N,t} - u_t) \cdot \nabla)\omega_t, \omega_{N,t} - \omega_t)| \leq |u_{N,t} - u_t| \, \|\nabla\omega_t\|_\infty |\omega_{N,t} - \omega_t|$$

$$\leq |\omega_{N,t} - \omega_t|^2 \|\nabla\omega_t\|_\infty. \tag{2.19}$$

The last step in the above estimate follows by the inequality

$$|u_{N,t} - u_t| < |\omega_{N,t} - \omega_t|. \tag{2.20}$$

Denoting by

$$M_p(T) = \sup_{0 \leq t \leq T} (1 + |\omega_t|_p^2), \tag{2.21}$$

which exists finite under our regularity assumption, we have (for p large enough $\|\nabla\omega_t\|_\infty \leq |\omega_t|_p$)

$$\frac{1}{2}\frac{d}{dt}|\omega_t - \omega_{N,t}|^2 \leq M_p(T)|\omega_t - \omega_{N,t}|^2 + M_p(T)N^{-p+1}. \tag{2.22}$$

Since

$$|\omega_0 - \omega_{N,0}|^2 \leq |(1 - P_N)\omega_0|^2 \leq M_p(T)N^{-p} \tag{2.23}$$

we conclude that

$$|\omega_t - \omega_{N,t}|^2 \leq (1 + T)M_p(T)N^{-p+1}, \qquad t \leq T. \tag{2.24}$$

Let us comment on estimate (2.24) which says that the Euler flow ω_t can be approximated, with high accuracy, by the solution of the trucated system $\omega_{t,N}$. For a fixed time T, we evaluate the number N which can be practically considered for a real simulation. If N is reasonably large, since p is arbitrary (provided that the initial condition is sufficiently smooth), we can choose p to minimize the error in the right-hand side of estimate (2.24), taking into account that $M_p(T)$ increases, at least exponentially, in p.

We must remark that estimate (2.23) is not very useful for estimating the error for large times. The a priori estimate on M_p gives an exponential (at least) growth in T which destroys the accuracy obtained for short times. However, this seems an intrinsic feature of the conservative system. We will discuss this important point in the last section.

5.3. Vortex Methods

In Chapter 4, Section 1, we have seen how the solution of the vortex model can be interpreted as a sort of weak solution of the Euler equation if the self-interaction of each vortex is neglected. We recall briefly the argument.

Let

$$\omega_t^N(dx) = \sum_{i=1}^N a_i \delta(x_i(t) - x)\, dx \tag{3.1}$$

be a probability measure describing the time evolution of the vorticity field according to the vortex motion

$$\frac{d}{dt}x_i(t) = \sum_{j=1, j \neq i}^{N} a_j K(x_i(t) - x_j(t)). \tag{3.2}$$

The point vortices are assumed to have the vorticity intensities $\{a_j\}, j = 1, \ldots, N, a_j > 0$, satisfying the normalization condition $\sum_{j=1}^{N} a_j = 1$. Then

$$\frac{d}{dt}\omega_t^N(f) = \omega_t^N(u \cdot \nabla f), \tag{3.3}$$

where f is a smooth function, $\omega^N(f) = \int \omega^N(dx)f(x)$, and u is given by the right-hand side of (3.2). Suppose now that at time zero

$$\omega_0^N \to \omega_0 \in L_\infty \cap L_1(\mathbb{R}^2) \qquad \text{as} \quad N \to \infty \tag{3.4}$$

in the sense of the weak convergence of measures, i.e.,

$$\lim_{N \to \infty} \omega_0^N(f) = \omega_0(f), \qquad f \text{ continuous and bounded}, \tag{3.5}$$

we expect that, denoting by ω_t the solution of the Euler equation with initial datum ω_0,

$$\omega_t^N \to \omega_t \qquad \text{as} \quad N \to \infty \quad \text{weakly} \tag{3.6}$$

if the Euler equation has the property that its solutions are weakly continuous with respect to the initial conditions. If condition (3.6) holds, the vortex flow provides a finite-dimensional approximation for the solution of the Euler equation in the plane.

In our discussion we are assuming vorticity profiles which are positive and normalized. This severe assumption is imposed for notational simplicity only: all the results can be easily extended to the general case with minor modifications. We now want to prove that the convergence (3.6) holds. We remark that this result is also interesting from a theoretical point of view. It says that the Euler evolution for the vorticity can be thought as the evolution of infinitely many vortices of infinitesimal intensity. Before proving (3.6) we need some machinery concerning the topology of the weak convergence of measures. Since we are going to compare two probability measures, ω_t^N and ω_t in the weak convergence topology, we want to introduce a corresponding metric on the space of the probability measures in $\mathbb{R}^2$.

What follows is general. Let M be a metric space with a bounded metric function $d: M \times M \to \mathbb{R}^+$. Denote by $\mathcal{M}(M)$ the space of the Borel probability measures on M. If $\mu_i \in \mathcal{M}(M)$, $i = 1, 2$, we denote by $C(\mu_1, \mu_2)$ the set of all joint representations of μ_1 and μ_2. A joint representation $P \in C(\mu_1, \mu_2)$ is a Borel probability measure on $M \times M$ satisfying

$$\int_{M \times M} P(dx_1, dx_2)f(x_i) = \int_M \mu_i(dx)f(x), \qquad i = 1, 2, \tag{3.7}$$

for all bounded measurable functions f. An example of joint representation is the product measure $\mu_1(dx_1)\mu_2(dx_2)$. Define

$$R(\mu_1, \mu_2) = \inf_{P \in C(\mu_1, \mu_2)} \int_{M \times M} P(dx_1, dx_2) d(x_1, x_2) \tag{3.8}$$

R is a distance on $\mathcal{M}(M)$ as follows by a direct inspection.

It can be proved that the topology induced by the metric (3.8) is equivalent to the topology of the weak convergence of the measures. This metric is called the Kantorovich–Rubistein (K–R) distance (sometimes called the Vasershtein distance). To give an idea of the meaning of this metric consider the case in which

$$\mu_k(dx) = N^{-1} \sum_{i=1}^{N} \delta(x_i^k - x) \, dx, \qquad k = 1, 2, \tag{3.9}$$

$x_i^k \in M$. Then

$$R(\mu_1, \mu_2) = \min_{\pi} N^{-1} \sum_{i=1}^{N} d(x_i^1, x_{\pi(i)}^2), \tag{3.10}$$

where the above minimum is taken over all the permutations of the indices 1, 2, …, N. In other words, in the above situation in which we want to compare two measures which are convex combinations of δ measures of equal masses, the distance between these two measures is obtained by choosing the most convenient way to sum the distance obtained by pairing the points on which the measures are supported. The proof of (3.10) is given in the Appendix. We notice that, formula (3.10) can be used to prove the equivalence between the weak convergence topology and the topology induced by the K–R distance (Exercise 5).

Let us come back to our original problem. Due to difficulties related to the divergence of the kernel K we consider, as a finite-dimensional approximating system, the regularized version

$$\frac{d}{dt} x_i^\varepsilon(t) = \sum_{j=1, j \neq i}^{N} a_j K_\varepsilon(x_i^\varepsilon(t) - x_j^\varepsilon(t)), \tag{3.11}$$

where K_ε is a C^∞ function satisfying $K_\varepsilon(0) = 0$, $\nabla \cdot K_\varepsilon = 0$, and differing from K only at short distances

$$K_\varepsilon(x - y) = K(x - y) \qquad \text{if} \quad |x - y| > \varepsilon. \tag{3.12}$$

Denoting by

$$\omega_t^{N, \varepsilon}(dx) = \sum_{i=1}^{N} a_i \delta(x_i^\varepsilon(t) - x) \, dx \tag{3.13}$$

we want to estimate $R(\omega_t^{N, \varepsilon}, \omega_t)$ where ω_t is the solution of the Euler equation with initial datum given by $\omega_0 \in L_\infty \cap L_1(\mathbb{R}^2)$. $\omega_0 \, dx$ is assumed to be a probability measure on $\mathbb{R}^2$. To specify the notion of the K–R distance in our context, we first need to introduce a bounded metric on $\mathbb{R}^2$ which is equivalent to the usual distance. We choose

$$d(x, y) = \begin{cases} |x - y| & \text{if } |x - y| < 1, \\ 1 & \text{otherwise.} \end{cases} \tag{3.14}$$

for which the metric R on $\mathcal{M}(\mathbb{R}^2)$ will be computed with the metric d on $\mathbb{R}^2$.

We are now in position to formulate the result.

Theorem 3.1. *Let $\omega_0 \, dx$ be a probability measure on $\mathbb{R}^2$ with $\omega_0 \in L_\infty \cap L_1(\mathbb{R}^2)$. Let $\omega_t \in L_\infty \cap L_1(\mathbb{R}^2)$ be the (weak) solution of the Euler equation with initial datum given by ω_0. Suppose that*

$$\lim_{N \to \infty} R(\omega_0, \omega_0^N) = 0, \tag{3.15}$$

where

$$\omega_0^N(dx) = \sum_{i=1}^N a_i \delta(x_i - x) \, dx. \tag{3.16}$$

Then for all sequences $\varepsilon = \varepsilon(N)$ for which

$$\lim_{N \to \infty} R(\omega_0, \omega_0^N) \exp[L_\varepsilon(T + e^{L_\varepsilon T})] = 0, \tag{3.17}$$

where

$$L_\varepsilon = \max\{H_\varepsilon, 2 \max K_\varepsilon\}, \tag{3.18}$$

$$H_\varepsilon = \text{Lipschitz constant of } K_\varepsilon,$$

we have

$$\lim_{N \to \infty} \sup_{0 \le t \le T} R(\omega_t, \omega_t^{N,\varepsilon}) = 0. \tag{3.19}$$

Remark. The above theorem asserts that, in order to have convergence at time t, the cutoff on scale ε must be removed quite gently with respect to the growth of the number of vortices N, as follows by an inspection of (3.10). If $R(\omega_0, \omega_0^{N,\varepsilon})$ is of order $N^{-\alpha}$ with $\alpha > 0$, then $\varepsilon(n)$ should be of order $(\log \log N)^{-1}$. As follows by the proof of the theorem, the error $R(\omega_t, \omega_t^{N,\varepsilon})$ is the sum of two terms, one essentially given by the left-hand side of (3.17), the other one of order ε^p for some small p, so that the convergence rate is not at all satisfactory. However, by general arguments we can see that the rate of convergence is not so horrible as can seem at first sight. Moreover, the accuracy can be considerably improved assuming more smoothness on the initial datum and choosing more carefully the cutoff. A more detailed analysis on the accuracy of the method will be presented in the next section.

We mention also that the original uncutoff problem also has a solution as we will discuss later.

PROOF. Consider the regularized Euler problem in the following weak form:

$$\frac{d}{dt} \omega_t^\varepsilon(f) = \omega_t^\varepsilon(u^\varepsilon \cdot \nabla f), \qquad \omega_0 \text{ given}, \tag{3.20}$$

where $u^\varepsilon = K_\varepsilon * \omega^\varepsilon$ and $f \in C^1(\mathbb{R}^2)$. The problem (3.20) has a unique solution for $\omega_0\, dx$ Borel probability measure on $\mathbb{R}^2$. The proof of the above statement follows easily by an application of the contraction mapping principle in $\mathcal{M}(\mathbb{R}^2)$ (see Exercises 1 and 2).

On the other hand, it can be proved that

$$\lim_{\varepsilon \to 0} R(\omega_t^\varepsilon, \omega_t) = 0, \tag{3.21}$$

where ω_t solves uniquely the Euler equation with initial datum ω_0 (see Exercise 3), and the above convergence holds for t belonging on compact sets. Actually, this is a way to prove an existence and uniqueness theorem for the two-dimensional Euler flow in $\mathbb{R}^2$.

The initial value (3.20) enjoys the following continuity property: if $\omega_{i,t}^\varepsilon$, $i = 1, 2$, are two solutions of the initial value problem (3.20) with initial data given by $\omega_{i,0}^\varepsilon \in \mathcal{M}(\mathbb{R}^2)$, then, for $t \le T$,

$$R(\omega_{1,t}^\varepsilon, \omega_{2,t}^\varepsilon) \le R(\omega_{1,0}^\varepsilon, \omega_{2,0}^\varepsilon) \exp[L_\varepsilon(T + e^{L_\varepsilon T})]. \tag{3.22}$$

The proof of estimate (3.22) is not hard and is left as an exercise to the reader (see Exercise 4).

The proof of the theorem is thus completed by the use of the triangle inequality, estimate (3.22), and condition (3.21)

$$R(\omega_t, \omega_t^{N,\varepsilon}) \le R(\omega_t, \omega_t^\varepsilon) + R(\omega_t^\varepsilon, \omega_t^{N,\varepsilon})$$

$$\le R(\omega_t, \omega_t^\varepsilon) + R(\omega_0, \omega_0^N) \exp[L_\varepsilon(T + e^{L_\varepsilon T}] \tag{3.23}$$

$$\square$$

5.4. Comments

The convergence result presented in Section 5.2 is essentially due to Boldrighini [Bol 79]. The content of Section 5.3 follows [MaP 82], [MaP 84]. This approach is based on the previous ideas of McKean [McK 69], Braun and Hepp [BrH 77], Neunzert [Neu 81], and Dobrushin [Dob 79].

As we remarked above, the convergence rate presented is so bad that the theorem discussed in Section 5.3 is not of practical interest from the point of view of numerical analysis. However, general probabilistic arguments based on the Central Limit Theorem (see [BrH 77]) can be used to provide a better investigation of the convergence rate. If the initial condition ω_0 is approximated by

$$\omega_0^N(dx) = \frac{1}{N} \sum_{i=1}^{N} \delta(x_i - x)\, dx, \tag{4.1}$$

where the x_i are chosen randomly and independently according to the distribution ω_0, the error at later times should be of order $1/\sqrt{N}$ with large probability.

The accuracy of the method can be considerably improved by an accurate choice of the regularizing cutoff and a better approximation at time zero. These methods are usually called "vortex blob methods" because the use of K_ε in place of K which is, in some way, as if to replace a point vortex with a blob of vorticity. We direct the reader to the following sequence of references [DeH 78], [Hal 87], [BeM 82]$_2$, [Cot 83], [Rav 83], [AnG 85], and [CaL 89]. In all of these references, the vortex blob method is proved to converge to the solution with an error decreasing polynomially in the number of computational elements N. ε, the scale length cutoff, is gently removed when N goes to infinity. The cutoff plays a technical role in proving the convergence theorem, because it prevents high velocities whenever two vortices stay at a short distance. Recently, Goodman, Hou, and Lowengroub [GHL 90] proved the convergence of the point vortex method without the use of any cutoff function (i.e., the problem stated at the beginning of Section 5.3). This result is interesting because it proves that a careful approximation of the initial condition allows the vortices to stay at a finite distance at later times. For a valuable review on the practical application of the vortex methods in simulating two-dimensional flows, we direct the reader to [Leo 80].

Also, the vorton system (see Chapter 4, Section 5) can be used for approximating the three-dimensional Euler equation (obviously for short times) following the same idea underlying the two-dimensional case. The vorton model was introduced for computational purposes by Beale and Majda [BeM 82]$_1$. Recently, the convergence of the pure point vorton approximation was proved in [HoL 90].

All these results (the two-dimensional as well as the three-dimensional case) give a very good estimate of the error, which is of the order $C(p)N^{-p}$, where p is practically arbitrary and $C(p)$ increases with p. However, $C(p)$ increases in time at least exponentially. This feature makes these results not very useful for simulations of many real flows on scales of times of practical interest (as for metereological purposes).

The stochastic Lagrangian picture, discussed in the last section of Chapter 2, suggests that we might take into account the effect of the viscosity by replacing the deterministic motion of the vortex system by a stochastic differential equation of Ito type

$$dx_i^\varepsilon(t) = \sum_{j=1, j \neq i}^{N} a_j K_\varepsilon(x_i^\varepsilon(t) - x_j^\varepsilon(t))\, dt + \sqrt{2\nu}\, dw_i, \qquad (4.2)$$

where $\{w_i\}_{i=1,\ldots,N}$ are N independent standard Brownian motins. We can prove that the measure valued stochastic process

$$\sum_{i=1}^{N} a_i \delta(x_i(t) - x)\, dx, \qquad (4.3)$$

obtained by solving (4.2), converges weakly, with probability one, to the deterministic flow ω_t^ν solution of the Navier–Stokes equation.

The idea of adding a random walk to the vortex motion to simulate viscous flows is due to Chorin [Cho 73]. For a rigorous proof of the convergence of the stochastic vortex blob method, see [MaP 82], [MaP 84], [Goo 87], and [Lon 88]$_1$. A proof of the convergence of the stochastic point vortex system has been obtained for suitably large values of the viscosity v [Osa 86].

Also, the three-dimensional Navier–Stokes flow can be approached, for short times, by a finite-dimensional stochastic system based on the vorton model ([EsP 89], [Lon 88]$_2$). We remark that the stochastic vortex system is a good approximation of the Navier–Stokes flow in the absence of a boundary. When obstacles are present in the fluid the situation is more complicated. The nonslip conditions $u = 0$ on the boundary produces vorticity. This means that any stochastic particle method must describe a particle production on the boundary (see [Cho 73] at level of general ideas) making the stochastic flow much more involved.

Appendix 5.1 (On K–R Distance)

Proof of equation (3.10). Any $P \in C(\mu_1, \mu_2)$ takes the form

$$P(dy) = \sum_{i;j=1}^{N} a_{i,j}\delta_{y_{i,j}}(dy), \tag{A1.1}$$

where $y_{i,j} = (x_i^1, x_j^2) \in M \times M$, y is the generic element of $M \times M$, and

$$a_{i,j} \geq 0, \tag{A1.2}$$

$$\sum_{j=1}^{N} a_{i,j} = \sum_{i=1}^{N} a_{i,j} = \frac{1}{N}.$$

The elements $a_{i,j}$ form a matrix A of positive elements for which the sum of any row and any column is $1/N$. Denote by $\mathcal{A}$ the set of all such matrices. Then $C(\mu_1, \mu_2)$ is in one-to-one correspondence with the set $\mathcal{A}$. The quantity

$$\int P(dx_1, dx_2)d(x_1, x_2) \tag{A1.3}$$

as a function of P corresponds to the hyperplane

$$\sum_{i;j}^{N} a_{i,j}d(x_i^1, x_j^2) \tag{A1.4}$$

which must be minimized in $\mathcal{A}$. This is a convex compact set in $\mathbb{R}^{N^2}$ so that the proof is achieved once we prove that the $N!$ elements of $\mathcal{A}$ defined, for a given permutation π, as

$$a_{i,j} = \delta_{j,\pi(i)}\frac{1}{N} \tag{A1.5}$$

are the only extremal elements of $\mathcal{A}$.

Denote by D the set of elements defined by (A1.5). Certainly D consists of extremal points. Let $A = \{a_{i,j}\} \notin D$ be an extremal point of $\mathscr{A}$. Then for some i_1, there exists j_1 and j_2 such that $0 < a_{i_1 j_2} < N^{-1}$ and $0 < a_{i_1 j_1} < N^{-1}, j_1 \neq j_2$. Moreover, there exists i_2 such that $0 < a_{i_2 j_2} < N^{-1}$ and so on. By iterating the procedure, we can construct a graph in the matrix A

$$a_{i_1 j_1} \to a_{i_1 j_2}$$
$$\downarrow$$
$$a_{i_2 j_2} \to a_{i_2 j_3} \cdots .$$

We easily realize that a graph contains a closed loop $l = a_{i_k j_k}, a_{i_k j_k + 1}, \ldots,$ $a_{i_{n-1} j_n}$ with positive different elements. Defining

$$A^{\pm} = \{a_{ij}^{\pm}\},$$

$$a_{ij}^{\pm} = a_{ij} \qquad \text{if} \quad a_{ij} \notin l,$$

$$a_{i_k j_k}^{\pm} = a_{i_k j_k} \pm \varepsilon \qquad \text{if} \quad a_{i_k j_k} \in l, \qquad \text{or} \qquad (A1.6)$$

$$a_{i_k j_k + 1}^{\pm} = a_{i_k j_k} \mp \varepsilon$$

in an alternate way along the loop

$$\varepsilon < \min_{a_{ij} \in l} a_{ij},$$

it is easily verified that $A^{\pm} \in \mathscr{A}$ and $A = \frac{1}{2}(A^+ + A^-)$. This contradicts the hypothesis of extremality. $\square$

Exercises

1. Given $\xi \in C([0, T]; M)$ where M is the space of the Borel probability measures on $\mathbb{R}^2$, solve (in the weak sense) the linear problem

$$\partial_t \omega + u \cdot \nabla \omega = 0,$$
$$u^\varepsilon = K^\varepsilon * \xi \tag{E.1}$$

by means of the solution of the ordinary differential equation

$$\frac{d}{dt} \phi_t(x) = u^\varepsilon(\phi_t(x)). \tag{E.2}$$

2. Given $\xi^i \in C([0, T]; M)$, $i = 1, 2$, denote by ω^i the solutions of the linear problem (E.1) and prove that, if T is small enough,

$$\sup_{0 < t < T} R(\omega_t^1, \omega_t^2) \leq \alpha \sup_{0 < t < T} R(\xi_t^1, \xi_t^2) \tag{E.3}$$

with $\alpha < 1$. Use the above fact to construct a global solution for the problem (3.20). (*Hint*: To prove (E.3) use the following fact. $P_t(dx, dy)$ defined by

$$\int P_t(dx, dy) f(x, y) = \int \omega_0(dx) f(\phi_t^1(x), \phi_t^2(x)), \tag{E.4}$$

(where $\phi_t^i(x)$, $i = 1, 2$, are the flows induced by the velocity fields $K^\varepsilon * \xi^i$) is a joint representation of ω_t^1 and ω_t^2.)

3. Prove (3.21). (*Hint*: Since $\|\omega_t^\varepsilon\|_p = \|\omega_0^\varepsilon\|_p$ for all $p = 1, \ldots, \infty$, it follows that $\lim_{\varepsilon \to 0} (K^\varepsilon - K) * \omega_t^\varepsilon = 0$.)

4. Prove (3.22). (*Hint*: Given a joint representation $P_0(dx, dy)$ of ω_0^i, $i = 1, 2$, construct a joint representation at later times

$$\int P_t(dx, dy) f(x, y) = \int P_0(dx) f(\phi_t^1(x), \phi_t^2(x)), \qquad \text{(E.5)}$$

where $\phi_t^1(x)$, $i = 1, 2$, are the flow lines solution of the problem (3.20) with initial data ω_0^i. Replace f in (E.5) by d, the metric function, and estimate.)

5. Prove the equivalence of the topology induced by the weak convergence of measures and the topology induced by the K–R distance. (*Hint*: Use (3.10).)

CHAPTER 6

Evolution of Discontinuities

This chapter is devoted to the study of some discontinuities arising in real fluids. Namely, we first study the so-called vortex sheet, that is, a situation in which the fluid is irrotational outside a manifold of codimension one. In other words, the vorticity is concentrated on a curve or on a surface, in two or three dimensions, respectively, and in these manifolds the velocity field is discontinuous.

In the final sections we study water waves, that is, the evolution of a density discontinuity.

6.1. Vortex Sheet

In many situations the vorticity has a tendency to be concentrated in a subset of the physical space of low dimensionality.

In Chapter 4 we widely discussed the case in which the vorticity is concentrated around points. In the present chapter, we want to investigate the so-called *vortex sheet*, that is, the situation in which the vorticity is concentrated as a δ function in sets of codimension one, a curve, or a surface for two-dimensional or three-dimensional flow, respectively.

There are many flows, occurring in nature, which can be schematized by a vortex sheet. If we observe a flow past an obstacle, the velocity field, for suitable values of the physical parameters, is approximately described by a potential flow outside some thin layers in which the vorticity is concentrated. The formation of such regions is due to the influence of the viscosity. More precisely, it is due to the flow–obstacle interaction, technically expressed by the vanishing of the velocity on the boundary and the consequent generation of large gradients of vorticity. These thin layers can be conveniently schematized by the vortex sheet.

We remark once more that the viscosity, which is responsible for the creation of the vorticity manifolds, is rather irrelevant for their evolution at a large Reynolds number and far from the obstacle. We will study mainly the two-dimensional case. We suppose that initially the vorticity is concentrated on a regular line of the plane x, y

$$y = \varphi(x) \tag{1.1}$$

and on this curve a vorticity density is given by the expression

$$\gamma = \gamma(x). \tag{1.2}$$

In other words, the vorticity distribution is

$$\omega_0(x, y)\, dx\, dy = \gamma(x)\delta(y - \varphi(x)). \tag{1.3}$$

The mathematical model expressing the time evolution of a vortex sheet is based on the following consideration. We first "assume" that a vortex sheet remains a vortex sheet at a later time, i.e.,

$$\omega(x, y, t)\, dx\, dy = \gamma(x)\delta(y - \varphi(x, t)). \tag{1.4}$$

Moreover, we hypothesize that the curve is convected by the velocity field generated by itself, as suggested by the Euler equation. However, there is an ambiguity because the velocity is discontinuous just on the sheet. We solve such a problem by defining the velocity on the sheet as the average of the upper and lower limit. Finally, the evolution of γ is given by the continuity equation expressing the conservation of the vorticity. As we will see later, such an evolution is compatible with the general laws of inviscid fluids, in the sense that it induces a weak solution of the Euler equation.

If we want to approach the vortex sheet dynamics by studying weak solutions of the Euler equation with initial data of the type (1.3), we face two problems.

(i) In Chapter 2 we proved the existence and uniqueness of weak solutions supposing that the vorticity was bounded, while here it is not so. Therefore we cannot use already known techniques.

(ii) Having solved a weak form of the Euler equation, we do not know a priori that the solutions we obtain are of type (1.4), although there is experimental evidence of this fact, at least for short times.

As a preliminary step, we proceed in a heuristic way in the research of the equation for the vortex sheet, according to the rules outlined above, and we discuss later a rigorous justification of such an equation. The model consists in the generalization to our case of the well-known property of the Euler equation which states that the vorticity is conserved along the particle paths. So a point of the line in which the vorticity is concentrated moves according to the velocity field computed at that point. At this stage, we need some care

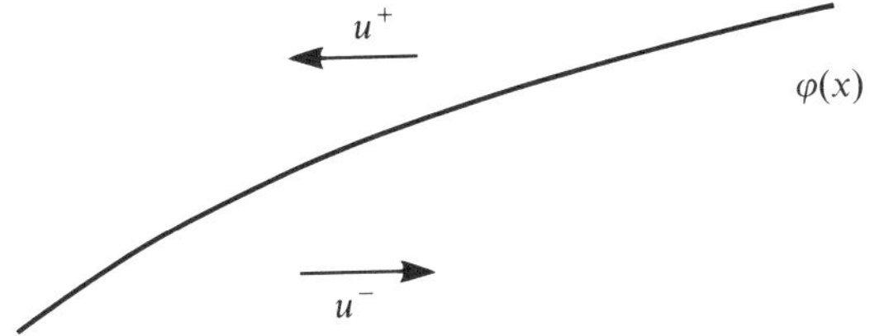

Figure 6.1

because this field has a discontinuity when we pass across the line. This is a problem deeply studied in the single layer potential theory ([CoH 37, Vol.2]) where are arrive at the following conclusion: the normal component of the velocity is continuous, while the tangential one has a jump equal to the vorticity at that point. This feature describes the physical meaning of the model we are considering: a fluid with a sudden jump in velocity (Fig. 6.1).

As we have seen, we want to calculate the velocity just at the discontinuity point $(x, \varphi(x))$. It is reasonable to suppose that the tangent component of the velocity $u_\tau(x, \varphi(x))$ is the average of the limit of the tangent velocity above u_τ^+ and below u_τ^-. The normal velocity is continuous across the curve

$$u_\tau(x, \varphi(x)) = \frac{u_\tau^+ + u_\tau^-}{2},$$

$$u_n(x, \varphi(x)) = u_n^+ = u_n^-. \tag{1.5}$$

We can write (1.5) by components. We use the relation between the vorticity and the velocity (Chapter 1, Section 2) to obtain $u(x, \varphi(x)) = (u_1, u_2)$

$$u_1 = -\frac{1}{2\pi} P \int \frac{\varphi(x) - \varphi(x')}{(x - x')^2 + (\varphi(x) - \varphi(x'))^2} \gamma(x') \, dx', \tag{1.6}_1$$

$$u_2 = \frac{1}{2\pi} P \int \frac{x - x'}{(x - x')^2 + (\varphi(x) - \varphi(x'))^2} \gamma(x') \, dx', \tag{1.6}_2$$

where $P \int$ means the Cauchy principal value.

Then we assume that φ and γ evolve under the action of the field (1.6). For φ we have

$$\frac{\partial}{\partial t} \varphi(x, t) = -u_1(x, t) \frac{\partial}{\partial x} \varphi(x, t) + u_2(x, t). \tag{1.7}_1$$

In fact, following the point $(x(t), \varphi(x(t))$ along the stream lines, we have

$$\frac{d}{dt} x(t) = u_1(x(t), t),$$

$$\frac{d}{dt} \varphi(x(t), t) = u_2(x(t), t). \tag{1.8}$$

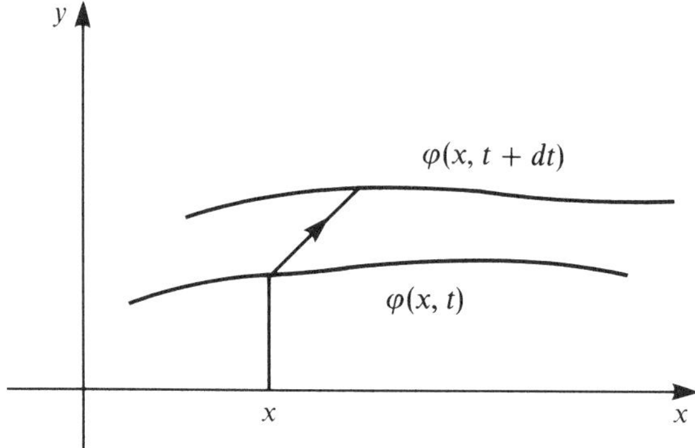

Figure 6.2

But

$$\frac{d}{dt}\varphi(x(t),\, t) = \partial_x\varphi(x(t),\, t)\frac{d}{dt}x(t) + \partial_t\varphi(x(t),\, t) \tag{1.9}$$

so that, using (1.8), we obtain $(1.7)_1$ (Fig. 6.2).

The vorticity contained in an arc of the curve φ is conserved during motion so that the continuity equation holds for γ

$$\frac{\partial}{\partial t}\gamma(x_1,\, t) + \frac{\partial}{\partial x_1}(\gamma(x_1,\, t)u_1(x_1,\, t)) = 0. \tag{1.7$_2$}$$

Until now we have written the vortex sheet curve using the Cartesian coordinates. In many situations this is not the most useful way to describe the curve; for instance, it can fail to describe it by a single value function of x. Other parametrizations of the vortex sheet curve are obviously possible. A special and useful parametrization is given in terms of the circulation variable for which (1.7) reduce to a simple complex variable equation, called the Birkhoff–Rott equation. We choose as a parameter the vorticity Γ contained in the region between a point of the curve and a reference fluid particle (on the curve).

We denote by $z = x_1 + ix_2$ the complex coordinate of a point of the plane. A point $z = z(\Gamma)$ moves under the action of the complex velocity field $u_1 + iu_2$. Expressing (1.6) in terms of $z(\Gamma)$ we readily arrive at

$$\frac{\partial}{\partial t}z^*(\Gamma,\, t) = -\frac{i}{2\pi}P\int_{-\infty}^{\infty}\frac{d\Gamma'}{z(\Gamma,\, t) - z(\Gamma',\, t)}, \tag{1.10}$$

where z^* indicates the complex conjugate of z: $z^* = x_1 - ix_2$. Observe that this parametrization is intrinsically Lagrangian in the sense that each element of the curve is parametrized once and forever at time zero (Fig. 6.3).

Let us now come back, for simplicity, to Cartesian coordinates. It is easy to see that (1.7) are in agreement with the weak form of the Euler equation

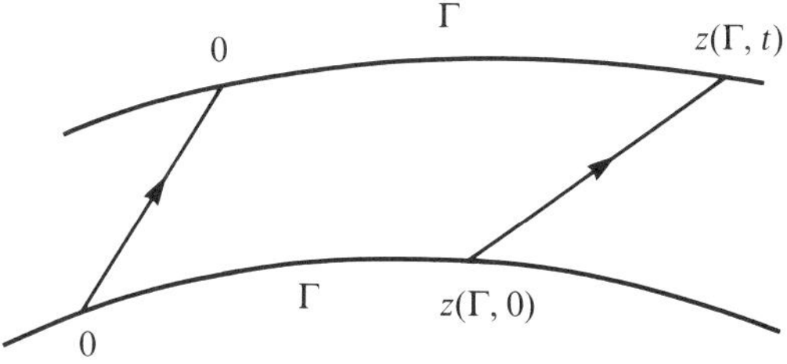

Figure 6.3

for the vorticity, at least at a formal level. In fact, choosing a regular enough test function, define

$$\omega_t(x, y)\, dx\, dy = \gamma(x, t)\delta(y - \varphi(x, t))\, dx\, dy \tag{1.11}$$

and

$$\omega_t(f) = \int \omega_t f\, dx_1, \tag{1.12}$$

we have

$$\frac{d}{dt}\omega_t(f)$$

$$= \int \frac{\partial}{\partial t}\{\gamma(x, t)\delta(y - \varphi(x, t))f(x, y)\}\, dx\, dy$$

$$= \int \left\{\left(\frac{\partial}{\partial t}\gamma(x, t)\right)f(x, \varphi(x, t))\right\}\, dx$$

$$+ \int \gamma(x, t)\left(\frac{\partial}{\partial \varphi(x, t)}f(x, \varphi(x, t))\right)\left(\frac{\partial}{\partial t}\varphi(x, t)\right) = (\text{using } (1.7))$$

$$= -\int \left\{\frac{\partial}{\partial x}(u_1\gamma)(x, t)f(x, \varphi(x_1, t))\, dx\right.$$

$$+ \int \gamma(x)\left(\frac{\partial}{\partial \varphi(x, t)}f(x, \varphi(x, t))\right)\left\{-u_1(x, t)\left(\frac{\partial}{\partial x}\varphi(x, t)\right) + u_2(x, t)\right\}\, dx.$$

$$\tag{1.13}$$

Finally, by an integration by parts,

$$\frac{d}{dt}\omega_t = \int \gamma(x, t)\left\{u_1\frac{\partial}{\partial x}f(x, \varphi) + u_2\frac{\partial}{\partial \varphi}f(x, \varphi)\right\}\, dx. \tag{1.14}$$

Thus we have proved:

Theorem 1.1. *Suppose* $(\varphi(\cdot, t), \gamma(\cdot, t))$ *is a solution of the system* (1.7) *with* φ *and* γ *smooth. Thus the measure* (1.11) *satisfies the Euler equation for the vorticity in the weak form.*

We will see that the converse of the statement of the above theorem can also be proved. Before discussing this point, let us introduce a weak formulation of the Euler equation for the velocity. It takes the form

$$(u, \nabla f) = 0, \tag{1.15}$$

$$\frac{d}{dt}(u, w) + \sum_{i=1}^{2} (u_i u, \partial_i w), \tag{1.16}$$

for any smooth function $f: \mathbb{R}^2 \to \mathbb{R}^1$ with compact support, and any time-independent divergenceless vector field $w: \mathbb{R}^2 \to \mathbb{R}^2$ with compact support. By definition

$$(u, w) = \int dx \, dy \sum_{i=1}^{2} \{u_i(x, y), w_i(x, y)\}. \tag{1.17}$$

The above formulation makes sense because, by virtue of the energy conservation, u_i is in $L_{2,\text{loc}}$.

Observe that (1.15) expresses the incompressibility condition and (1.16) the momentum conservation. These equations follow from the Euler equations (in the classical sense) by integrating by parts. We finally remark that the formulation (1.15)–(1.16) is weaker than the other weak form of the Euler equations we have already considered in Chapter 2. In fact, there the vorticity was supposed to be in $L_1 \cap L_\infty$. As a consequence of this, u was Hölder continuous. Here u is only $L_{2,\text{loc}}$.

In the present section we need such a weaker formulation for the following reason. In terms of vorticity we face a term like

$$\omega(u \cdot \nabla f) = (\text{curl } u, u \cdot \nabla f). \tag{1.18}$$

Now $\omega = \text{curl } u$ is a measure concentrated on the set of the discontinuity of u so that the right-hand side of (1.18) does not make sense.

We are now able to formulate the converse of Theorem 1.1.

Theorem 1.2. *Let u be a solution of (1.15) and (1.16). Suppose also that* curl u *is concentrated on a smooth curve $(x, \varphi(x, t))$ with intensity $\gamma(x, t)$ for $t \in [0, T]$. Then the pair (φ, γ) satisfies the vortex sheet equations (1.7).*

PROOF. Let η be such that $w = \nabla^\perp \eta$. Then denoting by

$$D_t^\pm = \{(x, y) | y \gtrless \varphi(x, t)\}, \tag{1.19}$$

we have

$$\int_{D_t^+} \sum_{i=1}^{2} u_i u \cdot \nabla^\perp \partial_i \eta = \int_{\partial D_t^+} \sum_{i=1}^{2} dl \, \partial_i \eta (u_i^+ u_\tau^+) - \int_{D_t^+} \nabla \eta \cdot \nabla^\perp E^+, \tag{1.20}$$

where u_n and $u_\tau^\pm$ denote the normal and (upper and lower) tangential component of $u^\pm$, dl is the infinitesimal arc length of ∂D_t, and finally

$$E^\pm = \tfrac{1}{2}(u^\pm)^2. \tag{1.21}$$

Equation (1.20) is a consequence of the following vector identity:

$$\operatorname{curl}(u_i u \partial_i \eta) = u_i u \nabla^\perp \partial_i \eta - \partial_i \eta u_i \operatorname{curl} u + \partial_i \eta (u \cdot \nabla^\perp) u_i. \tag{1.22}$$

By using the fact that $\operatorname{curl} u|_{D^+} = 0$, integrating by parts, and using (1.15), we easily obtain (1.20). Moreover, by using $\operatorname{curl} \nabla \eta = 0$, we have

$$\int_{D_t^+} \sum_{i=1}^2 u_i u \cdot \nabla^\perp \partial_i \eta = \int_{\partial D_t^+} \sum_{i=1}^2 dl\, \partial_i \eta (u_i^+ u_\tau^+) - \int_{\partial D_t^+} dl\, \partial_\tau \eta E^+. \tag{1.23}$$

Repeating the same argument in the lower semi-domain and summing, we arrive at the following identity:

$$\int_{\mathbb{R}^2} \sum_{i=1}^2 u_i u \cdot \nabla^\perp \partial_i \eta = -\int_{\partial D} dl\, \partial_\tau \eta (E^+ - E^-)$$

$$+ \int_{\partial D} \sum_{i=1}^2 dl\, \partial_i \eta (u_i^+ u_\tau^+ - u_i^- u_\tau^-) \tag{1.24}$$

The right-hand side of (1.24) is

$$\int_{\partial D} dl\, \partial_\tau \eta (u^+ - u^-) \cdot \bar{u} + \frac{1}{2} \int_{\partial D} dl\, \partial_n \eta (u_\tau^+ - u_\tau^-) u_n, \tag{1.25}$$

where

$$\bar{u} = \tfrac{1}{2}(u^+ + u^-).$$

Therefore, it is easy to realize that the momentum conservation can be expressed in the form

$$\frac{d}{dt}\omega(\eta) = \omega(\nabla \eta \cdot \bar{u}). \tag{1.26}$$

Equation (1.26) expresses the conservation of the vorticity and, in particular, the continuity equation $(1.7)_2$ and the fact that the curve $(x, \varphi(x, t))$ is convected through the average field $\bar{u}$. $\qquad\square$

The two theorems above assure the compatibility of the vortex-sheet dynamics, given by (1.7) or (1.10), with the motion of weak solutions for the Euler equation. In particular, Theorem 1.2 says that if we know a priori that a weak solution of the Euler equation (for the velocity) enjoys the property that the vorticity is concentrated on a smooth curve, then necessarily this curve is convected by the mean velocity $(u^+ + u^-)/2$ and the continuity equation holds. We remark that, defining

$$u(\lambda) = \lambda u^+ + (1 - \lambda)u^-, \qquad \lambda \in [0, 1], \tag{1.27}$$

the equation

$$\partial_t \varphi(x, t) + u_1(\lambda)\partial_x \varphi(x, t) = u_2(\lambda) \tag{1.28}$$

gives the same solution as $(1.7)_1$. In fact, since the normal component is continuous, the use of the convection velocity (1.27), in place of the usual mean velocity, gives rise only to an extra contribution of the velocity in the

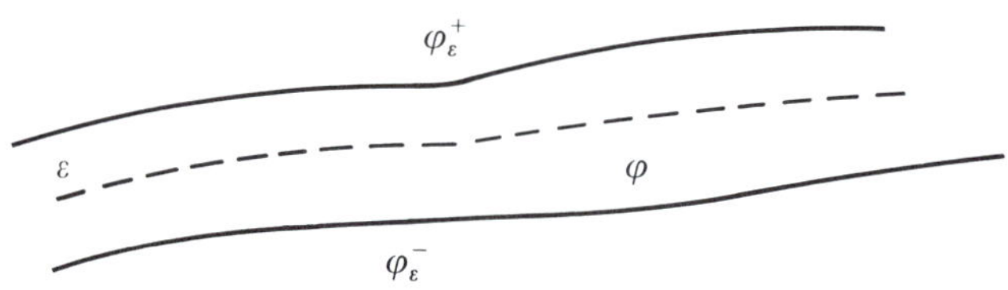

Figure 6.4

tangential direction, so that the shape of the curve is not affected. On the contrary, the choice of the mean velocity is absolutely necessary for the continuity equation $(1.7)_2$ to be consistent with the Euler equation. This feature can be understood at a physical level in the following way. Consider a very thin layer of vorticity given by

$$\omega_\varepsilon(x, y) = \frac{1}{\varepsilon} \chi_{\Lambda(\varepsilon)}(x, y), \tag{1.29}$$

where $\chi_{\Lambda(\varepsilon)}$ is the characteristic function of the set

$$\Lambda(\varepsilon) = \{(x, y)| \varphi_\varepsilon^-(x) < y < \varphi_\varepsilon^+(x)\} \tag{1.30}$$

φ_ε^- and φ_ε^+ are two functions which enclose φ at a distance approximately ε (Fig. 6.4).

The motion of this vortex layer, which maintains this structure at later times, is expected to approximate (in the sense of weak convergence of measures) the motion of the vortex sheet. Roughly speaking, φ_ε^+ and φ_ε^- are approximately convected by u^+ and u^-, respectively. Moreover, the continuity equation for the vortex layer dynamics is expressed by the following argument. Define

$$\gamma_\varepsilon(x, t) = \frac{\varphi_\varepsilon^+(x, t) - \varphi_\varepsilon^-(x, t)}{\varepsilon}. \tag{1.31}$$

Denoting by u^ε the velocity field generated by $\omega_\varepsilon(x, y, t)$, general arguments show that the curves $\varphi_\varepsilon^\pm$ are convected by u^ε. After simple calculations (making explicit use of the condition div $u^\varepsilon = 0$) we obtain

$$\partial_t \gamma_\varepsilon(x, t) = -\partial_x \left(\frac{1}{\varepsilon} \int_{\varphi_\varepsilon^-(x,t)}^{\varphi_\varepsilon^+(x,t)} dy\, u_1^\varepsilon(x, y, t) \right). \tag{1.32}$$

The term in the bracket on the right-hand side of (1.32) is approximately $\frac{1}{2}(u^+ + u^-)\gamma$.

The above arguments are heuristic and have been introduced to illustrate some features of the vortex sheet dynamics. A complete rigorous argument is not trivial and will be discussed in Section 6.3. However, it is worth underlining that a rigorous justification of the vortex sheet equation, assuming the Euler equation as the basic equation of the mathematical model of nonviscous fluids, can be given by proving continuity of the solutions of the Euler equation with respect to initial conditions close to data of the form (1.3). Namely, we would prove that solutions of the Euler equation initially close (with respect to the weak convergence of measure) to the data (1.3), remain close (in the same sense) to the vortex sheet solutions at later times. A similar problem was treated in Chapter 4 for a justification of the point vortex dynamics. A partial answer to this question, together with further considerations, will be given later in Section 6.3.

We briefly generalize the model in three dimensions. In this case, the vorticity is concentrated on a two-dimensional surface given by the equation

$$z = z(x, y, t). \tag{1.33}$$

The vorticity density $\Omega(x, y, t)$ is defined by the relation

$$\int \eta(x, y, z)\omega(x, y, z, t)\, dx\, dy\, dz = \int \eta(x, y, z(x, y, t))\Omega(x, y, t)\, dx\, dy, \tag{1.34}$$

where η is a test function.

The velocity of a point, exterior to the vortex sheet, is given by

$$u(r, t) = -\frac{1}{4\pi} \int \frac{r - r(x, y, t)}{|r - r(x, y, t)|^3} \wedge \Omega(x, y, t)\, dx\, dy, \tag{1.35}$$

where $r = (x, y, z)$ is the generic point and $(x, y) \to r(x, y, z)$ is the parametric equation of the surface given by (1.33).

We call $u^+(x, y, t)$ and $u^-(x, y, t)$ the two limits which are obtained when the point r tends to a point $r(x, y, t)$ in an arbitrary way to one or the other side of the vortex sheet. We define

$$U(x, y, t) = \frac{u^+(x, y, t) + u^-(x, y, t)}{2}$$

$$= -\frac{1}{4\pi} P \int \frac{r(x, y, t) - r(x', y', t)}{|r(x, y, t) - r(x', y', t)|^3} \wedge \Omega(x', y', t)\, dx'\, dy'. \tag{1.36}$$

Then the model of the vortex sheet is defined by the equations

$$\partial_t z = -U_1 \partial_x z - U_2 \partial_y z + U_3, \tag{1.37$_1$}$$

$$\partial_t \Omega + \partial_x(\Omega U_1) + \partial_y(\Omega U_2) = \Omega_1 \partial_x U + \Omega_2 \partial_y U. \tag{1.37$_2$}$$

Notice that in the three-dimensional case the velocity field also has a jump across the vortex sheet

$$[u] = u^+ - u^- \tag{1.38}$$

which satisfies the relations

$$[u] \cdot n = 0, \tag{1.39}$$

$$[u] \wedge n = 0, \tag{1.40}$$

where n is the normal vector.

6.2. Existence and Behavior of the Solutions

In this section we want to show how to construct a solution of the initial value problem associated with the vortex sheet equation. This task will be achieved only for short times. This is probably the best we can do, since the solution of the vortex sheet equation is expected, in general, to become singular after a finite time. Before discussing in a rigorous mathematical setup the initial value problem, it is useful to develop some classical heuristic considerations, based on the linear analysis of the vortex sheet equation, leading to the so-called Kelvin–Helmholtz instability.

From now on we limit ourselves to the simpler two-dimensional case. We suppose that initially the vorticity is uniformly concentrated in the x-axis. This is, of course, a stationary solution of the vortex sheet equation. Consider a small perturbation in the shape of a vortex sheet of intensity one. As consequence, the sheet is no longer stationary. For instance, consider the situation in the figure where a small bump is created around the origin (Fig. 6.5). A simple analysis shows that a nonzero average velocity is created on the top of the bump which starts to become distorted. On the other hand, the vorticity (initially constant) also moves and local concentrations of the vorticity are created. These contribute to create further rotations so that the disturbance may grow exponentially in time. This is what really happens at the level of linear analysis, as we will show.

Throughout this section we will consider the case of a small periodic disturbance of the flat profile. It is natural to study perturbatively the problem close to a known solution. The periodicity assumption is made for convenience. Consider the Birkhoff–Rott equation (1.10) and an initial condition of the type

$$z_0(\Gamma) = \Gamma + \varepsilon S_0(\Gamma), \tag{2.1}$$

where $S_0(\Gamma)$ is a 2π-periodic function of Γ. Notice that, in general, εS_0 is a perturbation of the flat sheet in shape as well as in vorticity intensity.

The equation for the perturbation is

$$\begin{aligned}
\varepsilon \partial_t S(\Gamma, t)^* &= \frac{1}{2\pi i} P \int d\Gamma' \frac{1}{\Gamma - \Gamma' + \varepsilon[S(\Gamma, t) - S(\Gamma', t)]} \\
&= \frac{1}{2\pi i} \sum_{n=0}^{\infty} (-\varepsilon)^n P \int d\Gamma' \frac{[S(\Gamma, t) - S(\Gamma', t)]^n}{(\Gamma - \Gamma')^{n+1}}.
\end{aligned} \tag{2.2}$$

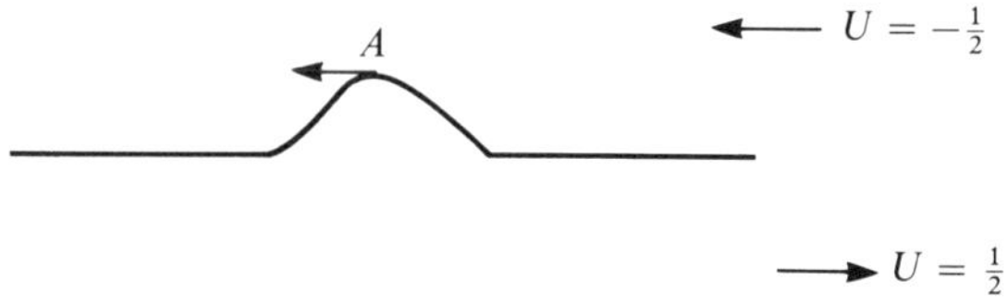

Figure 6.5

Furthermore we assume, for simplicity, that

$$S_0(-\Gamma) = -S_0(\Gamma) \tag{2.3}$$

so that the integral of s_0 vanishes. Moreover, the condition (2.3) is preserved at later times as follows by direct inspection of (2.2). The series expansion in the right-hand side of (2.2) is justified if $|\partial_\Gamma S|$ is sufficiently small. Since the term $n = 0$ in the right-hand side of (2.2) vanishes, the first-order (in ε) equation reduces to

$$\partial_t S(\Gamma, t)^* = -\frac{1}{2\pi i} P \int d\Gamma' \frac{S(\Gamma, t) - S(\Gamma', t)}{(\Gamma - \Gamma')^2}. \tag{2.4}$$

Obviously, the above equation is a linearization around the stationary flat profile.

Defining

$$S^+(\Gamma) = \sum_{k=1}^{\infty} s_k^\wedge \exp\{ik\Gamma\}, \tag{2.5}_1$$

$$S^-(\Gamma) = \sum_{k=-\infty}^{-1} s_k^\wedge \exp\{ik\Gamma\} = \sum_{k=1}^{\infty} s_{-k}^\wedge \exp\{-ik\Gamma\}. \tag{2.5}_2$$

Notice that the zero wave number component vanishes because of (2.3), so that $S = S^+ + S^-$. Therefore the integral in the right hand-side of (2.4) becomes

$$\frac{1}{2\pi i}\left(P \int d\xi \frac{S^+(\Gamma + \xi, t) - S^+(\Gamma, t)}{\xi^2} + P \int d\xi \frac{S^-(\Gamma + \xi, t) - S^-(\Gamma, t)}{\xi^2}\right). \tag{2.6}$$

Observe that S^+ can be analytically extended in the domain Im $\Gamma > 0$, since

$$S^+(\Gamma + i\rho) = \sum_{k \geq 1} s_k^\wedge \exp\{ik\Gamma\} \exp\{-k\rho\} \tag{2.7}$$

provides that mild conditions on $s_k^\wedge$ are verified, say $|s_k^\wedge| < C$.

Moreover, $S^+(\Gamma + i\rho)$ decays exponentially for $\rho \to \infty$. Thus

$$\frac{1}{2\pi i} P \int d\xi \frac{S^+(\Gamma + \xi, t) - S^+(\Gamma, t)}{\xi^2} = \tfrac{1}{2}\partial_\Gamma S^+(\Gamma, t). \tag{2.8}$$

The above integral has been obtained by the residue method replacing the integral on the real line with the integral in the infinite half-circle centered at the origin in the upper half-plane. The $\tfrac{1}{2}$ appearing in formula (2.8) is due to the fact that the only pole of the integrand is just on the integration path. A similar calculation holds for S^-. In conclusion, we have

$$\partial_t S(\Gamma, t)^* = \tfrac{1}{2}\partial_\Gamma S^+(\Gamma, t) - \tfrac{1}{2}\partial_\Gamma S^-(\Gamma, t). \tag{2.9}$$

In terms of Fourier coefficients (2.9) becomes

$$\frac{d}{dt} s_k^\wedge(t)^* = i\frac{k}{2} s_{-k}^\wedge(t), \tag{2.10}_1$$

$$\frac{d}{dt} s_{-k}^{\wedge}(t)^* = i\frac{k}{2} s_k^{\wedge}(t), \tag{2.10$_2$}$$

which implies

$$\frac{d^2}{dt^2} s_k^{\wedge}(t) = \frac{k^2}{4} s_k^{\wedge}(t), \tag{2.11$_1$}$$

$$\frac{d^2}{dt^2} s_{-k}^{\wedge}(t) = \frac{k^2}{4} s_{-k}^{\wedge}(t). \tag{2.11$_2$}$$

From (2.11), we finally obtain

$$s_k^{\wedge}(t) = \frac{s_k^{\wedge}(0) - i s_{-k}^{\wedge}(0)^*}{2} \exp\left\{\frac{kt}{2}\right\} + \frac{s_k^{\wedge}(0) + i s_{-k}^{\wedge}(0)^*}{2} \exp\left\{\frac{-kt}{2}\right\}, \tag{2.12$_1$}$$

$$s_{-k}^{\wedge}(t) = \frac{s_{-k}^{\wedge}(0) - i s_k^{\wedge}(0)^*}{2} \exp\left\{\frac{kt}{2}\right\} + \frac{s_{-k}^{\wedge}(0) + i s_k^{\wedge}(0)^*}{2} \exp\left\{\frac{-kt}{2}\right\}. \tag{2.12$_2$}$$

From the above expression we conclude that the vortex sheet dynamics is linearly unstable (Kelvin–Helmholtz instability). This is a well-known feature which is experimentally observed: a wake generated by an obstacle in a flow at a very high Reynolds number is very unstable. The sheet has the tendency to roll up and eventually breaks into complex configurations. Numerical simulations also confirm this behavior.

Let us now approach the problem of the construction of the solutions to the initial value problem associated to the vortex-sheet equation

$$\partial_t \varphi(x, t) + u_1 \partial_x \varphi(x, t) = u_2(x, t), \tag{2.13$_1$}$$

$$\partial_t \gamma(x, t) + \partial_x (u_1 \gamma)(x, t) = 0, \tag{2.13$_2$}$$

$$\varphi(x, 0) = \varphi_0(x), \qquad \gamma(x, 0) = \gamma_0(x), \tag{2.13$_3$}$$

$$u_1(x) = -\frac{1}{2\pi} P \int \frac{\varphi(x) - \varphi(x')}{(x - x')^2 + (\varphi(x) - \varphi(x'))^2} \gamma(x') \, dx', \tag{2.14$_1$}$$

$$u_2(x) = \frac{1}{2\pi} P \int \frac{(x - x')}{(x - x')^2 + (\varphi(x) - \varphi(x'))^2} \gamma(x') \, dx'. \tag{2.14$_2$}$$

There is no special reason for choosing the x and y parametrization for the vortex sheet equation. The considerations that follow are largely independent of the parametrization and apply as well to the Birkhoff–Rott equation.

We will show a local (in time) existence theorem for the system (2.13). There are good reasons for believing that there is an effective lack of regularity of the solution after a finite time. Thus the fact that the solution exists locally in time only arises from an intrinsic physical reason, and it is not due to the trivial fact that the curve described by φ roll up so that the manifold cannot be described by a single valued function φ. There are estimates on the critical time, which is the time in which the evolution ceases to be described by the Birkhoff–Rott equation due to the lack of regularity. These will be discussed in the next section.

We finally remark that we cannot exclude that weak solutions to the Euler equation, with an initial value given by a vorticity concentrated on a smooth curve as a delta function, can be constructed globally in time. What we are saying is that such solutions cannot be presumably described in terms of the vortex-sheet dynamics. Also this point will be further analyzed in the next section.

Coming back to the initial value problem (2.13), we first remark that the space of analytical functions is the natural ambient in which to seek solutions. In fact, the linear analysis shows that the Fourier components of the solutions grow, in general, as $\exp((k/2)t)$ (see (2.12)), so that the initial conditions must decay at least exponentially in k if we want smooth solutions. On the other hand, the construction of distributional solutions for the vortex-sheet equation is meaningless because the velocity field (see (2.14)) is a Cauchy principal value integral, so that some smoothness on the solutions is required.

The basic tool for our analysis is the following theorem of the Cauchy–Kowalevski type:

Theorem 2.1. *Let $\{X_\rho\}_{\rho \geq 0}$ be a scale of Banach space, i.e., a family of Banach spaces satisfying $X_{\rho'} \subset X_\rho$, $\|\cdot\|_\rho \leq \|\cdot\|_{\rho'}$ for $\rho < \rho'$, where $\|\cdot\|_\rho$ denotes the norm in X_ρ. Consider the Cauchy problem*

$$\xi(t) = \xi_0 + \int_0^t dx \, F(\xi(s)), \tag{2.15}$$

where $F: \{\xi \mid \xi \in X_{\rho'}, \|\xi\|_{\rho'} < R\} \to X_\rho$ is a continuous mapping for some $R > 0$, $\rho < \rho' < \rho_0$.

Moreover, there exists a constant C_R such that for all $\xi, \xi_1, \xi_2 \in X_{\rho'}$ with $\|\xi\|_{\rho'}, \|\xi_1\|_{\rho'}, \|\xi_2\|_{\rho'} < R$

$$\|F(\xi)\|_\rho \leq C_R \frac{\|\xi\|_{\rho'}}{\rho' - \rho}, \tag{2.16}$$

$$\|F(\xi_1) - F(\xi_2)\|_\rho = C_R \frac{\|\xi_1 - \xi_2\|_{\rho'}}{\rho' - \rho}. \tag{2.17}$$

Then, for $\xi_0 \in X_{\rho_0}$ with $\|\xi\|_{\rho_0} < R_0 < R$, there exists a unique continuous solution $\xi(t) \in X_\rho$ satisfying $\|\xi\|_\rho < R$ for $t \in [0, a(\rho_0 - \rho)]$ and a is chosen sufficiently small.

PROOF. See Appendix A.1.

To apply the above theorem to our case let us introduce the real Banach space of all 2π periodic functions with a norm given by

$$\|\varphi\|_\rho = \sum_{k \in \mathbb{Z}} |\varphi^\wedge(k)| e^{|k|\rho}. \tag{2.18}$$

We denote by X_ρ such a Banach space. Notice that if $\varphi \in X_\rho$ it may be analytically extended in the complex plane in the region $|\text{Im } z| < \rho$, and continuously up to $|\text{Im } z| = \rho$.

Suppose now that

$$\|\partial_x \varphi\|_\rho^2 < \tfrac{1}{2}. \tag{2.19}$$

Under this hypothesis we want to give a bound on u_1 and u_2 defined by (2.14). We have

$$u_1(x) = -\frac{1}{2\pi} P \int d\alpha \, \frac{\varphi(x) - \varphi(x + \alpha)}{\alpha^2 + [\varphi(x) - \varphi(x + \alpha)]^2} \gamma(x + \alpha)$$

$$= -\frac{1}{2\pi} P \int \frac{d\alpha}{\alpha^2} \, \frac{\varphi(x) - \varphi(x + \alpha)}{1 + [\varphi(x) - \varphi(x + \alpha)/\alpha]^2} \gamma(x + \alpha)$$

$$= -\frac{1}{2\pi} \sum_{n=0}^{\infty} (-1)^n P \int \frac{d\alpha}{\alpha} \gamma(x + \alpha) \left[\frac{\varphi(x) - \varphi(x + \alpha)}{\alpha} \right]^{2n+1}. \tag{2.20}$$

The above series is absolutely convergent by virtue of condition (2.19), and the obvious bound $\|\partial_x \varphi\| = \sup_{0 \le x \le 2\pi} |\partial_x \varphi(x)| \le \|\partial_x \varphi\|_\rho$.

Remark. It is not difficult to show that the sum in (2.20), originally in the integral sign, can be interchanged.

Taking the Fourier transform in (2.20) we obtain

$$\hat{u}_1(k) = -\frac{1}{2\pi} \sum_{n=0}^{\infty} (-1)^n \sum_{k_0 \in \mathbb{Z}} \sum_{k_1 \in \mathbb{Z}} \cdots \sum_{k_{2n+1} \in \mathbb{Z}} \delta\left(\sum_{i=0}^{2n+1} k_i, \, k \right),$$

$$P \int \frac{d\alpha}{\alpha} \gamma^\wedge(k_0) \exp(i\alpha k_0) \prod_{j=1}^{2n+1} \frac{1 - \exp(ik_j \alpha)}{\alpha} \varphi^\wedge(k_j), \tag{2.21}$$

after recalling that the Fourier transform of $\varphi(x + \alpha)$ is $e^{ik\alpha} \varphi^\wedge(k)$ and where $\delta(i, j) = 0$ if $i \ne j$ and $\delta(i, i) = 1$. Moreover,

$$P \int \frac{d\alpha}{\alpha} \exp(i\alpha k_0) \prod_{j=1}^{2n+1} \frac{1 - \exp(ik_j \alpha)}{\alpha}$$

$$= P \int \frac{d\alpha}{\alpha} \exp(i\alpha k_0) \prod_{j=1}^{2n+1} (-ik_j) \int_0^1 dt_j \, \exp(ik_j t_j \alpha)$$

$$= \prod_{j=1}^{2n+1} (-ik_j) \int_0^1 dt_1 \cdots \int_0^1 dt_{2n+1} \, P \int \frac{d\alpha}{\alpha} \exp\left(i\alpha \sum_{j=1}^{2n+1} k_j t_j \right) \exp(i\alpha k_0). \tag{2.22}$$

The last integral can be explicitly calculated. By the residue theorem

$$P \int \frac{d\alpha}{\alpha} e^{i\alpha A} = \pi i \, \text{sgn } A. \tag{2.23}$$

Thus the right-hand side of (2.22) is

$$(-1)^n \pi \prod_{j=1}^{2n+1} k_j \int_0^1 dt_1 \cdots \int_0^1 dt_{2n+1} \, \mathrm{sgn}\left(\sum_{j=1}^{2n+1} k_j t_j + k_0\right) \qquad (2.24)$$

and the above expression can be bounded by

$$\pi \sum_{j=1}^{2n+1} |k_j|. \qquad (2.25)$$

From (2.21) we finally have

$$|\hat{u}_1(k)| \le \frac{1}{2} \sum_{n \ge 0} \sum_{k_0 \in \mathbb{Z}} \sum_{k_1 \in \mathbb{Z}} \cdots \sum_{k_{2n+1} \in \mathbb{Z}} \delta\left(\sum_{i=0}^{2n+1} k_i, k\right) |\gamma^\wedge(k_0)|\right) \prod_{j=1}^{2n+1} |k_j| \, |\varphi^\wedge(k_j)| \qquad (2.26)$$

from which

$$\|u_1\|_\rho = \sum_{n \ge 0} e^{\rho |k|} |\hat{u}_1(k)| \le \frac{1}{2} \sum_{n \ge 0} \sum_{k_0 \in \mathbb{Z}} \sum_{k_1 \in \mathbb{Z}} \cdots \sum_{k_{2n+1} \in \mathbb{Z}} \exp(\rho |k_0|) |\gamma^\wedge(k_0)|$$

$$\times \prod_{j=1}^{2n+1} \exp(\rho |k_j|) |\varphi^\wedge(k_j)| \, |k_j|$$

$$= \frac{1}{2} \sum_{n \ge 0} \|\gamma\|_\rho \|\partial_x \varphi\|_\rho^{2n+1} = \frac{1}{2} \frac{\|\gamma\|_\rho \|\partial_x \varphi\|_\rho}{1 - \|\partial_x \varphi\|_\rho^2} \le \|\gamma\|_\rho \|\partial_x \varphi\|_\rho. \qquad (2.27)$$

(For the last step we have used $\|\partial_x \varphi\|_\rho^2 < \frac{1}{2}$.)

Similarly, we obtain

$$\|u_2\|_\rho \le \frac{1}{2} \frac{\|\gamma\|_\rho \|\partial_x \varphi\|_\rho + \|\gamma - 1\|_\rho}{1 - \|\partial_x \varphi\|_\rho^2} \le \|\gamma\|_\rho \|\partial_x \varphi\|_\rho + \|\gamma - 1\|_\rho. \qquad (2.28)$$

In a completely analogous way we can obtain Lipschitz estimates on the velocity fields. More precisely, denoting $u_i(\eta, \sigma)$, $i = 1, 2$, the velocity field generated by the curve $y - \eta(x) = 0$ and the vorticity intentisy $\sigma = \sigma(x)$, we have

$$\|u_1(\varphi, \gamma) - u_1(\eta, \sigma)\|_\rho \le 2(\|\partial_x \varphi - \partial_x \eta\|_\rho \|\gamma\|_\rho + \|\partial_x \varphi\|_\rho \|\gamma - \sigma\|_\rho), \qquad (2.29)_1$$

$$\|u_2(\varphi, \gamma) - u_2(\eta, \sigma)\|_\rho \le 2(\|\partial_x \varphi - \partial_x \eta\|_\rho \|\gamma\|_\rho + \|\gamma - \sigma\|_\rho). \qquad (2.29)_2$$

We are now in a position to apply Theorem 2.1. Defining

$$\xi = \begin{pmatrix} \varphi \\ \mu \end{pmatrix}, \qquad (2.30)$$

where $\mu = \gamma - 1$, and

$$F(\xi) = \begin{pmatrix} -u_1 \partial_x \varphi + u_2 \\ -(1 + \mu)\partial_x u_1 - u_1 \partial_x \mu \end{pmatrix}, \qquad (2.31)$$

where $u_1 = u_1[\varphi, \mu]$ and $u_2 = u_2[\varphi, \mu]$ are understood as functionals of φ and $\mu = \gamma - 1$, according to (2.14).

Setting

$$\|\xi\|_\rho = \|\varphi\|_\rho + \|\partial_x\varphi\|_\rho + \|\mu\|_\rho, \tag{2.32}$$

by (2.27) and (2.28) we obtain

$$\|u_1\partial_x\varphi + u_2\|_\rho \leq \|u_1\|_\rho\|\partial_x\varphi\|_\rho + \|u_2\|_\rho$$

$$\leq (1 + \|\mu\|_\rho)\|\partial_x\varphi\|_\rho^2 + \|\mu\|_\rho(1 + \|\partial_x\varphi\|_\rho) + \|\partial_x\varphi\|_\rho$$

$$\leq \frac{1}{\sqrt{2}}\|\partial_x\varphi\|_\rho + \tfrac{1}{2}\|\mu\|_\rho + \left(1 + \frac{1}{\sqrt{2}}\right)\|\mu\|_\rho + \|\partial_x\varphi\|_\rho$$

$$\leq \text{const.}\ \|\xi\|_\rho, \tag{2.33}$$

$$\|\partial_x(u_1\partial_x\varphi + u_2)\|_\rho \leq \text{const.}\frac{\|\xi\|_{\rho'}}{\rho' - \rho}. \tag{2.34}$$

This follows by the general inequality

$$\|f'\|_\rho = \sum_k e^{\rho|k|}|f^\wedge(k)|\,|k| \leq \sup_k |k|e^{-(\rho'-\rho)|k|}\sum_k e^{\rho'|k|}|f^\wedge(k)|$$

$$\leq \|f\|_{\rho'}\frac{1}{\rho' - \rho}. \tag{2.35}$$

Finally, for $\|\mu\|_{\rho'} < R$ and $\|\partial_x\varphi\|_{\rho'} < R$, we have

$$\|(1 + \mu)\partial_x u_1 - u_1\partial_x\mu\|_\rho \leq C_R\frac{\|\partial_x\varphi\|_{\rho'} + \|\mu\|_{\rho'}}{\rho' - \rho} \leq C_R\frac{\|\xi\|_{\rho'}}{\rho' - \rho}. \tag{2.36}$$

In conclusion

$$\|F(\xi)\|_\rho \leq C_R\frac{\|\xi\|_{\rho'}}{\mu' - \mu} \tag{2.37}$$

provided that $\|\xi\|_{\rho'} < R \leq 1/\sqrt{2}$. The Lipschitz estimate follows the same lines.

Therefore we have proved the following theorem:

Theorem 2.2. *Consider an initial datum* (γ_0, φ_0) *for the initial value problem associated with* (2.13). *Suppose that*

$$\|\varphi_0\|_{\rho_0} + \|\partial_x\varphi_0\|_{\rho_0} + \|1 - \gamma\|_{\rho_0} < R_0 < \frac{1}{\sqrt{2}} \tag{2.38}$$

for some $R_0 > 0$ *and* $\rho_0 > 0$.

Then there exists a unique solution $\varphi(t)$, $\gamma(t)$ *in* X_ρ, $\rho < \rho_0$, *of the initial value problem in the integral form, satisfying for* $R > R_0$ *the estimate*

$$\|\varphi\|_\rho + \|\partial_x\varphi\|_\rho + \|1 - \gamma\|_\rho < R \tag{2.39}$$

up to time $t = a(\rho_0 - \rho)$, *for some* a *depending on* R_0 *and* R.

6.3. Comments

The well-posedness of the initial value problem for the vortex-sheet equation was discussed in the space of analytic functions. In this framework, a short time result was obtained in [SSB 81]. For the Cauchy–Kowalevski theorem there is a wide literature. We mention [Nir 72], [BaG 77], [Nis 77], [KaN 79], and [Caf 90]. In the last reference is discussed a simplified version of such type of theorems, which is well suited for the vortex-sheet problem. An existence and uniqueness result for the vortex-sheet problem, global in time, was proved in [DuR 86] and [DuR 88] for a special (nontrivial) class of initial conditions.

The ill-posedness of such a problem has been studied in [DuR 86], [DuR 88], [Ebi 88], and [CaO 89]. In this last paper, the authors construct a solution of the linearized problem, (2.9), of the following form:

$$S(\Gamma, t) = \varepsilon \left\{ \left(1 - \exp\left(-\frac{t}{2} - i\Gamma \right) \right)^{1+\nu} - \left(1 - \exp\left(-\frac{t}{2} + i\Gamma \right) \right)^{1+\nu} \right\}. \quad (3.1)$$

For $t > 0$, S is a well-defined analytical function. However, for $t = 0$ and Γ small,

$$S(\Gamma, 0) \cong 2i\varepsilon \; \mathrm{Im} \, \exp\left\{ i\frac{\pi}{2}(1 + \nu) \right\} \Gamma^{1+\nu}. \quad (3.2)$$

Therefore

$$\frac{\partial^2 Z}{\partial \Gamma^2} = \frac{\partial^2 S}{\partial \Gamma^2} \cong \mathrm{const.} \; \Gamma^{\nu-1} \quad (3.3)$$

so that if $\nu \in (0, 1)$, Z has a singularity in the second derivative, a cusp appears.

Thus we can invert the time and obtain a solution of the linearized equation, which is analytic at the initial time $t_0 < 0$, exhibiting a singularity at time zero. Notice that $|t_0|$ may be chosen arbitrarily small. Also the initial condition can be chosen arbitrarily small as follows by the fact that $s \to 0$ as $t \to \infty$ in (3.1). Of course, we must wait longer to see the singularity if the initial datum is small enough. Caflish and Orellana [CaO 88] were able to show that the same behavior occurs when the full equation is considered. If ε is small enough, the nonlinear terms do not change the qualitative behavior of the solution.

Heuristic and numerical arguments ([Moo 79], [Moo 84], [MBO 82], and [Kra 86]) show that, for periodic initial data S of suitable size ε, the critical time ε, that is, the time in which the first singularity occurs, is $t_c = O(\log \varepsilon)$. Actually, a critical time of this order of magnitude can be found rigorously

Theorem 3.1 ([CaO 86]). *Let $0 < k < 1$, $0 < \varepsilon < 1$, and $\rho > 0$. Let $s(\Gamma, 0)$ be analytical in the strip $|\mathrm{Im} \, \Gamma| < \rho$ such that*

$$\sup_{|\mathrm{Im} \, \Gamma| < \rho} \{ |s(\Gamma, 0) + |\partial_\Gamma s(\Gamma, 0)| \} < \varepsilon \quad (3.4)$$

for ε sufficiently small. Then (1.10) *has an analytic solution for time t,*

$$0 \le t \le T = k \min(2\rho, \varepsilon^{-1}). \tag{3.5}$$

Moreover, $k \to 1$ as $\varepsilon \to 0$.

Notice that the existence time T becomes larger as the initial datum gets smaller. A natural question is to see whether the above estimate of the critical time is satisfactory. To see this, we apply Theorem 3.1 to the case of an initial perturbation of a sinuisoidal form

$$s(\Gamma, 0) = i\frac{\varepsilon'}{2} \sin 2\Gamma. \tag{3.6}$$

Using the expression

$$\sin 2\Gamma = \frac{\exp(i2\Gamma) - \exp(-i2\Gamma)}{2i}$$

(3.4) gives

$$\text{const. } \varepsilon' \exp(2\bar{\rho}) < \varepsilon. \tag{3.7}$$

Hence

$$\exp(2\rho) < \frac{\varepsilon}{\varepsilon'} \text{ const.}$$

that is,

$$2\rho < \ln \varepsilon - \ln \varepsilon' + \text{const.}, \qquad \varepsilon, \varepsilon' < 1. \tag{3.8}$$

To maximize the time in (3.5) it is convenient to choose $\varepsilon \gg \varepsilon'$ but such that $2\rho > |\ln \varepsilon|$. Thus

$$T \cong k|\ln \varepsilon|, \tag{3.9}$$

where $k \to 1$ as $\varepsilon \to 0$.

We do not give here the proof of Theorem 3.1. We mention only that it is based on an approximate equation for the vortex sheet due to Moore which we are going to derive. We rewrite (1.10) as

$$\frac{\partial}{\partial t}s^*(\Gamma, t) = B[s] = B[s^+] + B[s^-] + D[s^+, s^-], \tag{3.10}$$

where

$$B[s] = \frac{i}{2\pi}P\int_{-\infty}^{\infty} \frac{d\xi}{\xi + s(\Gamma + \xi) - s(\Gamma)}. \tag{3.11}$$

$S^{\pm}$ are defined by (2.5) and D is defined implicitly by (3.10). The Moore approximation consists in neglecting the term D. The remaining terms can be handled by the complex variable manipulation as in the previous section. We obtain

$$B[s^+] = \tfrac{1}{2} \operatorname*{res}_{\xi \to 0} \frac{s^+(\Gamma + \xi) - s^+(\Gamma)}{\xi + s^+(\Gamma + \xi) - s^+(\Gamma)} \frac{1}{\xi}$$

$$= \tfrac{1}{2}\frac{\partial_\Gamma s^+(\Gamma)}{1 + \partial_\Gamma s^+(\Gamma)}. \tag{3.12}$$

Similarly with $B[s^-]$. In conclusion, we obtain

$$\frac{\partial}{\partial t}\{(s^+)^* + (s^-)^*\} = \frac{1}{2}\left\{\frac{\partial_\Gamma s^+(\Gamma)}{1 + \partial_\Gamma s^+(\Gamma)} - \frac{\partial_\Gamma s^-(\Gamma)}{1 + \partial_\Gamma s^-(\Gamma)}\right\}. \tag{3.13}$$

Equation (3.13) is called the Moore equation. In [CaO 86] it is proved that (3.13) is a good approximation for (1.10) when the initial data are small. More precisely, for any t, $0 \le t \le T$, where T is defined in Theorem 3.1, for any solution of the approximating equation (3.13), there exists nearly a solution of the full equation (1.10). Good control of the solutions of the Moore equation, combined with the use of the Cauchy–Kowalevski theorem, allow us to achieve the proof of Theorem 3.1.

Numerical simulations show that a periodic vortex sheet rolls up into a wound spiral in a time of the order of the critical time. The vortex blob method can be used (exactly as for smooth solutions of the two-dimensional Euler equation as discussed in Chapter 5, [Kra 86]). Caflisch and Lowengroub [CaL 89] proved the convergence of the vortex blob and point vortex method for the vortex-sheet problem.

Another numerical algorithm based on contour dynamics is often used. Consider a weak solution of the two-dimensional Euler equation of the form

$$\omega_t^\varepsilon(x, y) = \varepsilon^{-1}\chi_{\Lambda\varepsilon(t)}(x, y), \tag{3.14}$$

where

$$\Lambda_\varepsilon(t) = \{(x, y)|\varphi^-(x, t) < y < \varphi^+(x, t)\} \tag{3.15}$$

and $\chi_{\Lambda\varepsilon(t)}$ is the characteristic function of the set $\Lambda_\varepsilon(t)$. We assume $\varphi^\pm$ 2π-periodic functions and

$$\int_0^{2\pi} \{\varphi^+(x, t) - \varphi^-(x, t)\}\, dx = \varepsilon. \tag{3.16}$$

Relation (3.16) holds for all time provided it is satisfied at time zero by the vorticity conservation law. Then the profiles φ^+ and φ^- are expected to be a good approximation of the vortex-sheet dynamics. For numerical implementation of this idea, see [BaS 89], [ShB 88], and [ShB 90].

A convergence proof of the above algorithm can be found in [BeP 92]. More precisely we can prove the following theorem:

Theorem 3.2 ([BeP 92]). *Suppose initially*

$$\|\varphi_\varepsilon^+\|_{\rho_0} + \|\partial_x\varphi_\varepsilon^+\|_{\rho_0} + \|\varphi_\varepsilon^-\|_{\rho_0} + \|\partial_x\varphi_\varepsilon^-\|_{\rho_0} + \|\mu_\varepsilon\|_{\rho_0} < R_0 < \tfrac{1}{4}, \tag{3.17}$$

where

$$\mu_\varepsilon = \eta_\varepsilon - 1, \qquad \gamma_e = \frac{\varphi_e^+ - \varphi_\varepsilon^-}{\varepsilon}, \tag{3.18}$$

(for $\|\cdot\|_{\rho 0}$, see definition (2.18)).

Then there exists $a > 0$ for which, for $t \in [0, a(\rho_0 - \rho)]$,

$$\lim_{\varepsilon \to 0} \omega_\varepsilon(\cdot, t) = \omega(\cdot, t) \tag{3.19}$$

in the sense of weak convergence of measures and

$$\omega(dx, dy, t) = \delta(x - \varphi(x, t))\gamma(x, t) \tag{3.20}$$

and φ and γ are the unique solution to (2.13) in X_ρ.

Notice that the above theorem provides a rigorous derivation of the vortex-sheet equation in terms of solutions of the Euler equation for bounded data. In the previous theorem the vortex-sheet curve was approximated in terms of analytical contours. This seems necessary for the following reason. Consider the flat sheet of constant intensity, $\varphi = 0$, $\gamma = 1$. This is the only analytical solution of the initial value problem. We now approximate the sheet by

$$\varphi^\pm = \begin{cases} \pm\varepsilon/2 & \text{if } x \in (-\infty, \alpha(\varepsilon)) \cup [0, +\infty), \\ 0 & \text{otherwise.} \end{cases} \tag{3.21}$$

Then $\omega_\varepsilon(x, y)\, dx\, dy = \varepsilon^{-1}\chi_{\Lambda\varepsilon(0)}(x, y)\, dx\, dy$ converge weakly in the sense of the measure, provided that $\alpha(\varepsilon) \to 0$ as $\varepsilon \to 0$. However, if we compute the second component of the velocity field at time zero at the origin, we find

$$-\frac{1}{2\pi\varepsilon}\int_0^{\alpha(\varepsilon)} dx \int_{-\varepsilon/2}^{\varepsilon/2} dy\, \frac{x}{x^2 + y^2} = -\frac{1}{\pi}\int_0^{\alpha(\varepsilon)/\varepsilon} d\xi\, \arctan\left(\frac{1}{2\xi}\right). \tag{3.22}$$

By (3.22) we conclude that when $\alpha(\varepsilon)/\varepsilon \to \text{const.}$ the field is not vanishing and it may even diverge if $\alpha(\varepsilon)/\varepsilon \to \infty$. Notice also that the above example may be easily modified by choosing suitable sequences $\varphi_\varepsilon^\pm$ belonging to C^∞.

As a consequence, we do not expect that the solutions of the regularized Euler problem converge to the (analytical) stationary solution of the vortex-sheet problem. This is somehow not very surprising. There is some numerical evidence of more than one solution to the vortex-sheet problem with the same initial datum (see [Pul 89] and reference quoted therein). For instance, beside the stationary flat profile solution, we can exhibit the two-branches (obviously not analytical) solution of the type sketched in Fig. 6.6.

In this situation we expect the existence (even for short times) of many weak solutions of the Euler equation (for the velocity, see (1.16)), only one of these being the solution of the vortex-sheet problem in the case of analyticity at time zero. Actually, there are recent results proving a global existence theorem for the weak solution of the Euler equation with vortex sheet as initial datum. However, among these solutions, obtained by compactness

Figure 6.6

methods, it is difficult to isolate those which correspond to the vortex sheets. Notice that such a difficulty exists even before the critical time is reached. In this context a natural question arises. Is there a physical prescription selecting the vortex-sheet solution (the analytical one) among all the weak solutions of the Euler equations? It is reasonable to conjecture that the analytic solution would be characterized, for a short time, as the vanishing viscosity limit of the Navier–Stokes solution which can be uniquely and classically constructed, globally in time, for positive viscosity coefficients, with initial datum given by an analytical profile. No results are known in this direction apart from the simple case of the flat profile with constant intensity. In this case the nonlinear terms disappear, the Navier–Stokes equation reduces to the heat equation, and it is easy to see that the previous conjecture holds.

An analogous, but less singular problem, is that of the time evolution of a vortex patch, i.e., the evolution of the characteristic function of a simply connected bounded set (see Chapters 2 and 3). Suppose that the boundary of such a set is a regular curve. It is natural to see whether the regularity is preserved in time. From the Euler equation we obtain an evolution equation for the boundary, parametrized by $[0, 2\pi) \to x(t, s)$,

$$\partial_t x(t, s) = \frac{1}{2\pi} \int_0^{2\pi} \partial_\sigma x(t, \sigma) \log |x(t, s) - x(t, \sigma)| \, d\sigma. \tag{3.23}$$

It has been conjectured ([Maj 86], [CoT 88]), that singularities in the curve can occur in a finite time. This is also based on some numerical experiences. However, it has recently proved a theorem showing regularity in the class $C^\alpha([0, 2\pi); \mathbb{R}^2)$ globally in time (see [Che 91] and also [BeC 92]).

6.4. Spatially Inhomogeneous Fluids

Up to now we have studied incompressible homogeneous fluids, that is, fluids in which the density does not depend on space and time. However, there are some physically interesting situations concerning incompressible fluids in which the density can no longer be assumed spatially constant. For instance, a river flowing into the sea is an example of a mixture of two fluids with different densities (due to the different salt concentration), each of them incompressible.

In general, our fluid is described by velocity and density fields $u = u(x, t)$ and $\rho = \rho(x, t)$. We assume the incompressibility conditions

$$\nabla \cdot u = 0. \tag{4.1}$$

As a consequence of the continuity equation we have a density that satisfies

$$\frac{\partial}{\partial t}\rho + u \cdot \nabla \rho = 0 \tag{4.2}$$

which means that the density is carried out by the fluid particles, that is, the density is constant along a particle path.

Finally, the Newton law gives us

$$\rho D_t u = -\nabla p + \rho F, \tag{4.3}$$

where F is an external (given) force per unit mass.

Equations (4.1), (4.2), (4.3) form a system of partial differential equations which are hard to deal with. In fact, as regards the initial value problem (in the whole space, for simplicity) in the three-dimensional case, we observe that it is more difficult than the case $\rho = 1$. Thus we cannot hope to have existence and uniqueness of the solutions for all times. We do add no further comments. Furthermore, the lack of conservation of vorticity makes difficult even the solvability in the main part of the Cauchy problem in the planar case. Actually, this problem is still unsolved, as far as we know.

In the two-dimensional case, (4.1), (4.2), (4.3) give, for $F = 0$,

$$\rho^2 D_t \omega = -\nabla p \cdot \nabla \rho. \tag{4.4}$$

Differentiating (4.2)

$$D_t(\partial_i \rho) = -(\partial_i u) \cdot \nabla \rho. \tag{4.5}$$

In (4.4), to control ω, we need to control ∇p. On the other hand, in (4.5), to control ∇p, we need to control the right-hand side of (4.5) which is quadratic in ω and $\nabla \rho$, assuming that ∇u is of the same order of ω. Thus, without using more sophisticated arguments, we are led to an inequality of the type $(d/dt)Y \le Y^2$ for a suitable function $Y = Y(t)$ controlling the growth of ω and $\nabla \rho$. $Y(t)$ may explode in finite time. Of course, we cannot exclude that a deeper geometric analysis of the vector fields we are considering could lead to a priori estimates preventing singularities. We are simply saying that a global existence and uniqueness theorem (if any) does not follow by the arguments developed up to now.

6.5. Water Waves

A case of particular interest in the applications is when the density takes initially (and hence at any further time) two values only. In other words, we want to approach the problem in which there is a fluid with two phases (described by different values of the density) separated by a regular surface. More precisely, we consider two fluids of density ρ and 0 (say, for instance, water and air), and focus our attention on the "heavy" fluid.

The homogeneous heavy fluid of constant density ρ is moving in the half-space D: $D = \{(x, y, z) \in \mathbb{R}^3 | z \ge -h, h > 0\}$ under the action of a gravitational field. It occupies only a part of D and we suppose that at the equilibrium the fluid stays in the region $z \le 0$ (for instance, we can think of water in a basin of deepness h). We confine ourselves to the study of the simplest irrotational case.

The Euler equation can be written

$$\rho(\partial_t + u\nabla\cdot)u = -\nabla p - \rho g n, \tag{5.1}$$

where g is the gravitational constant and n is the unit vector in the z direction. Since the motion is irrotational, a function φ exists such that

$$u = \nabla\varphi. \tag{5.2}$$

The incompressibility condition implies that φ is a harmonic function

$$\Delta\varphi = 0. \tag{5.3}$$

Therefore by (5.1) we have

$$\rho\nabla[\partial_t\varphi + \tfrac{1}{2}(\nabla\varphi)^2 + gz] = -\nabla p. \tag{5.4}$$

By integration

$$\rho[\partial_t\varphi + \tfrac{1}{2}(\nabla\varphi)^2 + gz] = -(p - p_0) + \text{const.}, \tag{5.5}$$

where p_0 is the atmospheric pressure acting on the free surface of the fluid. The constant depends only on time and so it can be absorbed in φ.

We now discuss the boundary conditions. As usual, we impose that the velocity must be tangent to the boundary $z = -h$, that is,

$$\partial_z\varphi = 0 \quad \text{if} \quad z = -h. \tag{5.6}$$

We now want to find the equation of the free surface of the fluid. We write this surface in the form

$$z = S(x, y, t). \tag{5.7}$$

The time evolution of S can be derived following the same arguments used in deriving the equation of a vortex sheet. Actually, the point $(x, y, S(x, y, t))$ is convected by the velocity field $u = \nabla\varphi$. Therefore

$$\frac{d}{dt}x(t) = \partial_x\varphi(x(t), y(t), z(t), t),$$

$$\frac{d}{dt}y(t) = \partial_y\varphi(x(t), y(t), z(t), t), \tag{5.8}$$

$$\frac{d}{dt}S(x(t), y(t), t) = \partial_z\varphi(x(t), y(t), z(t), t).$$

Since

$$\frac{d}{dt}S(x(t), y(t), t) = \partial_t S(x(t), y(t), t) + (\partial_x\varphi\partial_x S)(x(t), y(t), t)$$

$$+ (\partial_y\varphi\partial_y S)(x(t), y(t), t) \tag{5.9}$$

we conclude that

$$\partial_t S + \partial_x\varphi\partial_y S + \partial_y\varphi\partial_y S = \partial_z\varphi. \tag{5.10}$$

Finally, we must relate the jump of pressure on the interface with the shape of S. The jump is related to the different nature of the two fluids (say, air and water) and gives rise to a force called *surface tension*. It is a phenomenological fact that the energy of the interface is proportional to its surface measure. From this it can be proved that, for small deformation, the following holds (Laplace formula):

$$p - p_0 = -T\left(\frac{\partial^2}{\partial x^2}S + \frac{\partial^2}{\partial y^2}S\right),\tag{5.11}$$

where T is a constant called the surface tension coefficient. A derivation of this formula is given in Appendix 6.2. For the moment we consider the relation (5.11) which is valid in linear approximation. The general case will be discussed later.

Summarizing, our initial boundary value problem is

$$\Delta\varphi = 0 \qquad \text{in the domain} \quad \{-h < z < S(x, y, t)\},\tag{5.12}$$

$$\partial_t S + \partial_x\varphi\,\partial_x S + \partial_y\varphi\,\partial_y S = \partial_z\varphi \qquad \text{for} \quad z = S(x, y, t),\tag{5.13}$$

$$\partial_t\varphi + \tfrac{1}{2}[(\partial_x\varphi)^2 + (\partial_y\varphi)^2 + (\partial_z\varphi)^2] + gS = T\rho^{-1}\left(\frac{\partial^2}{\partial x^2}S + \frac{\partial^2}{\partial y^2}S\right)$$

$$\text{for} \quad z = S(x, y, t),\tag{5.14}$$

$$\partial_3\varphi = 0 \qquad \text{for} \quad z = -h,\tag{5.15}$$

$$\varphi(x, 0) = \varphi_0 \qquad \text{and} \qquad S(x_1, x_2, 0) = S_0.\tag{5.16}$$

From now on, for simplicity, we consider initial data that do not depend on y. Then the problem becomes

$$\left(\frac{\partial^2}{\partial x^2} + \frac{\partial^2}{\partial y^2}\right)\varphi = 0 \qquad \text{in} \{-h < z < S(x, t)\},\tag{5.17}$$

$$\partial_t S + \partial_x\varphi\,\partial_x S = \partial_z\varphi \qquad \text{for} \quad z = S(x, t),\tag{5.18}$$

$$\partial_t\varphi + \tfrac{1}{2}[(\partial_x\varphi)^2 + (\partial_z\varphi)^2] + gS = T\rho^{-1}\frac{\partial^2}{\partial x^2}S \qquad \text{for} \quad z = S(x, t),\tag{5.19}$$

$$\partial_3\varphi = 0 \qquad \text{for} \quad z = -h,\tag{5.20}$$

$$\varphi(x, z, 0) = \varphi_0(x, z) \qquad \text{and} \qquad S(x, 0) = S_0(x).\tag{5.21}$$

As we have already said the surface tension term in (5.14) takes into account the interaction between the two fluids. In many cases, and in particular in the situation of water/air in which we are interested, this effect is negligible. So we study first the case $T = 0$ and afterwards discuss possible modifications when the surface tension cannot be neglected.

Systems (5.17)–(5.21) are very complicated and it is difficult to say something about the general solution, without making use of further approximations. These will be discussed in the next section. First, we underline the difficulty of the problem we are considering. This is a Laplace problem in a domain whose boundary $z = S(x, t)$ is unknown. S is the solution of the nonlinear equation (5.18). In fact, even knowing φ, the term $\partial_z \varphi$ is calculated in the point $(x, S(x, t))$, yielding a nonlinearity in the right-hand side of (5.18). Equation (5.18) is coupled with another nonlinear equation for the trace of φ on the interface, which depends on the solution of the Laplace problem, again through $\partial_z \varphi$. This problem is reminiscent of the vortex-sheet problem treated in Section 6.2. Thus we hope to get the same kind of short time result. Actually, we can prove a short-time result, in the framework of the Cauchy–Kowalevski type of theorem, as for the vortex-sheet problem, for initial data which are a small perturbation of the trivial solution. We do not give the details here but simply address the reader to the [Nal 69], [Ovs 71], [Ovs 74], [Ovs 76], [Shi 76], and [ShR 76].

For the present time we investigate the stability problem for the stationary solution $S = 0$, $\varphi = 0$. The norm we use for such a problem is too weak to control the behavior of the solutions so that the result is valid up to times for which smooth solutions exist. The stability we claim is related to the energy conservation which we are going to illustrate. Consider first the case $T = 0$. The total energy takes the form (for a periodic channel of period L)

$$E = \int_0^L dx \int_{-h}^{S(x,t)} dz \left\{ \tfrac{1}{2} u^2(x, z, t) + gz \right\}. \tag{5.22}$$

Differentiating formally the above expression we get

$$\frac{d}{dt} E = \int_0^L \int_{-h}^{S(x,t)} dz\, u(x, t) \cdot \partial_t u(x, t)$$

$$+ \int_0^L dx \left\{ \tfrac{1}{2} u^2(x, z, t) + gS(x, t) \right\} \partial_t S(x, t). \tag{5.23}$$

The first term in the right-hand side of (5.21) is

$$= -\int_0^L dx \int_{-h}^{S(x,t)} dz \left\{ u \cdot (u \cdot \nabla) u + u \cdot \nabla p + gu \cdot n \right\}$$

$$= -\int_0^L dx \int_{-h}^{S(x,t)} dz \left\{ \nabla[\tfrac{1}{2} u^2 u + pu + gzu] \right.$$

$$= \int_\Sigma d\sigma \left[\tfrac{1}{2} u^2 + p + gx_3 \right] u \cdot v, \tag{5.24}$$

where Σ is the upper boundary, $d\sigma$ is the line element on it and, finally, v is the inward normal. Since p is constant on the surface, by the Green lemma and the incompressibility condition, we have

$$\int_\Sigma d\sigma\, u \cdot v = \int_{\Lambda(t)} dx\, \nabla \cdot u = 0, \tag{5.25}$$

where $\Lambda(t) = \{(x, z) | 0 < x < L, -h < z < S(x, t)\}$.

We get

$$\frac{d}{dt}E = \int_{\Sigma} d\sigma \left[\tfrac{1}{2}u^2 + gz\right]u \cdot v + \int_0^L dx \left[\tfrac{1}{2}u^2 + gS\right]\partial_t S. \qquad (5.26)$$

Consider now $u \cdot v$. We have

$$v = \frac{1}{[1 + (\partial_x S)^2]^{1/2}}(\partial_x S, -1) \qquad (5.27)$$

from which

$$u \cdot v = \frac{\partial_x \varphi \partial_x S - \partial_z \varphi}{[1 + (\partial_x S)^2]^{1/2}}. \qquad (5.28)$$

Finally, using (5.18)

$$u \cdot v = -\frac{\partial_t S}{[1 + (\partial_x S)^2]^{1/2}}. \qquad (5.29)$$

Inserting this expression in (5.26) we have $(d/dt)E = 0$.

We make expression (5.22) explicit

$$E = \int_0^L dx \int_{-h}^{S(x,t)} dz \,\tfrac{1}{2}u^2(x, z, t) + \tfrac{1}{2}g \int_0^L dx \, S^2(x, t) - \tfrac{1}{2}gh^2 L. \qquad (5.30)$$

Thus in the norm

$$\int_0^1 dx \int_{-h}^{S(x,t)} dz \,\tfrac{1}{2}u^2(x, z, t) + \tfrac{1}{2}g \int_0^L dx \, S^2(x, t) \qquad (5.31)$$

is conserved and stability is achieved in this norm.

If $T > 0$, we must add the term arising from the surface tension to the energy

$$E = \int_0^L dx \int_{-h}^{S(x,t)} dz \,\{\tfrac{1}{2}u^2(x, z, t) + gz\} + \frac{T}{2} \int_0^L dx \,(\partial_x S)^2. \qquad (5.32)$$

Differentiating as above

$$\frac{d}{dt}E = \int_{\Sigma} d\sigma \,(p - p_0)u \cdot v + T \int_0^L dx \,(\partial_x S)\frac{\partial^2 S}{\partial x \, \partial t}. \qquad (5.33)$$

By using (5.11)

$$\frac{d}{dt}E = T \int_{\Sigma} d\sigma \left(\frac{\partial^2 S}{\partial x^2}\right)u \cdot v + T \int_0^L dx \,(\partial_x S)\frac{\partial^2 S}{\partial x \, \partial t}. \qquad (5.34)$$

By using (5.29) and integrating by parts

$$\frac{d}{dt}E = T \int_{\Sigma} d\sigma \left(\frac{\partial^2 S}{\partial x^2}\right)(\partial_t S) + T \int_0^L dx \,(\partial_x S)\frac{\partial^2 S}{\partial x \, \partial t} = 0. \qquad (5.35)$$

In this case, the natural norm to investigate the stability is the conserved quantity

$$\frac{1}{2}\int_0^L dx \int_{-h}^{S(x,t)} dz\, u^2(x, z, t) + \frac{g}{2}\int_0^L dx\, S^2(x, t) + \frac{T}{2}\int_0^L dx\, [\partial_x S(x, t)]^2.$$

(5.36)

Notice that here we have a better control on the solutions through the L_2 norm of the derivatives of the interface.

We now spend a few words in discussing the surface tension for finite deformations. Physical arguments show that the energy E related to the surface tension is proportional to the area of the separation surface of fluids (see [Bat 67]):

$$E = T\int_0^L dx\left[\left\{1 + \left(\frac{\partial}{\partial x}S\right)^2\right\}^{1/2} - 1\right].$$

(5.37)

It is easy to calculate the related force by variational methods (see Appendix 6.2). It is

$$(p - p_0) = -T\partial_x\left\{\frac{\partial_x S}{[1 + (\partial_x S)^2]^{1/2}}\right\}.$$

(5.38)

As we can see, the Laplace formula (5.11) is an approximation that is valid for small values of $\partial_x S$. This correction is not so important when we deal with the evolution of small water waves. Vice versa it becomes important when dealing with large deformations. In the equilibrium case and in the presence of boundaries we face very interesting problems of "minimal surface" that require a variational technique.

When dealing with small waves a natural problem arises. Starting with a small initial datum, so that the linear approximation at time zero is reasonable, we want a control on the solution of the full equation to garantee the reasonableness of the approximation. An indication in this direction can be obtained again by using the energy. However, what we really need is a pointwise control on the smallness of $\partial_x S$ while we are able to control the integral norm only. The energy is defined

$$E = \int_0^L dx \int_{-h}^{S(x,t)} dz\, [\tfrac{1}{2}u^2 + gz] + T\int_0^L dx\, [\sqrt{1 + (\partial_x S)^2} - 1].$$
(5.39)

The proof of energy conservation follows the previous lines.

To obtain stability we have to derive an upper and a lower bound of the second term on the right hand side of (5.39). By the obvious inequality

$$\sqrt{1 + a^2} \le 1 + a.$$

(5.40)

We have

$$\int_0^L dx\, (\sqrt{1 + (\partial_x S)^2} - 1) \le \int_0^L dx\, |\partial_x S|.$$

(5.41)

On the other hand

$$(1 + a^2)^{1/2} - 1 \geq \begin{cases} a^2(\sqrt{2} - 1) & \text{if } 0 \leq a \leq 1, \\ a(\sqrt{2} - 1) & \text{if } a \geq 1. \end{cases} \qquad (5.42)$$

Hence

$$\int_0^L dx \, [\sqrt{1 + (\partial_x S)^2} - 1] \geq \int_0^L dx \, \chi(|\partial_x S| \leq 1)(\partial_x S)^2(\sqrt{2} - 1)$$

$$+ \int_0^L dx \, \chi(|\partial_x S| \geq 1)|\partial_x S|(\sqrt{2} - 1), \qquad (5.43)$$

where χ is the characteristic function. By the Cauchy–Schwarz theorem

$$\int_0^L dx \, \chi(|\partial_x S| \leq 1)(\partial_x S)^2(\sqrt{2} - 1)$$

$$= \int_0^L dx \, \chi^2(|\partial_x S| \leq 1)(\partial_x S)^2(\sqrt{2} - 1)$$

$$\geq (\sqrt{2} - 1)L^{-1}\left(\int_0^L dx \, \chi(|\partial_x S| \leq 1)|\partial_x S|\right)^2. \qquad (5.44)$$

In conclusion

$$\left(\int_0^L dx \, \chi(|\partial_x S| \leq 1)|\partial_x S|\right)^2$$

$$\leq \left(\frac{L}{\sqrt{2} - 1}\right)\int_0^L dx \, [\sqrt{1 + (\partial_x S)^2} - 1]$$

$$\leq \left(\frac{L}{\sqrt{2} - 1}\right)[E(t) + \tfrac{1}{2}gh^2L] = \left(\frac{L}{\sqrt{2} - 1}\right)[E(0) + \tfrac{1}{2}gh^2L]$$

$$= \left(\frac{L}{\sqrt{2} - 1}\right)\left[\frac{1}{2}\int_0^L dx \int_{-h}^{S(x,0)} dz \, u^2(x, z, 0)\right.$$

$$+ \frac{g}{2}\int_0^L dx \, S^2(x, 0) + T \int_0^L dx \, |\partial_x S(x, 0)|\right]$$

$$\leq \left(\frac{L}{\sqrt{2} - 1}\right)\left[\frac{1}{2}\int_0^L dx \int_{-h}^{S(x,0)} dz \, u^2(x, z, 0)\right.$$

$$+ \frac{g}{2}\left(\int_0^L dx \, |\partial_x S(x, 0)|\right)^2 + T \int_0^L dx \, |\partial_x S(x, 0)|\right]. \qquad (5.45)$$

Here we assumed that the mean value of the perturbation is vanishing

$$\frac{1}{L}\int_0^L dx \, S(x, 0) = 0. \qquad (5.46)$$

In the last step in (5.45) we have used the inequality

$$\int_0^L dx\, S^2(x) = \int_0^L dx \left[\int_{x_0}^\xi d\eta\, \partial_\eta S(\eta) \right]^2 \le \int_0^L dx \left[\int_{x_0}^x d\xi\, |\partial_\xi S(\xi)| \right]^2$$

$$\le L \left(\int_0^L dx\, |\partial_x S(x, 0)| \right)^2, \tag{5.47}$$

where x_0 denotes the point in which $S(x_0) = 0$. (It exists by (5.46).)

Then in (5.45) we have estimated

$$\int_0^L dx\, \chi(|\partial_x S| \le 1)|\partial_x S|. \tag{5.48}$$

In a similar fashion we can estimate

$$\int_0^L dx\, \chi(|\partial_x S| > 1)|\partial_x S| \tag{5.49}$$

and

$$\int_0^L dx \int_{-h}^{S(x,t)} dz\, \tfrac{1}{2}u^2 \tag{5.50}$$

in terms of continuous functions of the L_1 norm of $\partial_x S$ and the kinetic energy at time zero. This achieves the stability property for the full equation in the norm

$$\int_0^L dx \int_{-h}^{S(x)} dz\, \tfrac{1}{2}u^2(x, z) + \int_0^L dx\, |\partial_x S|. \tag{5.51}$$

6.6. Approximations

The equations for the free boundary are very complicated indeed so that they should be simplified, whenever possible, neglecting some unimportant terms. The equations we obtain are simpler to treat and often give a mathematical model in good agreement with the experimental data. From a mathematical point of view, we should also discuss the relation between these equations and the "true" ones introduced in the previous sections. In general, we can show that, if we chose two inital data sufficiently close such that the approximation holds, then the solutions of the two equations remain close for a finite amount of time during which the existence of the solutions is ensured. When the time becomes large, the neglected terms could produce relevant effects and the two solutions could become very different. (Even dealing with a single equation, two initially close data perform trajectories which diverge, in general, exponentially in time. Even worse is the situation wherein we also perturb the equation.) To prevent this we would need some very strong stability property not only for the approximate solutions but also, and especially, for the "true" ones. These stability properties would ensure the validity of the approximation. Although we do not have such theorems, some of the approximations seem to remain valid for a long time.

The simplest approximation we can do is the *linear one*. It consists in neglecting the quadratic term in the Euler equation, $(u \cdot \nabla)u$, with respect to the linear term $\partial_t u$. We discuss now the physical meaning of this approximation by heuristic arguments. We anticipate immediately that the linearization will produce a wave motion of wavelength λ, period τ, and amplitude a. Assuming that such a wave motion takes place, then a particle of fluid during time τ moves with a quantity of order a so that its velocity is almost $u \approx a\tau^{-1}$. On the other hand, the velocity varies, during time τ, on a scale of λ, so that

$$\partial_t u \approx a\tau^{-2} \quad \text{and} \quad (u \cdot \nabla)u \approx a^2\tau^{-2}\lambda^{-1}. \tag{6.1}$$

The statement $(u \cdot \nabla)u \ll \partial_t u$ is equivalent to $a \ll \lambda$. In conclusion the linearization is expected to be valid whenever $a \ll \lambda$.

We now write the linearized equations of motion. Equations (5.18), (5.19) become, neglecting quadratic terms,

$$\frac{\partial}{\partial t} S = \frac{\partial}{\partial z} \varphi, \tag{6.2}$$

$$\partial_t \varphi + gS = \frac{T}{\rho} \frac{\partial^2}{\partial x^2} S \quad \text{for} \quad z = S(x, t). \tag{6.3}$$

Deriving (6.3) with respect to time and inserting it in (6.2) we have

$$\frac{\partial^2}{\partial t^2} \varphi + g\frac{\partial}{\partial z}\varphi - \frac{T}{\rho} \frac{\partial}{\partial z} \frac{\partial^2}{\partial x^2}\varphi = 0 \quad \text{for} \quad z = S(x, t). \tag{6.4}$$

Supposing the oscillations to be small enough, (6.4) can be considered valid if $z = 0$ rather than $z = S(x, t)$, so that

$$\frac{\partial^2}{\partial t^2} \varphi + g\frac{\partial}{\partial z}\varphi - \frac{T}{\rho} \frac{\partial}{\partial z} \frac{\partial^2}{\partial x^2}\varphi = 0 \quad \text{for} \quad z = 0 \tag{6.5}$$

that together with (5.17)

$$\left[\frac{\partial^2}{\partial x^2} + \frac{\partial^2}{\partial z^2} \right] \varphi = 0 \tag{6.6}$$

and the initial and boundary conditions (5.20), (5.21) constitute the linear problem of the gravity waves.

We look for some periodic solution

$$\varphi = \cos(kx - \omega t)f(z). \tag{6.7}$$

(Obviously, $\tau = 2\pi\omega^{-1}$ is the period and $\lambda = 2\pi k^{-1}$ is the wavelength.) Inserting (6.7) in (6.6) we have

$$\frac{\partial^2}{\partial z^2} f(z) - k^2 f(z) = 0 \tag{6.8}$$

whose general solution is

$$f(z) = A \exp(kz) + B \exp(-kz). \tag{6.9}$$

We impose the boundary conditions (5.20)

$$Ak \exp(-kh) - Bk \exp(kh) = 0 \tag{6.10}$$

so that

$$B = A \exp(-2kh). \tag{6.11}$$

In conclusion we have

$$\varphi = A[\exp(kz) + \exp(-2kh)\exp(-kz)]\cos(kx - \omega t). \tag{6.12}$$

Inserting expression (6.12) in (6.5) we obtain

$$-\omega^2[1 + \exp(-2kh)] + gk[1 - \exp(-2kh)] + \frac{T}{\rho}k^3[1 - \exp(-2kh)] = 0, \tag{6.13}$$

that is,

$$\omega^2 = gk\frac{1 - \exp(-2kh)}{1 + \exp(-2kh)} + \frac{T}{\rho}k^3\frac{1 - \exp(-2kh)}{1 + \exp(-2kh)}. \tag{6.14}$$

(A relation between ω and k is called a "dispersion relation.")

It is easy now to find the expression for S by using (6.2). Obviously it will also be of the form (6.7). It is also easy to find the velocity field corresponding to the velocity potential (6.12)

$$u_1 = -Ak[\exp(kz) + \exp(-2kh)\exp(-kz)]\sin(kx - \omega t),$$
$$u_2 = Ak[\exp(kz) - \exp(-2kh)\exp(-kz)]\cos(kx - \omega t). \tag{6.15}$$

A subsequent integration gives the parametric equation of a particle path. As an example we consider the case $h = \infty$. Moreover, we assume z essentially constant $z = z_0$ in the expression of the velocity field (6.15). Then

$$x_1 - x_0 = -\frac{Ak}{\omega}\exp(kz_0)\cos(kx - \omega t),$$
$$x_3 - x_0 = -\frac{Ak}{\omega}\exp(kz_0)\sin(kx - \omega t). \tag{6.16}$$

In conclusion, the particles of the fluid describe a uniform circular motion around the point (x_0, z_0) of radius $Ak/\omega \exp(kz_0)$, decreasing to the interior in an exponential way.

We now discuss the role of surface tension which is present in the dispersion relation (6.14). First, we consider the case in which $h = \infty$. Then

$$\omega^2 = gk + \frac{T}{\rho}k^3. \tag{6.17}$$

Hence, for

$$k \ll \sqrt{\frac{\rho g}{T}} \quad \text{(long waves)}, \tag{6.18}$$

the surface tension can be neglected and the wave has a gravitational origin
only

$$\omega^2 = gk. \tag{6.19}$$

On the contrary, when

$$k \gg \sqrt{\frac{\rho g}{T}} \qquad \text{(short waves)}, \tag{6.20}$$

the gravitational effect can be neglected and so

$$\omega^2 = \frac{T}{\rho} k^3 \tag{6.21}$$

(capillary waves).

A similar discussion can be made for finite h when $kh \ll 1$. In this case

$$\omega^2 = ghk^2 + \frac{T}{\rho} k^4 h \tag{6.22}$$

and

$$k \ll \sqrt{\frac{g\rho}{T}} \quad \text{implies} \quad \omega^2 = ghk^2 \qquad \text{(gravity waves)}, \tag{6.23}$$

$$k \gg \sqrt{\frac{g\rho}{T}} \quad \text{implies} \quad \omega^2 = \frac{T}{\rho} hk^4 \qquad \text{(capillary waves)}. \tag{6.24}$$

In general, both effects are present and we speak of capillo–gravity waves.

Until now we have discussed special wave solutions. In general, the solution will be a superposition of wave motions. We consider initial data which
produce a wave packet. As is well known, this packet moves with a velocity
given by the expression

$$U = \frac{d\omega}{dk} \qquad \text{(group velocity)}. \tag{6.25}$$

From (6.14)

$$U = \frac{1}{2}\sqrt{\frac{g}{k}} \qquad \text{if} \quad h = \infty \quad \text{and} \quad T = 0, \tag{6.26}$$

$$U = \sqrt{gh} \qquad \text{if} \quad hk \ll 1 \quad \text{and} \quad T = 0, \tag{6.27}$$

$$U = \frac{3}{2}\sqrt{\frac{Tk}{\rho}} \qquad \text{if} \quad h = \infty \quad \text{and} \quad k \gg \sqrt{\frac{g\rho}{T}}, \tag{6.28}$$

$$U = k\sqrt{\frac{Th}{\rho}} \qquad \text{if} \quad hk \ll 1 \quad \text{and} \quad k \gg \sqrt{\frac{g\rho}{T}}. \tag{6.29}$$

There are some physical phenomena in which the linear approximation is
too drastic, because the nonlinear terms cannot be neglected. A famous example is given by the Korteweg–de Vries equation (from now on denoted as
K–deV) in which a genuine nonlinear effect happens: the phenomenon of
solitary waves. We briefly discuss this fact. It is well known, through experi-

mental observation, that a shallow channel can produce a wave packet which moves with constant velocity and remains equal in form during time. (For more details on the following argument, see [DEG 82].) This fact conflicts with the linear equations that produce a dispersion (different phase velocities for different wave numbers) and hence an enlargement of the wave packet. Only the nonlinear terms could prevent the dispersion of the wave packet.

We discuss now which approximation allows us to obtain the K–deV equation from (5.17)–(5.21). At present we neglect surface tension which, being a linear term, does not give rise to great complications.

The K–deV equation arises when we consider long waves, whose length is very large with respect to the deepness h. Moreover, we require that the amplitude be small with respect to h. We introduce the two parameters μ, σ

$$\mu = \left(\frac{h}{\lambda}\right)^2 \qquad (\lambda = \text{wavelength}), \tag{6.30}$$

$$\sigma = \frac{\text{amplitude of the perturbation}}{h}, \tag{6.31}$$

and we study the approximation in which μ and σ are of the same order of magnitude. The K–deV equation arises following a wave during the motion and in considering only the lower orders in μ, σ, as we shall later see.

It is convenient to write (5.17)–(5.21) in a dimensional form. We introduce a characteristic length λ which we choose equal to the wavelength of a wave solution of the linearized equations and the characteristic time t_0 which we choose as $t_0 = \lambda/U$ where U is given by $U = (gh)^{1/2}$ (see (6.27)). We take as new variables

$$x' = x\lambda^{-1}, \qquad t' = tt_0^{-1}, \qquad z' = zh^{-1} + 1. \tag{6.32}$$

The unknown quantity S must also be rescaled. Calling a a characteristic quantity, we define

$$S' = Sa^{-1}. \tag{6.33}$$

Using these new variables, (5.19) becomes

$$\left(\frac{\sqrt{hg^{-1}}}{\lambda a}\right)\partial_{t'}\varphi + \frac{1}{2ga\lambda^2}\left(\frac{\partial\varphi}{\partial x'}\right)^2 + \frac{1}{2h^2 ga}\left(\frac{\partial\varphi}{\partial z'}\right)^2 + S' = 0. \tag{6.34}$$

A reasonable rescaling for φ is given by

$$\varphi' = \varphi\sqrt{hg^{-1}}\,\frac{1}{a\lambda}. \tag{6.35}$$

Hence (6.34) becomes

$$\frac{\partial\varphi'}{\partial t'} + S' + \tfrac{1}{2}\sigma\left(\frac{\partial\varphi'}{\partial x'}\right)^2 + \tfrac{1}{2}\frac{\sigma}{\mu}\left(\frac{\partial\varphi'}{\partial z'}\right)^2 = 0 \qquad \text{for} \quad z' = 1 + \sigma S', \tag{6.36}$$

where $\mu = h^2/\lambda^2$ and $\sigma = a/h$.

With this scaling the other equations become

$$\left(\mu\frac{\partial^2}{\partial x'^2} + \frac{\partial^2}{\partial z'^2}\right)\varphi' = 0, \tag{6.37}$$

$$\mu\sigma\frac{\partial\varphi'}{\partial x'}\frac{\partial S'}{\partial x'} + \mu\frac{\partial S'}{\partial t'} = \frac{\partial\varphi'}{\partial z'} \qquad \text{for} \quad z' = 1 + \sigma S, \tag{6.38}$$

with the boundary condition

$$\frac{\partial\varphi'}{\partial z'} = 0 \qquad \text{if} \quad z' = 0. \tag{6.39}$$

We proceed formally. By developing φ' in series of powers z'

$$\varphi'(x', z', t') = \sum_{n=0}^{\infty} (z')^n \rho_n(x', t'). \tag{6.40}$$

Putting (6.40) in (6.37) we obtain the recurrence relation

$$\mu\left(\frac{\partial^2\rho_n}{\partial x'^2}\right) + (n+2)(n+1)\rho_{n+2} = 0. \tag{6.41}$$

Using the boundary condition (6.39) we see that $\rho_1 = 0$ and so all odd terms vanish. The even terms depend on $\rho_0(x', t')$

$$\varphi'(x, (t')) = \sum_{j=0}^{\infty} (-1)^j \mu^j \frac{(z')^{2j}}{(2j)!} \frac{\partial^{2j}\rho_0}{\partial(x')^{2j}}. \tag{6.42}$$

Define

$$W(x', t') = \frac{\partial}{\partial x'}\rho_0(x', t') \tag{6.43}$$

then, inserting (6.42) in (6.36), derivating with respect to x', we obtain

$$S_x + W_t - \mu\partial_t\partial_x\left[\frac{(1+\sigma S)^2}{2}W_x\right] + \sigma\tfrac{1}{2}\partial_x[W]^2$$

$$+ \sigma\mu\left[-W\frac{(1+\sigma S)^2}{2}W_{xx}\right] + \tfrac{1}{2}\sigma\mu\partial_x[(1+\sigma S)^2 W_x^2] + O(\mu^2) = 0, \tag{6.44}$$

where, for notational simplicity, we have neglected the $'$ and we denote the derivative as $\cdot_t$ or $\cdot_x$. We will use this convention until the end of this section.

Equation (6.38) becomes

$$S_t + (1 + \sigma S)W_x - \frac{\mu}{6}[(1 + \sigma S)^3 W_{xxx}]$$

$$+ \sigma S_x\left[W - \frac{\mu}{2}(1 + \sigma S)^2 W_{xx}\right] + O(\mu^2) = 0. \tag{6.45}$$

We study these two last equations. The simplest approximation follows by putting $\mu = 0$ and $\sigma = 0$. We have

$$S_x + W_t = 0, \tag{6.46}$$

$$S_t + W_x = 0, \tag{6.47}$$

which imply the classical wave equation

$$S_{xx} - S_{tt} = 0. \tag{6.48}$$

If we put only $\mu = 0$, (6.44) and (6.45) reduce to

$$S_x + W_t + \sigma W W_x = 0, \tag{6.49}$$

$$S_t + W_x + \sigma(S W_x + S_x W) = 0, \tag{6.50}$$

which are called the nonlinear equations of the shallow waters. Equations (6.44) and (6.45) give

$$S_x + W_t - \frac{\mu}{2} W_{xxt} + \sigma W W_x = 0, \tag{6.51}$$

$$S_t + W_x + \sigma(S W_x + S_x W) - \frac{\mu}{6} W_{xxx} = 0, \tag{6.52}$$

when considering first order in μ or in σ.

We now follow the evolution of the progressive wave solution of the linearized equation looking at very large times. This suggests the change of variables

$$\xi = x - t, \tag{6.53}$$

$$\tau = \varepsilon t, \tag{6.54}$$

where ε is a parameter which gives a measure of the time rescaling. Of course,

$$\partial_x = \partial_\xi, \tag{6.55}$$

$$\partial_t = -\partial_\xi + \varepsilon \partial_\tau, \tag{6.56}$$

and (6.51), (6.52) become

$$S_\xi - W_\xi + \varepsilon W_\tau = \frac{\mu}{2}(-\partial_\xi + \varepsilon \partial_\tau) W_{\xi\xi} - \sigma W W_\xi, \tag{6.57}$$

$$-S_\xi + \varepsilon S_\tau + W_\xi = -\sigma(S W_\xi + S_\xi W) + \frac{\mu}{6} W_{\xi\xi\xi}. \tag{6.58}$$

We sum the two equations

$$\varepsilon(S_\tau + W_\tau) = -\frac{\mu}{3} W_{\xi\xi\xi} - \sigma[W W_\xi + S W_\xi + S_\xi W] + \frac{\varepsilon\mu}{2} W_{\xi\xi\tau}. \tag{6.59}$$

We now consider the case in which, at order zero in μ, σ, ε, there is only a progressive wave. So in this order $\partial_\xi S = \partial_\xi W$, and hence $S = W + O(\mu, \sigma, \varepsilon) + \text{const}$. We choose the initial conditions in such a way that this constant vanishes. In conslusion, we have

$$S_\tau + \tfrac{3}{2}\sigma^* S S_\xi + \frac{\mu^*}{6} S_{\xi\xi\xi} = 0 \qquad \text{(K–deV equation)}, \qquad (6.60)$$

where $\mu^* = \varepsilon^{-1}\mu$ and $\sigma^* = \varepsilon^{-1}\sigma$, and we have neglected the term $O(\varepsilon)$, which means that we are looking at large times.

We obtain similar results if we follow a regressive wave. Putting $\xi = x + t$, $\tau = \varepsilon t$, we have

$$S_\tau - \tfrac{3}{2}\sigma^* S S_\xi - \frac{\mu^*}{6} S_{\xi\xi\xi} = 0. \qquad (6.61)$$

Until now we have neglected the surface tension. However, qualitatively, this approximation is not very drastic because to take the surface tension into account we need to modify only the magnitude of μ^*.

Writing the K–deV equation in terms of the physical variables x, t defined as

$$x = \xi + \tau, \qquad (6.62)$$

$$t = \tau, \qquad (6.63)$$

we obtain

$$S_t + S_x + \tfrac{3}{2}\sigma^* S S_x + \frac{\mu^*}{6} S_{xxx} = 0. \qquad (6.64)$$

This is a nonlinear partial differential equation. It has been extensively studied and gives rise to an integrable system which has been completely investigated using the spectral transform technique. For a wide discussion, see [DEG 82]. We observe here that it explains the existence of solitary waves. A very old and well-known famous example is the following. Consider values of μ^* and σ^* such that (2.64) reads

$$S_t + S_x + 12 S S_x + S_{xxx} = 0 \qquad (6.65)$$

then the expression

$$S = \tfrac{1}{4} \frac{A^2}{\cosh^2 \tfrac{1}{2}[Ax - (A + A^2)t + \delta]} \qquad (6.66)$$

is a solution for any δ. It represents a wave packet moving with velocity $(1 + A^2)$ and remaining constant in form.

Another important equation, the Boussinesq equation, can be obtained using the same approximation used in the K–deV equation, but considering initial conditions in which the progressive and regressive waves are both present.

The discussion of this section is at a formal level. It is possible to make it rigorous at least for a short time. See [KaN 86]. For a mathematical justification of the shallow-water equations (6.49) and (6.50) (for short times), see [KaN 79].

Appendix 6.1 (Proof of a Theorem of the Cauchy–Kowalevski Type)

We give the proof of Theorem 2.1. To construct the solution, we use the classical iteration scheme

$$\xi^n(t) = \xi_0 + \int_0^t ds\, F(\xi^{n-1}(s)),$$

$$\xi^{(0)}(t) = \xi_0. \tag{A1.1}$$

We assume inductively that

$$\|\xi^\kappa(t)\|_\rho \le R \qquad \text{for} \quad t < a_k(\rho_0 - \rho), \qquad k < n, \tag{A1.2}$$

and we prove that the same is true for $k = n$. Here a_k denotes a decreasing sequence of positive numbers which will be chosen later.

Setting

$$\eta^n = \xi^n - \xi^{n-1} \tag{A1.3}$$

we have

$$\eta^n(t) = \int_0^t ds\, F(\xi^{n-1}(s)) - F(\xi^{n-2}(s)). \tag{A1.4}$$

By the inductive hypothesis we can use (2.17) so that

$$\|\eta^n(t)\|_\rho = C_R \int_0^t ds\, \frac{\|\eta^n(t)\|_{\rho(s)}}{\rho - \rho(s)} \tag{A1.5}$$

with

$$\rho(s) = \tfrac{1}{2}\left(\rho_0 + \rho - \frac{s}{a_{n-1}}\right). \tag{A1.6}$$

Denoting

$$M_n = \sup_{\substack{0 < \rho < \rho_0 \\ 0 \le t < a_n(\rho_0 - \rho)}} \left(1 - \frac{t}{a_n(\rho_0 - \rho)}\right) \|\xi^n(t)\|_\rho \tag{A1.7}$$

we get

$$\|\eta^n(t)\|_\rho \le C_R M_{n-1} \int_0^t ds\, \frac{1}{\rho(s) - \rho}\, \frac{1}{1 - \dfrac{s}{a_{n-1}(\rho_0 - \rho(s))}}$$

$$= 2a C_R M_{n-1} \int_0^t ds\, \frac{a(\rho_0 - \rho) + s}{[a(\rho_0 - \rho) - s]^2}$$

$$\le 4a_{n-1} C_R M_{n-1} \frac{a_{n-1}(\rho_0 - \rho)}{a(\rho_0 + \rho) - t}. \tag{A1.8}$$

From one side we obtain

$$M_n \le 4a_{n-1} C_R M_{n-1} \le (4a_0 C_R)^n, \tag{A1.9}$$

where $a_0 > a_1$ is given by

$$a_0 = \min\left(\frac{1}{16C_R}, \frac{R - R_0}{8C_R}\right). \tag{A1.10}$$

On the other hand, by estimate (A1.8),

$$\|\xi^n(t)\|_\rho \leq \|\xi_0\|_\rho + \sum_{k=1}^{n} \|\eta^k(t)\|_\rho$$

$$\leq \|\xi_0\|_\rho + \sum_{k=1}^{n} 4a_{k-1} C_R M_{k-1} \frac{a_{k-1}}{a_{k-1} - a_k}$$

$$\leq \|\xi_0\|_\rho + \sum_{k=1}^{n} 8a_0 C_R 2^{-k}. \tag{A1.11}$$

The last step is a consequence of (A1.9) (A1.10) and the choice

$$a_n = a_{n-1} - a_0 2^{-n}. \tag{A1.12}$$

Therefore

$$\|\xi^n(t)\|_\rho \leq \|\xi_0\|_\rho + 2a_0 C_R \leq R. \tag{A1.13}$$

Finally, setting $a = \lim_n a_n = a_0/2$, for $t < a(\rho_0 - \rho)$ we have $\eta^n(t) \to 0$ exponentially and

$$\xi(t) = \lim_n \xi^n(t) \tag{A1.14}$$

satisfies the bound

$$\|\xi(t)\|_\rho \leq R. \tag{A1.15}$$

$\square$

Appendix 6.2 (On Surface Tension)

By reasonable physical arguments (see [Bat 67]) we state that the energy E of the surface separating two different fluids (say air/water) depends linearly on the measure of the interface. In two dimensions

$$E = T \int_0^L dx \,\{\sqrt{1 + (\partial_x S)^2} - 1\}, \tag{A2.1}$$

where T is a constant depending on the nature of the fluids (called the surface tension coefficient).

From now on, for the sake of simplicity, we assume periodicity with period L. Consider a deformation of $S(x)$ by adding an arbitrary smooth function $\gamma(x)$

$$S_1(x) = S(x) + \gamma(x). \tag{A2.2}$$

The energy of the deformed surface is

$$E_1 = T \int_0^L dx \,\{\sqrt{1 + (\partial_x S + \partial_x \gamma)^2} - 1\}. \tag{A2.3}$$

The first order in γ gives

$$E_1 - E = T \int_0^T dx \, \frac{\partial_x S \partial_x \gamma}{\sqrt{1 + (\partial_x S)^2}}$$

$$= \text{(by integration by parts)}$$

$$= -\int_0^L dx \, \gamma \frac{\partial}{\partial x} \frac{\partial_x S}{\sqrt{1 + (\partial_x S)^2}}. \tag{A2.4}$$

We compare this expression with the infinitesimal work of the pressure jump in the interface

$$dL = \int_0^L dx \, \gamma(p - p_0). \tag{A2.5}$$

We have

$$\int_0^L dx \, \gamma \left\{ \frac{\partial}{\partial x} \frac{\partial_x S}{\sqrt{1 + (\partial_x S)^2}} + (p - p_0) \right\} = 0. \tag{A2.6}$$

By the arbitrariness of γ we have

$$p - p_0 = -\frac{\partial}{\partial x} \frac{\partial_x S}{\sqrt{1 + (\partial_x S)^2}}. \tag{A2.7}$$

Neglecting $\partial_x S$ with respect to 1, we finally obtain (5.11).

CHAPTER 7

Turbulence

In this chapter we describe turbulent flows. More specifically, we will try to illustrate the statistical theories from both a phenomenological and mathematical point of view and the transition regime from laminar to turbulent flows.

7.1. Introduction

The spirit of this chapter is essentially different from that of all the others. The reason is that there is no complete mathematical theory describing turbulent flows, so that any exposition must be necessarily, at least partially, of a phenomenological nature. Moreover, turbulent phenomena are very complicated and not yet completely understood from a physical point of view, so that there is no general agreement on what "turbulence" is. In fact, the term turbulence describes very different fluid behaviors: from discussions with colleagues we have the impression that there are as many notions of turbulence as there are individuals working on the subject.

We will try to treat this topic by discussing those few points which we believe to be relevant and reasonably well understood. In the absence of a complete mathematical theory, it is not very meaningful to discuss in great detail the existing rigorous results. Although having their own mathematical interest, they are at best similar to pieces of a mosaic which is far from being complete. Given this situation we choose to present the matter discursively, addressing the reader to the existing literature for proofs and further details.

Let us start by agreeing on a (possible, but rough and imprecise) definition of a turbulent flow. It often happens in practice that real fluids, for certain values of the characteristic parameters, behave chaotically. Take, for instance, a gas flowing through a pipe at high speed, a jet, or a quick flow past an obstacle. In all these cases, preparing the system in the same way at different times and measuring the same physical quantities, say the velocity in a certain point, we can possibly find different values. In other words, we can make the following ideal experiment. Putting a large number N of identical systems under the same conditions, we obtain N different outputs from the same measuring process. If so, we have only a statistical way to describe this phenomenology.

According to this definition, a turbulent motion is not only complex but also needs, for intrinsic reasons, a statistical description. These two aspects of a turbulent motion, complexity of the flow and its statistical nature, both play an important role in understanding the essence of the phenomenon. The relation between these two aspects is not yet fully clarified as we will see later. For the moment, let us come back to a possible strategy for a statistical description of a flow. The situation under consideration resembles somewhat that occurring in statistical mechanics. Consider, for example, a gas in a box. A state of such a system is, in principle, described by a huge number of positions and momenta of the molecules constituting the gas. However, such information, even if disposable, is not of practical interest. In most of the cases, we are interested in the collective behavior of the molecules. Namely, we want to know a few averaged quantities by means of which to construct the thermodynamics. Also the result of a hypothetical explicit integration of the Newton equation governing the motion of the system is of relative interest, for the purpose of understanding the relaxation toward the thermal equilibrium or for the study of stationary nonequilibrium states, say, for instance, those describing heat fluxes.

In statistical mechanics the words "microscopic state" (or microstate) is often used to indicate a complete description of the system (in terms of positions and momenta of the molecules in our example), and the words "macroscopic state" (or macrostate) to indicate a description of the system in terms of the few parameters (temperature, density, etc., in our case) yielding a complete description of the system from a macroscopic or thermodynamical point of view. Thus we can say that statistical mechanics provides a drastic reduction of the number of degrees of freedom of the system.

The basic assumption of equilibrium statistical mechanics is the introduction, a priori, of a probability measure on the phase space of the system, called the Gibbs measure, by means of which we compute the averages. In our example, N molecules in a box of volume V, the Gibbs measure depends on two macroscopic parameters, the temperature, and the density N/V and, obviously, on the interaction among the particles. Explicitly, the Gibbs measure μ takes the form

$$\mu(x_1, \ldots, x_N, v_1, \ldots, v_N)\, dx_1, \ldots, dx_N, dv_1, \ldots, dv_N$$

$$= \frac{e^{-\beta H}\, dx_1, \ldots, dx_N, dv_1, \ldots, dv_N}{Z(\beta, N)}, \qquad (1.1)$$

where H is the Hamiltonian of the system, β is proportional to the inverse temperature, $x_1, \ldots, x_N, v_1, \ldots, v_N$ are the position and velocities of our molecules, and $Z(\beta, N)$ is a normalization factor.

The main idea is that the Gibbs measure gives more weight to those points of the phase space in which the system is more likely found. In other words, the averages with respect to the measure (1.1) would replace the time averages, which, in principle, could be computed by solving the dynamical problem.

Another interesting measure which is also introduced to postulate the equilibrium distributions of the microscopic states is the microcanonical ensemble, which is defined by

$$\mu(dx_1, \ldots, dx_N\, dv_1, \ldots, dv_N) = \Omega(E)^{-1}\delta(H - E)\, dx_1, \ldots, dx_N, dv_1, \ldots, dv_N,$$

(1.2)

where

$$\Omega(E) = \int \delta(H - E)\, dx_1, \ldots, dx_N, dv_1, \ldots, dv_N$$

$$= \int \frac{d\sigma}{|\mathrm{grad}\, H|}.$$

(1.3)

Here $d\sigma$ denotes the surface measure on the manifold $H = E$. The last step follows by the Liouville theorem (see [Tho 72], for instance).

The measure (1.2) essentially says that all the points on the surface $H = E$ have the same probability. The usual reason for justifying the measure (1.3) is the postulate that most of the orbits of our physical system are dense in the surface $H = E$, so that time averages can be replaced by averages computed by means of the microcanonical measure (ergodic hypothesis). A complete rigorous justification of the introduction of the measure (1.2) (the Gibbs ansatz), from the first principles of mechanics has not been fully given. This problem is related to the solution of the ergodic problem (i.e., to prove the equivalence between the time and microcanonical averages) for mechanical systems, and a discussion on this interesting and delicate point is much beyond the purpose of the present book. However, we can also survive without solving the ergodic problem: the Gibbs ansatz works in practice. In fact, statistical mechanics gives predictions which are generally in agreement with the experiments, and its methodology can be applied to a large variety of situations so that the success of the theory is, somehow, a justification of the theory itself.

We address the reader to [Hua 63], [Tho 72], and [Rue 69] (increasing as regards the mathematical exposition) for the basic notions of statistical mechanics. We only mention that the two measures (1.1) and (1.2) are equivalent in the thermodynamic limit which is $N \to \infty$, $V \to \infty$, and $N/V \to$ const.

Nonequilibrium statistical mechanics is much less understood. There is no equivalent of the Gibbs ansatz for calculating the macroscopic observables in stationary problems (say the heat flux flowing through a gas between two walls at different temperatures). Thus, to deal with these problems, we have to combine statistical and dynamical considerations with procedures which are ad hoc for the problem under consideration.

After this long digression let us come back to turbulence and consider stationary problems by which we mean the following. By performing measurements on the turbulent flow we find, in many cases, that the quantities of interest, say the mean velocity in a point, the mean square, and so on, do not depend on time. In this case, we say that the turbulence is stationary. For a stationary turbulent flow one hope quite naturally arises, that there exists, at least in a large variety of cases, an a priori measure which allows us to compute the time averages of the quantity of interest, without solving the dynamical problem. This is in analogy with the Gibbs ansatz in equilibrium statistical mechanics. Obviously, there is no reason to believe that there is a universal behavior in all stationary turbulent flows. Moreover, even admitting a strict analogy with statistical mechanics concepts, it should be reasonable to conceive stationary flows which are of nonequilibrium type. Thus the identification of the equilibrium parameters, like the turbulent temperature and density, should be preliminary to establish this distinction.

It is probably too naive to hope that there exists a single probability measure defined on the space of the configurations of the system describing a large class of stationary turbulent flows, however, the possibility is appealing and makes it worth describing attempts in this direction. We are not able to give a conclusive positive answer to this challenging problem and the theory, if any in the future, seems quite far from our present knowledge.

On the other hand, a phenomenological theory, based on dimensional arguments due to Kolmogorov, predicts a rather universal behavior for turbulent flows, at least in suitable regimes, so that one of the most basic problems in turbulence theory is to give a rigorous justification of this or other similar theories. This, in view of the fact that the Kolmogorov predictions are in rather good agreement with the experiments. We will describe phenomenological approaches and rigorous attempts in Sections 7.3, 7.6, and 7.4, 7.5, respectively. The nonstationary problems are, of course, even more difficult. There exists a huge heuristic literature on the subject on which we do not try to address the reader.

Up to now we have discussed the statistical nature of a fully developed turbulent flow. As we have often remarked, this nature has not yet been understood. However, recent progress has been made on the problem of understanding why turbulence develops. We mean the so-called onset of turbulence or the transition from laminar to turbulent flows. Here we arrive at the other important feature of a turbulent flow: the complexity of the motion. Actually, the general theory of dynamical systems provides a convincing explanation of many features accompanying the transition from a laminar to a turbulent regime. This is basically the content of Section 7.2.

A final remark concerning statistical theories. Obviously, the a priori measure, describing the statistical nature of a turbulent flow, if it exists, should be a consequence of the nature of the dynamics describing real fluids. So dynamical systems arguments applied to fluid dynamical equations should give insights on what the probability measure describing the fully developed turbulence is. However, this problem is even more difficult to approach than the ergodic problem in statistical mechanics. In fact, to prove that the time averaging is equivalent to the average on ensembles, it is necessary to have an idea about the concept of "averaging on the ensembles." In other words, we first have to construct interesting invariant measures for the Euler or Navier–Stokes flows. This is the reason why we posed the problem of the research of an a priori measure and regard the problem of the onset and fully developed turbulence as essentially distinct.

Coming back to the onset of the turbulence problem, we will not describe this topic in great detail for two reasons. First, there exists a large literature on the subject to which we will systematically address the reader and second, because in this book we have chosen to deal with conservative systems only. Thus, following the same philosophical attitude of the rest of the book, we want to outline, even for turbulence, those features which are mostly connected with the nonlinear or conservative part of the Navier–Stokes equation. Actually, viscosity plays an important role in turbulence, basically to set the turbulent regime going and, as the reader will see, most of the considerations in Section 7.2 concern dissipative systems. We found it useful to insert this section for conceptual completeness. However, once the turbulence is established, possible universal features depend essentially on the nonlinear part of the Navier–Stokes equation only, or, in other words, on the Euler flow. To this we concentrate most of our efforts.

7.2. The Onset of Turbulence

In this section we illustrate the mechanism to produce turbulent flows. As we said, since the phenomenon is essentially dissipative, we give only a quick and qualitative account of the basic ideas.

As we mentioned in the previous section, viscosity plays a role in getting turbulent flows going, so that we are forced to replace the mathematical model used up to now, the Euler equation, by a new one, which takes into account dissipative effects. This model is based on a new equation called the Navier–Stokes equation. We write it without additional comments.

$$(\partial_t + u \cdot \nabla)u = -\nabla p + \nu \Delta u + f, \qquad (2.1)_1$$

$$\nabla \cdot u = 0, \qquad (2.1)_2$$

ν is a positive constant called the (kinematical) viscosity coefficient and f is an external, given force per unit volume.

If the fluid is contained in a volume Λ, we must also specify the boundary conditions which are, usually, the perfect adherence on the walls

$$u = 0 \qquad \text{on } \partial\Lambda. \qquad (2.2)$$

One of the most important features of the Navier–Stokes equation is the similitude law which we are going to illustrate.

Suppose that the problem at hand exhibits a characteristic length and velocity, L and U, respectively. Introducing the new dimensionless variables

$$v = u/U, \qquad r = x/L, \qquad \tau = t/T, \tag{2.3}$$

where

$$T = L/U, \tag{2.4}$$

the Navier–Stokes equation becomes

$$(\partial_t + v \cdot \nabla_r)v = -\nabla_r \pi + R^{-1}\Delta_r v + F, \tag{2.5}_1$$

$$\nabla_r \cdot v = 0, \tag{2.5}_2$$

where ∇_r means the gradient with respect to r,

$$\pi = p/U^2, \qquad F = Lf/U^2, \tag{2.6}$$

and finally

$$R = LU/v. \tag{2.7}$$

For example, for problem (2.1), we could put $L = \operatorname{diam} \Lambda$ and a natural scale for the velocity could be obtained by f (which has the dimensions of an acceleration), say $U^2/L = \max |f|$. R is called the Reynolds number and, as we will see later, plays the role of a control parameter.

The reason for introducing the Reynolds number is that two fluids in two domains of the same form, under the same (rescaled) F and with the same Reynolds number, behave identically. This is in agreement with the adimensional form of the Navier–Stokes equation. This is the similitude law for which a real airplane "should" behave as a smaller model of it, under suitable conditions yielding the same Reynolds number. Actually, on this simple observation is based the possibility to build small and manageable models of real large objects such as an airplane, a harbor, a river, and so on.

In many situations also the details of the external force F are not important, so that the only parameter responsible for the qualitatitive behavior of the flow is R. This statement must be taken with some care. Often a single parameter is not enough to characterize the kind of the flow completely, however, in most practical situations, we observe drastic changes in the behavior of the flow by varying the Reynolds number, so that it is certainly interesting to investigate this feature.

If the Reynolds number is sufficiently small the asymptotic behavior of the flow for $t \to \infty$ is trivial. There exists a unique attractive solution v^*. We can prove that such a solution is globally attractive in two dimensions (v^* attracts any orbit) and locally attractive in three dimensions (v^* attracts any orbit starting from an initial datum sufficiently close to v^*). This is what is really observed in practice: when R is small we see a stationary, stable solution (after a short and uninteresting transient), whatever the initial state for the system.

It is an experimental fact that at a larger Reynolds number this stationary solution disappears (or better, as we will see in the sequel, this stationary solution loses its stability property, so that it can no longer be observed). However, due to the dissipativity of the Navier–Stokes equation, a single attractive solution can be replaced by a more general attractive set, called an attractor. We will later give a more precise definition of an attractor. For the moment, we want only to underline that we are interested in the geometry and the dimension of the Navier–Stokes attractor as a measure of the complexity of the motion. Before trying to describe some general features of the motion associated to the Navier–Stokes equation, it is worth giving some basic ideas from the theory of dissipative ordinary differential systems, which give some insight on what happens when the Reynolds number increases.

Consider an ordinary differential system in $\mathbb{R}^n$

$$\frac{d}{dt} x(t) = F_R(x(t)), \tag{2.8}$$

where $F: \mathbb{R}^+ \times \mathbb{R}^n \to \mathbb{R}^n$ is a one-parameter smooth vector field. R plays the role of the Reynolds number in the Navier–Stokes equation. We suppose that for all $R > 0$ the origin is a stationary solution of (2.8), i.e.,

$$F_R(0) = 0 \tag{2.9}$$

and for $R < R_0$ the origin is asymptotically stable (see Chapter 3). Moreover, to fix the ideas, we focus our attention on the simplest nontrivial case, $n = 2$. If R_0 is a critical value for which something happens, that is, the asymptotic stability property of the origin is lost, there are two possibilities. Consider the linearization around 0 of the vector field F_R

$$A_{ij}^R = \frac{\partial (F_R)_i}{\partial x_j}. \tag{2.10}$$

The supposed asymptotic stability ensures, for $R < R_0$, that Re $\lambda_i(R) < 0$, where $\lambda_1(R)$ and $\lambda_2(R)$ are the eigenvalues of A^R, which are either real or conjugate because A^R has real entries. At $R = R_0$ we mentioned two possibilities. They are:

(i) $\lambda_1(R) < \lambda_2(R)$ are both real and when $R \to R_0$, $\lambda_2(R)$ crosses the imaginary axis (at the origin) in the complex plane.

(ii) The eigenvalues are complex conjugate $\lambda_0(R) \pm i\mu(R)$. When $R \to R_0$ they simultaneously cross the imaginary axis.

There is a third possibility: $\lambda_1(R_0) = \lambda_2(R_0) = 0$. This is somehow an exceptional event (nongeneric) so that it will not be taken into consideration.

The expected behavior for the two situations is the following:

(i) The lack of stability of the origin for $R > R_0$ makes the origin repulsive in the direction associated to λ_2. On the other hand, the change in the sign of an eigenvalue has a local character (for a genuine nonlinear system, i.e., for a

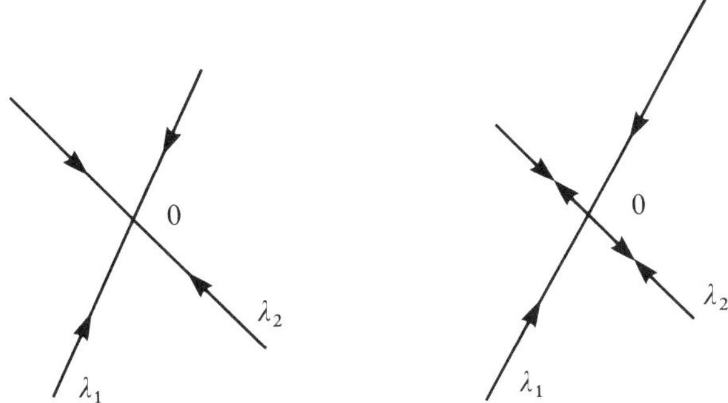

Figure 7.1

system in which there are nonvanishing higher terms in the development of the vector field F_R around the origin) so that we cannot expect, in general, a global change of the system. Thus, roughly speaking, in the λ_2 direction the repelling tendency of the origin might combine with the attracting longer distance effect and create two new stationary solutions (Fig. 7.1). For $R = R_0$ we are in the presence of a "bifurcation." By bifurcation we mean a sudden change of the stability character of a given solution. The effect we have already described can be summarized by the bifurcation diagram (Fig. 7.2). This kind of bifurcation is called "fork bifurcation" for obvious reasons.

(ii) In this case, the so-called Hopf bifurcation can occur (Fig. 7.3). For $R > R_0$ the two competing effects of the repelling character of the origin and the attractivity at larger distances might give rise to a periodic orbit. The bifurcation diagram is shown in Fig. 7.4. We do not give the proof of these bifurcations nor the exact hypothesis under which they can occur. At this level we want only to convince the reader of the possibility of the occurrence of such features.

By increasing R, a stable periodic orbit might bifurcate. For instance, it might double its period at some critical value $R = R_1$ (Fig. 7.5). For $n > 2$, other types of bifurcations are also possible. For example, think of a periodic

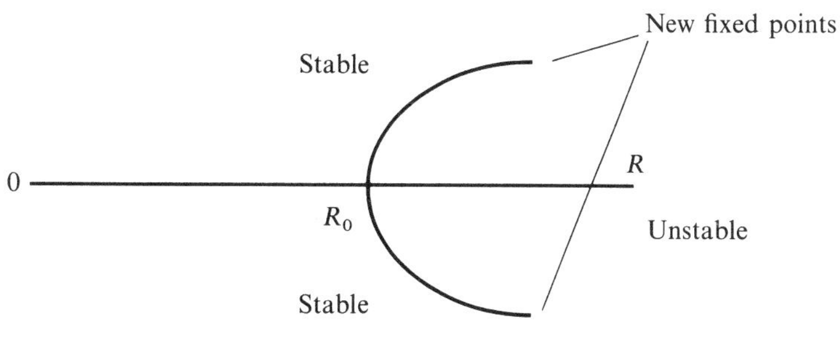

Figure 7.2

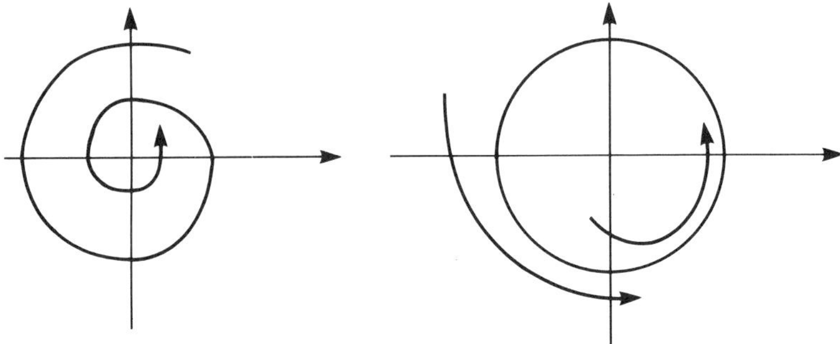

Figure 7.3

orbit which, continuously with R, has the tendency to invade a manifold of dimension two (Fig. 7.6). It can eventually bifurcate at $R = R_1$ into a quasi-periodic motion on a two-dimensional torus. We recall that a j-dimensional quasi-periodic motion is a solution of the type

$$x(t) = g(\omega_1 t, \ldots, \omega_j t), \qquad (2.11)$$

where g is a 2π-periodic function in each variable and the frequencies are rationally independent. Of course, it is possible, at least topologically, that a two-dimensional quasi-periodic motion might bifurcate into higher-dimensional quasi-periodic motions at other critical values of R.

Thus we have roughly discussed some kind of bifurcations which might occur in the study of an ordinary differential system. We have seen how the dimensionality of the attracting set can increase with R (stationary solution to a periodic orbit, to a two-dimensional quasi-periodic motion, to higher-dimensional quasi-periodic motions, and so on), as well as the complication of the geometry of an attracting set. However, we anticipate that the complicated nature of the motion on an attracting set must not be confused with a possible high dimensionality of such a set. We can conceive of a high-dimensional attractor with a simple structure, for instance, a j-dimensional quasi-periodic motion with very large j, and a complicated attractor of low dimensionality. Moreover, the previous phenomenology might induce the

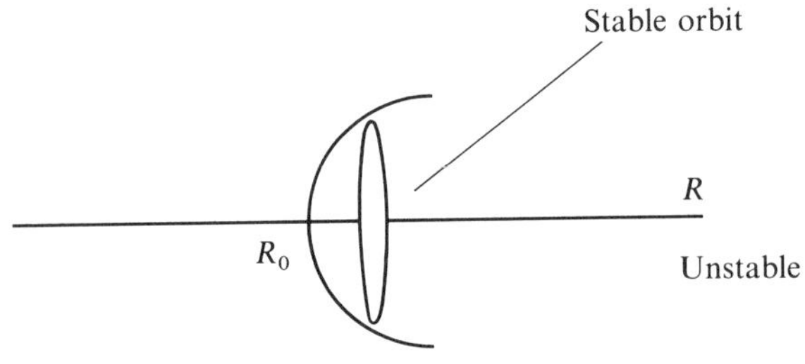

Figure 7.4

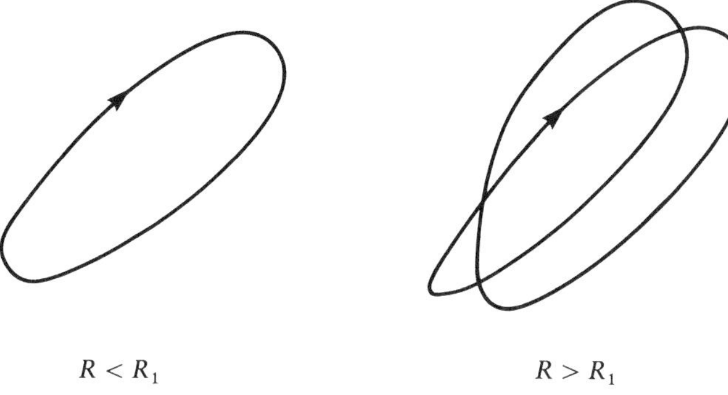

Figure 7.5

reader to believe that an explanation of the transition to a turbulent regime is a chain of bifurcations leading to a sequence of quasi-periodic motions with increasing dimensionality. As we will see in the sequel this does not seem to be the case: the onset of turbulence refuses to be explained in such simple terms.

Let us now come back to fluid dynamics to see how the type of bifurcations we have discussed can occur in real fluids. Consider, for instance, the Couette–Taylor flow. Two coaxial cylinders (very long to be schematized as cylinders of infinite length), with radii $r_1 < r_2$, rotate with angular velocities $-\omega$ and ω. Inside there is a viscous fluid. A plane stationary solution can be found by expressing the Navier–Stokes equation in polar coordinates. Assuming the velocity field u tangent to any coaxial cylinder enclosed between the two and depending only on the distance r from the axis, denoting by $v = v(r)$ its modulus, we have

$$\frac{dp}{dr} = \frac{v^2}{r}, \tag{2.12$_1$}$$

$$\frac{d^2v}{dr^2} + \frac{1}{r}\frac{dv}{dr} - \frac{v}{r^2} = 0. \tag{2.12$_2$}$$

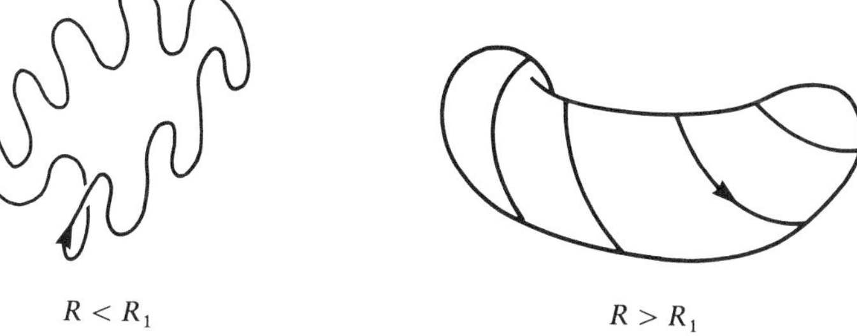

Figure 7.6

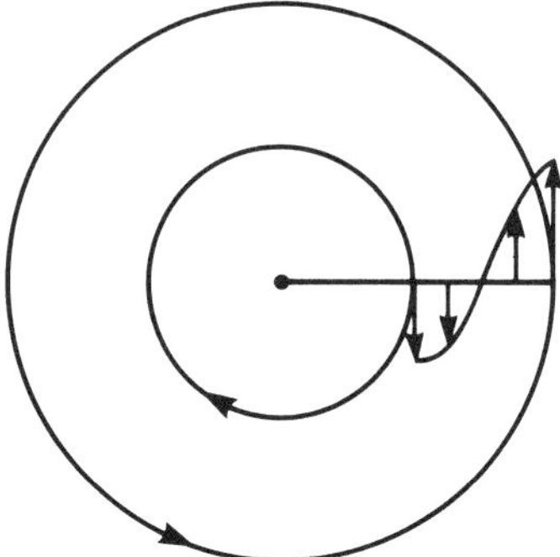

Figure 7.7

(Notice that we assume unities for which $v = 1$, r_1 and r_2 will be fixed so that ω determines the Reynolds number.) From $(2.12)_2$ we get $v(r) = \alpha r + \beta/r$ where α and β must be determined by the boundary conditions

$$-\omega r_1 = \alpha r_1 + \frac{\beta}{r_1}, \tag{$2.13)_1$}$$

$$\omega r_2 = \alpha r_2 + \frac{\beta}{r_2}. \tag{$2.13)_2$}$$

Finally

$$v(r) = \omega r \frac{r_2^2 + r_1^2}{r_2^2 - r_1^2} - \frac{\omega r_2^2 r_1^2}{r_2^2 - r_1^2} \frac{2}{r}. \tag{2.14}$$

Such a solution is really observed for ω sufficiently small (Fig. 7.7).

Increasing ω, such a solution loses stability and a new stationary solution appears. Such a solution is no longer planar. The velocity field is tangent to parallel coaxial tori. See a schematization of the cross section of the two cylinders in Fig. 7.8. Actually, the possible stationary solutions are infinitely many, differing for a vertical translation only. These tori are called Taylor cells. This kind of bifurcation (stationary solution to a stationary solution) is followed, by increasing ω, by a bifurcation of stationary solution to a periodic orbit. In fact, at a critical value of ω the cells begin to oscillate with a given frequency. Afterward, new frequencies occur and through a sequence of bifurcations the motion becomes more and more complicated, the cells break down, and the motion appears as chaotic.

An attempt to give an explanation for this behavior was given by Landau [LaL 68]$_1$. The Landau theory consisted in conjecturing that the main bifurcations are from invariant tori to invariant tori of higher dimensions. We are talking of invariant tori in the space of the states of the system thus the motion is (after an unimportant transient) quasi-periodic. Therefore, following Landau, the transition to a turbulent flow is the appearance of an increas-

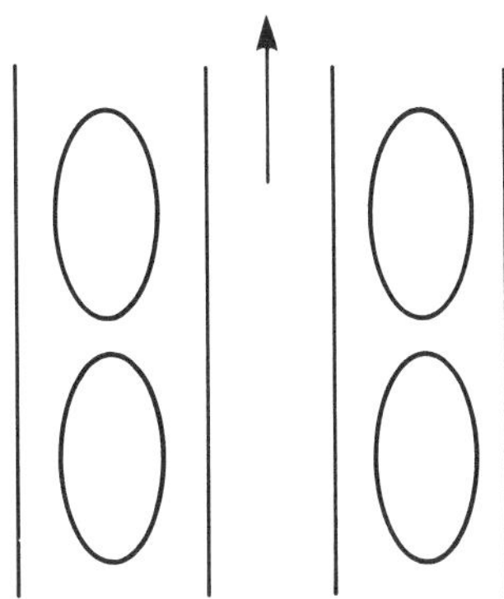

Figure 7.8

ing number of frequencies, and a measure of its complexity is given by the number of frequencies we need to describe the motion.

As remarked by Ruelle and Takens [RuT 71] and others, the notion of quasi-periodicity does not seem adequate to describe a turbulent motion for the following two reasons.

(a) Quasi-periodicity is not a "robust" property.

This means the following. Consider a two-dimensional torus T^2 and the dynamical system

$$\frac{d\phi}{dt} = \omega + \varepsilon f(\phi), \tag{2.15}$$

where $\phi \in T^2$, $\omega = (\omega_1, \omega_2)$ are rationally independent frequencies, $\varepsilon \geq 0$, and f is a smooth vector field on T^2. For $\varepsilon = 0$ the motion is quasi-periodic while if $\varepsilon > 0$ the motion is no longer quasi-periodic for an arbitrary f. On the other hand, the Navier–Stokes equation is an approximate equation and it is natural to assume reasonable, for a fluid, those properties which are stable with respect to small perturbations of the second member of the equation describing it.

(b) A quasi-periodic motion does not give decay of time correlations.

Given two observables, i.e., two functions f and g defined on the space of the velocity fields, define

$$\langle f \rangle = \lim_{T \to \infty} \frac{1}{T} \int_0^T f(t)\, dt, \tag{2.16}$$

where $f(t)$ is the time evolution of the observable f and

$$C_t(fg) = \langle f(t)g \rangle - \langle f \rangle \langle g \rangle. \tag{2.17}$$

It is a fact experimentally observed (but not analytically proven) that

$$C_t(fg) \to 0 \qquad \text{as} \quad t \to \infty. \tag{2.18}$$

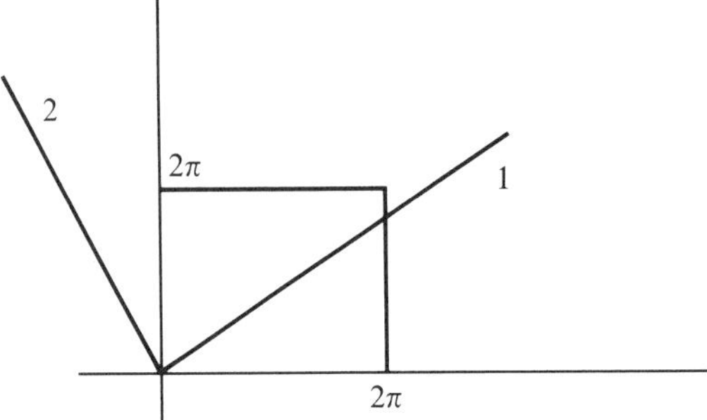

Figure 7.9

However, property (2.18) is not fulfilled by a quasi-periodic motion. We know from ergodic theory that a quasi-periodic motion is ergodic but not mixing (see [ArA 68]). The mixing property is exactly expressed by condition (2.18), and thus if we want a theory of the turbulence in which time correlations decay we must abandon the notion of quasi-periodicity.

According to Ruelle and Takens [RuT 71] (see also [Rue 87] and references quoted therein), we look for dynamical systems enjoing the property of being robust and exhibiting time decay of the correlations. Systems of this type exist and are called hyperbolic or Axiom A. The Ruelle and Takens theory conjectures the occurrence of finite-dimensional hyperbolic systems as attractors at high Reynolds numbers. Here we do not give a precise definition of a hyperbolic or Axiom A system (see [Rue 78]), we only present a simple example to give an idea of the kind of motion.

Consider the torus $T^2 = [0, 2\pi)^2$. Consider also the action of the matrix

$$S = \begin{pmatrix} 1 & 1 \\ 1 & 2 \end{pmatrix}, \tag{2.19}$$

S has as eigenvalues $(3 \pm \sqrt{5})/2$. Consider the (nonlinear!) transformation in T^2: $x \to Sx$ (mod 2π). Denote this map A (Fig. 7.9). The motion contracts along direction 1 and expands along direction 2. It is easy to figure out the chaoticity of such a motion: two points, x and y, which are very close initially, after repeated applications of the map A, will have very different histories. Actually, T^2 contains two dense one-dimensional manifolds (stable and unstable manifolds) that are the eigendirections of the matrix S, one attracting and the other repelling (Fig. 7.10).

We conclude here the dynamical system analysis of the turbulence. We address the reader to [Rue 84], [Rue 87], and [GuH 83] for a deeper and more rigorous analysis of the topic. We can just say that the transition to chaos mechanism has been relatively well clarified from a logical point of view and a first notion of turbulence (what we call here the onset of turbulence) has been achieved. However, many points must be investigated. In particular, the following questions are quite natural:

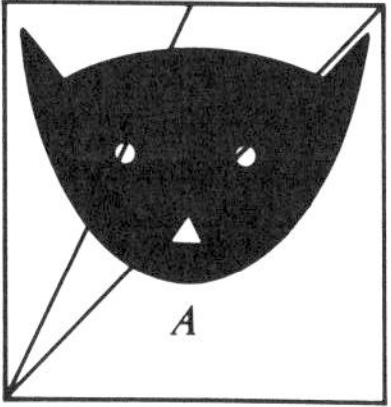

Figure 7.10

(1) Do the Navier–Stokes equations possess attractors?
(2) If so, do they have finite dimension?
(3) (After a positive answer to questions (1) and (2).) Is the motion on the attractor hyperbolic?

The first two questions have a positive answer. We discuss briefly this point and address the reader to the Temam book [Tem 88] where these arguments are treated in great detail.

For simplicity, we consider the case of a two-dimensional viscous fluid moving in a flat torus $T^2 = [0, 2\pi] \times [0, 2\pi]$. Then we assume the Navier–Stokes equation (2.1) with u, p, and f 2π-periodic functions. Moreover, we suppose, without the sake of generality

$$\int_{T^2} f(x) = \int_{T^2} u(x) = 0. \tag{2.20}$$

It is well known that, in the space H of all divergence-free vector fields u, with finite energy (that is, a Hilbert space equipped with norm $\|u\| = (\int_{T^2} u(x)^2)^{1/2}$) the initial value problem associated to the Navier–Stokes equation possesses a unique solution belonging to $C([0, T]; H)$ (see, e.g., [Tem 84] and references therein). Then we can consider the semigroup $S(t)$ defined by

$$S(t)u_0(x) = u(x, t), \qquad u_0 \in H, \tag{2.21}$$

where $u(x, t)$ is the solution of the Navier–Stokes problem associated to the initial value u_0. A global attractor $A \subset H$ is an invariant set

$$S(t)A = A \tag{2.22}$$

which has the property to attract any orbit

$$\text{dist}(S(t)u_0, A) \to 0 \qquad \text{as} \quad t \to \infty \tag{2.23}$$

and it is also compact.

It is easy to show the existence of an absorbing ball in H, i.e., a ball which is invariant and attracts all the orbits. This is actually a consequence of the energy identity

$$\frac{1}{2}\frac{d}{dt}\|u\|^2 = v(u, \Delta u) + (u, f). \tag{2.24}$$

Here, and from now on, we will assume f independent of time. Thanks to the obvious inequalities

$$-(u, \Delta u) \geq \|u\|^2, \tag{2.25}$$

$$(u, f) \leq \|u\| \, \|f\|, \tag{2.26}$$

we arrive at

$$\frac{d}{dt} \|S(t)u_0\| \leq -\nu \|S(t)u_0\| + \|f\| \tag{2.27}$$

and finally

$$\|S(t)u_0\| \leq e^{-\nu t} \|u_0\| + \frac{1}{\nu} \|f\| (1 + e^{-\nu t}). \tag{2.28}$$

From the estimate (2.28) we conclude that any ball in H of radius $R > \|f\|/\nu$ is an invariant attracting set.

The above argument may be improved to get the existence of an attracting ball in the Sobolev space V (which is the set of all divergence-free vector fields with the first derivative square-summable) and this is enough to prove the existence of a global attractor (see [Tem 88]). We give an idea of this fact. From the energy equality (2.24), integrating with respect to time, we obtain

$$\tfrac{1}{2}(\|u(t + r)\| - \|u(t)\|) + \nu \int_t^{t+r} |u(\tau)|_1^2 \, d\tau \leq \int_t^{t+r} d\tau \, \|u(\tau)\| \, \|f\|, \quad (2.29)$$

where $|\ |_1$ is defined in Chapter 2, Section 1, and r is fixed. By the boundedness of $\|u(t)\|$ we obtain the boundedness of $\int_t^{t+r} |u(\tau)|_1^2 \, d\tau$. This argument can be improved to get a bound on $|u(t)|_1$ and the attractivity of a ball in V.

The successive step is the analysis of such an attractor. This set could have a very complicated geometric structure. An estimate of the "size" of this set and, in a sense, of its complexity, is a measure of its Hausdorff dimension. We now give the definition. Let M be a metric space and let $N \subset M$ be a subset of M. Given two positive numbers, $d, \varepsilon > 0$, denote

$$\mu(N, d, \varepsilon) = \inf \sum_{i \in I} r_i^d, \tag{2.30}$$

where the infimum is taken over all the covering of N by a family of balls $\{B_i\}_{i \in I}$ of radii $r_i < \varepsilon$. The number

$$\mu(N, d) = \lim_{\varepsilon \to 0} \mu(N, d, \varepsilon) \tag{2.31}$$

is called the d-dimensional Hausdorff measure of the set N. It is easy to realize that if $\mu(N, d') < +\infty$ for some d', then $\mu(N, d) = 0$ for all $d > d'$. Hence there exists a $d_0 \geq 0$ such that $\mu(n, d) = 0$ for $d > d_0$ and $\mu(N, d) = +\infty$ for $d < d_0$. Such a d_0 is called the Hausdorff dimension of the set N.

Coming back to our problem, the compactness of the attractor suggests that it should have a finite Hausdorff dimension. Actually, we can prove that the Hausdorff dimension of the Navier–Stokes attractor is finite and, usually, increases with the Reynolds number, going to infinity as $R \to \infty$. This fact is interpreted as an increase of the complexity of the attractor. However, it is not always true that the dimension of the attractor increases whenever $R \to \infty$, as the following example shows.

Choose

$$f = iAv \cos y, \tag{2.32}$$

where from now on we denote by (x, y) the two components on T^2 and i is the versor in the x direction. A is a constant. The corresponding stationary state is

$$u^*(x, y) = iA \cos y. \tag{2.33}$$

We choose A related to the size of f, as a control parameter, say the Reynolds number of the problem.

We want to show that the energy variation

$$\tfrac{1}{2}\|u - u^*\|^2 = E \tag{2.34}$$

vanishes as $t \to \infty$. To this purpose it is convenient to introduce the vorticity variation

$$\tfrac{1}{2}\|\omega - \omega^*\|^2 = N, \tag{2.35}$$

where

$$\omega^*(x, y) = A \sin y \tag{2.36}$$

is the stationary vorticity field.

Writing the Navier–Stokes equation for the vorticity, and computing the time derivatives of E and N, we find

$$\frac{d}{dt} E = A \int v_1 v_2 \sin y - v \int \sum_{i,j=1}^{2} (\partial_i v_j)^2, \tag{2.37}$$

$$\frac{d}{dt} N = A \int v_1 v_2 \sin y - v \int \sum_{i=1}^{2} (\partial_i \delta)^2, \tag{2.38}$$

where

$$v = u - u^* \qquad \text{and} \qquad \delta = \omega - \omega^*. \tag{2.39}$$

Hence

$$\frac{d}{dt}(N - E) = v \int \sum_{i,j=1}^{2} (\partial_i v_j)^2 - v \int \sum_{i=1}^{2} (\partial_i \delta)^2. \tag{2.40}$$

It is easy to see, by the continuity equation and by the use of the explicit Fourier expansion, that the right-hand side of (2.40) is bounded by $-4v(N-E)$, which is also a positive quantity. By the Gronwall inequality we conclude that $N - E$ vanishes exponentially as $t \to \infty$. However, we want to show that N and E vanish separately. We find another differential inequality from (2.37). In order to evaluate the right-hand side of (2.37), we develop v_1 and v_2 in Fourier series

$$v_1 = a \cos y + b \sin y + \varphi(x, y),$$

$$v_2 = c \cos x + d \sin x + \psi(x, y). \tag{2.41}$$

where φ and ψ depend only on higher wave numbers. Actually, by the continuity equation, at the lowest order, v_1 depends only on y while v_2 depends only on x.

Inserting (2.41) in (2.37), using the orthogonality property of the trigonometric functions and the Cauchy–Schwarz inequality, we have that

$$\frac{d}{dt} E \leq A(\|\psi\|^2\|v_1\|^2 + \|\varphi\|^2\|v_2\|^2) - v \int \sum_{i,j=1}^{2} (\partial_i v_j)^2. \tag{2.42}$$

On the other hand, by an explicit Fourier development, we can prove that

$$\int \sum_{i,j=1}^{2} (\partial_i v_j)^2 \geq 2E \tag{2.43}$$

and

$$\|\varphi\| \leq 2(N - E),$$
$$\|\psi\| \leq 2(N - E). \tag{2.44}$$

Finally, from (2.42)

$$\frac{d}{dt} E \leq 4A(N - E)^{1/2}E^{1/2} - 2vE \tag{2.45}$$

from which we easily get $E \to 0$ as $t \to \infty$. For this example of the absence of turbulence we have followed [Mar 86]. (See also [Mar 87] and [CFT 88].)

Let us come back to the very central question about the nature of the Navier–Stokes attractor and the motion on it. Unfortunately, very little is known. The finite Hausdorff dimensionality of the attractor could induce us to believe that, for a fixed Reynolds number R, everything goes as if the (nontransient) relevant motion were taking place on a manifold, a point of which is determined by few parameters, which are the relevant degrees of freedom of the system. However, this picture is too optimistic. No result concerning the smoothness of the attractor was known until now and so the capture of the relevant degrees of freedom seems far from present knowledge.

There are attempts to study the Navier–Stokes equation taking into account only a finite number of modes (see [BoF 79] and [FrT 85] for more recent results and references), however, there is no reason to believe that a finite number of Fourier modes, evolved according to the truncated Navier–Stokes equation, are enough to determine the long-time behavior of the solutions of the Navier–Stokes equation itself.

7.3. Phenomenological Theories

When the Reynolds number diverges the dimension of the Navier–Stokes attractor is also expected to diverge. The motion is chaotic and strongly unstable, as follows by experimental observations. In these circumstances, as we said in the Introduction, the notion of a single observable loses its meaning and a statistical approach seems more appropriate. The velocity profile of the flow becomes a random field with respect to a probability measure, expressing which kind of profiles we are more likely to observe. Moreover, we are no longer interested in a single solution of the Navier–Stokes equation, but rather in the time evolution of probability measures on the space of the velocitiy profiles.

The simpler case to study is a stationary situation. We mean the following. We have a fluid contained in a bounded domain, for simplicity, we choose a three-dimensional torus $V(T) = [-T, T]^3$. The fluid is performing a (stationary) turbulent motion. From a mathematical point of view, this means that there exists a probability measure μ on the space of the velocity profiles H, for which $u = u(x)$ is a stationary (in time) random field.

Let us be a little more precise. Let H be the subspace of $L_2(T^3)^3$ of the divergenceless vector fields. Suppose that μ_T is a Borel probability measure on H with the property

$$\int \mu_T(du)F(u(t)) = \int \mu_T(du)F(u), \tag{3.1}_1$$

where $u(t)$ is the solution of the Cauchy problem for the Navier–Stokes equation with the initial value given by u, and F is a continuous bounded function on H. We also suppose that μ_T is translationally invariant

$$\int \mu_T(du)f(u(x + r)) = \int \mu_T(du)f(u(x)), \qquad x \in [-T, T]^3, \tag{3.1}_2$$

for all bounded continuous real-valued functions f defined on $\mathbb{R}^3$. We are not very much concerned about the fact that unique smooth solutions of the Navier–Stokes equation are not known to exist, for all times and the large Reynolds number. As we will see, our analysis is so rough and preliminary that this seems a minor problem.

Equations (3.1) express the space–time invariance of the measure μ_T which we require, since we assume to deal with stationary turbulence. Physically speaking, this means that, although the results of the same experiment can change, the probability of an event is constant in time and does not depend on the place where the experiment is performed.

It is clear that all the relevant information concerning the turbulent phenomenon is in the measure μ_T. However, in contrast to the equilibrium statistical mechanics, in which the relevant invariant measures are given by the Gibbs ansatz, in fluid dynamics we do not know how to produce invariant measures for the Navier–Stokes (or Euler) flow so that we cannot go much further with our analysis. However, some conclusions can be drawn by means of purely dimensional arguments. Let us first introduce two basic quantities

$$\|u\|_{T^2} = \frac{1}{(2T)^3}\int_{V(T)} |u|^2\,dx, \qquad (|u|_{1,T})^2 = \frac{1}{(2T)^3}\sum_{i=1}^{3}\int_{V(T)} |\nabla u_i|^2\,dx, \tag{3.2}$$

$$E_T = \frac{1}{2}\int \mu_T(du)\,\|u\|_T^2, \qquad \varepsilon_T = \nu\int \mu_T(du)(|u|_{1,T})^2. \tag{3.3}$$

E_T and ε_T denote the mean energy per unit volume and the mean dissipation energy rate per unit volume, respectively. Since we do not want a dependence on the size T of the torus $V(T)$ (we want basically to describe situations which are spatially homogeneous and hence quite far from the boundary of the volume containing the fluid) we take (formally) the limit $T \to \infty$. Denoting by E and ε the limits of E_T and ε_T, respectively, thanks to the homogeneity of the limiting measure (we are assuming the existence of such a limiting measure, denoted by μ, enjoying the property of being translationally invariant), we have the expressions

$$E = \frac{1}{2} \int \mu\,(du)\,u^2(x), \qquad \varepsilon = v \int \mu\,(du)\,|\nabla u(x)|^2. \tag{3.4}$$

Notice once more that the above expressions do not depend on x.

Another quantity of interest is the following object, called the velocity autocorrelation tensor,

$$R_{i,j}(r) = \int d\mu\, u_i(x)u_j(x + r), \qquad r \in \mathbb{R}^3, \quad i, j = 1, 2, 3, \tag{3.5}_1$$

and its Fourier transform

$$\phi_{i,j}(k) = (2\pi)^{-3/2} \int_{\mathbb{R}^3} dr\, e^{-ikr} R_{i,j}(r), \qquad k \in \mathbb{R}^3. \tag{3.5}_2$$

Finally, define

$$E(s) = \sum_{i=1}^{3} \int_{|k|=s} d\sigma\,(k)\,\phi_{i,j}(k), \tag{3.6}$$

where $d\sigma(k)$ is the surface element on the sphere $|k| = s$. Expression (3.6) is the called energy spectrum and plays an important role in the theory of fully developed turbulence.

An easy consequence of the previous definitions is the following identity:

$$E = \frac{1}{2}\sum_i R_{i,i}(0) = \frac{1}{2}\sum_i \int dk\, \phi_{i,i}(k) = \frac{1}{2}\int_0^\infty ds\, E(s). \tag{3.7}$$

How is the energy distributed among the various s? Is a universal behavior conceivable for the function $E(s)$?

Strictly speaking, the answer to the second question is certainly no. To see this, consider the Navier–Stokes equation in terms of the Fourier transform. A slight modification of the arguments seen in Chapter II, Section 5, yields

$$\frac{d}{dt}\hat{u}_t(k) = i\{kp^\wedge(k) - \int_{\mathbb{R}^3} dh\, \hat{u}_t(k - h) \cdot h\hat{u}_t(h))\} - vk^2\hat{u}_t(k) + f^\wedge(k),$$

$$k \cdot \hat{u}_t(k) = 0, \tag{3.8}$$

where $f^\wedge$ is the Fourier transform of the given external force.

From the structure of (3.8) we realize that the energy is dissipated through

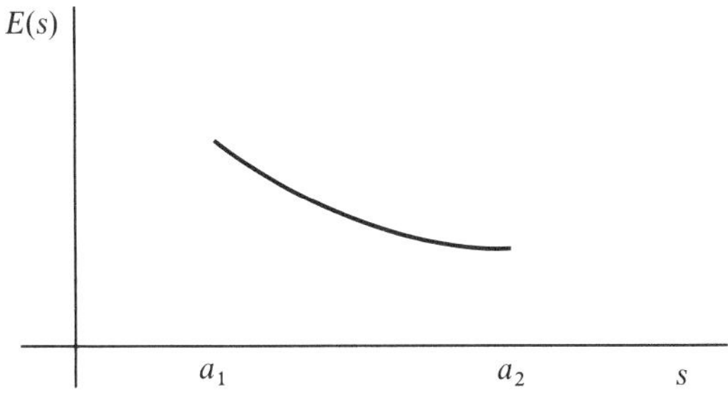

Figure 7.11

the viscosity term at large k, which means at small scales. To make the system alive, we give energy to the system through the external force f which can be assumed, as much as possible, simple. For instance, we can assume $f^\wedge(k)$ different from zero for a finite number of k's. Thus we inject energy in the system at large scales (small k) and this is dissipated at small scales (large k).

As we know from the theory of the Euler equation, the inertial term in brackets preserves the energy and is responsible for the energy transfer from large to small scales. If we plot an energy spectrum, as actually observed in the experiments, we find behavior of the type as shown in Fig. 7.11. The spectrum below a_1 is sensitive to the external force f, while over a_2 the dissipative term is dominant. a_1 and a_2 are approximately defined as those numbers for which from 0 up to a_1 almost all the energy is localized while from a_2 up to infinity almost all the energy dissipation is localized

$$\int_0^{a_1} E(s)\, ds \cong \sum_{s=0}^{\infty} E(s)\, ds = 2E, \tag{3.9}$$

$$\int_{a_2}^{\infty} s^2 E(s)\, ds \cong \int_0^{\infty} s^2 E(s)\, ds = \varepsilon/v. \tag{3.10}$$

The form of the spectrum between a_1 and a_2 is essentially described by the conservative (inertial) term. For this reason it is called "inertial subrange." This part of the spectrum seems universal enough, not depending on the details of the system at hand. The shape of this part of the spectrum can be conjectured by dimensional arguments. This is the well-known Kolmogorov law [Kol 41]. We give the argument.

By using definitions (3.2) and the Navier–Stokes equation, we establish the dimensions of v and ε:

$$v = [V][L], \qquad \varepsilon = [V]^3[L]^{-1}, \tag{3.11}$$

where $[v]$ and $[L]$ denote velocity and length dimensions. From this we get typical length and velocity, l and v, respectively,

$$l = (v^3/\varepsilon)^{1/4}, \qquad v = (v\varepsilon)^{1/4}. \tag{3.12}$$

The dimensions of $E(s)$ are

$$E(s) = [V]^2[L] \tag{3.13}$$

therefore, *if $E(s)$ behaves like a power*, in unity v and l it must be

$$E(s) = (sl)^\alpha v^2 l \tag{3.14}$$

since sl is a pure number. Finally expressing everything in terms of v and ε

$$E(s) = s^\alpha (v^3/\varepsilon)^{\alpha/4} (v\varepsilon)^{1/2} (v^3/\varepsilon)^{1/4}. \tag{3.15}$$

We now make the essential hypothesis that *in the inertial subrange $E(s)$ does not depend on v*. From this and (3.15) we get $\alpha = -\frac{5}{3}$ and the celebrated universal law

$$E(s) = \text{const. } s^{-5/3}. \tag{3.16}$$

The Kolmogorov law is in relatively good agreement with the experiments. However, the experiments devoted to establishing the behavior of the energy spectrum are delicate and there are those who believe that corrections to the $\frac{5}{3}$ law are needed. The same type of arguments can be used to determine the energy spectrum in dimension two. The result is that $E(s) \approx s^{-3}$.

From the point of view of mathematical physics we need a characterization of the measure (3.1) in terms of the mathematical model describing incompressible fluids. The following sections will be devoted to this important point. We finally mention that some rigorous connections between the phenomenological theories presented in this section and the dynamical system approach discussed in Section 7.2 have been established in [CFT 85], [CFM 85], and [CFT 88].

7.4. Statistical Solutions and Invariant Measures

We have already emphasized that a single state of a fluid may be an inadequate description of a turbulent flow: we need ensembles of states with a statistical prescription. In more precise mathematical language, we are led to introduce a probability measure on the space of all possible fields associated with the fluid. Moreover, instead of considering a single evolution of the velocity field (according to the Euler or Navier–Stokes equations), we want to investigate the evolution of probability measures.

Consider, for simplicity, a fluid in a d-dimensional torus $\Lambda = [-\pi, \pi]^d$, $d = 2, 3$. All possible velocity profiles with bounded energy are the divergence-free elements of $L_2(\Lambda)^d$. We denote this space by H. On H (better on a σ-algebra of sets to be specified later) we define a probability measure μ_0. Suppose now that the solutions of the Euler equation give rise to a flow in H

$$T_t u(x) = u(x, t), \tag{4.1}$$

where $u(x, t)$ is the solution of the Euler equation with initial datum given by $u = u(x)$. Strictly speaking, we face many difficulties in defining such a flow since we do not know if the Cauchy problem is well defined in H. Actually, we do not need the flow be defined in all H, but only in a sufficiently large subset. For instance, in dimension two, the existence theory ensures that the flow T is well defined in $H \cap C^1$. In three dimensions, however, the known existence theory is too poor to go any further nor is the situation much better for the Navier–Stokes equation. On the other hand, since our future considerations based on the existence of such a flow will be mostly descriptive, we simply ignore this problem. The mathematically oriented reader not agreeing with this procedure can restrict himself to the two-dimensional case, and think of T as defined in a suitable subspace of H.

Now let μ_t denote the time evolved measure, defined from μ_0 as usual, by

$$\mu_t(A) = \mu_0(T_{-t}A), \tag{4.2}$$

where A is a measurable set in H and

$$T_{-t}A = \{u \mid u(x, t) \in A\}. \tag{4.3}$$

Our next target is to derive an evolution equation for μ_t. First we consider the same problem for an ordinary differential system. For instance, put $H = \mathbb{R}^n$ and replace the Euler equation by

$$\frac{d}{dt}u = F(u), \tag{4.4}$$

where F is a smooth vector field in $\mathbb{R}^n$. Accordingly, define the semiflow T as the solution of the Cauchy problem

$$\frac{d}{dt}T_t u = F(T_t u),$$

$$T_0 u = u. \tag{4.5}$$

Given a Borel probability measure μ_0, the evolved measure μ_t is defined by (4.2) or, equivalently, by

$$\int \mu_t\,(du)\,f(u) \equiv \mu_t(f) = \int \mu_0\,(du)\,f(T_t u) \tag{4.6}$$

for all bounded $f \in C_1(\mathbb{R}^n)$. Finally, by a simple calculation we obtain

$$\frac{d\mu_t(f)}{dt} = \mu_t(\nabla f \cdot F). \tag{4.7}$$

In order to translate the above analysis to our partial differential context we should replace ∇f by a functional derivative. However, we will follow a simpler way. Denote by

$$u^\wedge = \{\hat{u}(k) \mid k \in \mathbb{Z}^d\} \tag{4.8}$$

the sequence of the Fourier transforms of u. Then

$$H = \{\hat{u}|\textstyle\sum |\hat{u}(k)|^2 < +\infty, k \cdot \hat{u}_t(k) = 0\} \tag{4.9}$$

and the Euler equation can be seen as

$$\frac{d}{dt}\hat{u}_t(k) = ikp^{\wedge}(k) + B_k(\hat{u}), \tag{4.10}_1$$

$$k \cdot \hat{u}_t(k) = 0, \tag{4.10}_2$$

where $p^{\wedge}(k)$ denotes the Fourier transform of the pressure, $(4.10)_2$ expresses the incompressibility condition, and

$$B_k(\hat{u}) = i \sum_{h \in \mathbb{Z}^d} \hat{u}_t(k - h) \cdot h\hat{u}_t(h). \tag{4.11}$$

We notice that, by virtue of $(4.10)_2$, the pressure can be easily eliminated in $(4.10)_1$. In fact, (4.10) are equivalent to

$$\frac{d}{dt}\hat{u}_t(k) = B_k^{\perp}(\hat{u}), \tag{4.12}$$

where $B_k^{\perp}(\hat{u})$ is the projection of $B_k(\hat{u})$ into the subspace orthogonal to k.

Once having established the Euler equation in a convenient form, we construct a suitable set of test functions. Consider the set $C(\Lambda_M)$ of all cylindrical functions based on $\Lambda_M \equiv [-M, M]^d \cap \mathbb{Z}^d$, $M \in \mathbb{Z}^+$, as the set of all functions

$$u^{\wedge} \to f(u^{\wedge}) \tag{4.13}$$

of the type

$$f(u^{\wedge}) = \phi(u_M^{\wedge}), \tag{4.14}$$

where

$$u_M^{\wedge} = \{\hat{u}(k)|k \in \Lambda_M\} \tag{4.15}$$

denotes the restriction of $\hat{u}$ to Λ_M and $\phi \in C_1 (\mathbb{C}^{(2M+1)d})$.

Putting

$$C = \bigcup_{M=1}^{\infty} C(\Lambda_M) \tag{4.16}$$

we see that, if $f \in C$, f depends on $u^{\wedge}$ only through a finite number of $u^{\wedge}(k)$'s. For this reason we choose C as the family of test functions. The analogue of (4.7) is

$$\frac{d\mu_t(f)}{dt} = \mu_t\left(\sum_{k \in \mathbb{Z}^d} \frac{\partial f}{\partial \hat{u}(k)} B_k^{\perp}(\hat{u})\right). \tag{4.17}$$

Notice that the sum appearing in the right-hand side of (4.17) contains a finite number of elements since the derivatives of f vanish but for a finite number of k's.

This choice of C, making natural the extension of (4.7) to our context, also suggests the σ-algebra on which to define μ_t. We first introduce the roughest topology making the elements of C continuous. This topology is equivalent to the weak L_2 convergence in H. We then require that μ_t be defined on the σ-algebra of the Borel sets on H equipped with this topology. Thus we have defined the concept of statistical solutions for the Euler equation as a map $t \to \mu_t$, Borel probability measure valued, satisfying (4.17).

A comment on the family C is needed. We introduced it simply as a matter of convenience. However, the choice of the set of test functions in the study of statistical solutions of a partial differential equation should be dictated by physical considerations only. The test functions are the observables of the physical problem under consideration, those functions whose averages are the relevant physical objects. Our choice seems reasonable: we allow as observables all smooth functions of $\hat{u}(k)$, with k arbitrarily large. Such types of equations for the study of statistical solutions have been introduced by Hopf [Hop 52].

We do not go far in the study of (4.17). Its analysis is even more difficult than that of the Euler equation itself. The existence of the solutions is known, by (4.2), whenever the existence of the flow T is ensured. The uniqueness of the solutions is a more delicate problem which can be achieved (in a natural class of flows $t \to \mu_t$) as a consequence of the regularity properties of T. Of course, it is not known for the Euler and Navier–Stokes dynamics in three dimensions, globally in time.

There is wide heuristic literature concerning evolution equations for the fluctuations of the velocity field (which, in this context, is a random field) with respect to its mean value. Due to the nonlinearity of the fluid dynamical equations, the exact expression of the time derivative of a momentum of the type

$$\mu_t(|u(x) - \mu_t(u(x))|^m)$$

involves higher-order momenta. This leads to the so-called "closure problem": in order to make these equations suitable, we have to truncate this hierarchical structure on the basis of a suitable ansatz on the measure μ_t. The problem is closely related to that encountered in nonequilibrium statistical mechanics in dealing with the **BBKGY** hierarchy. Here the closure problem is achieved by exploiting asymptotic regimes in which the statistical independence of relevant random variables is expected to hold, as for the Boltzmann–Grad or weak coupling limits, yielding the Boltzmann or the Master equation. In turbulence we do not know of any physically significant assumption, even at a heuristic level, which makes the treatment of (4.17) or any equivalent equation easier. In conclusion, the study of the nonstationary turbulence seems especially difficult and our discussion ends here.

Coming back to the equilibrium problem, let us recall once again that the central problem in the theory of turbulence is to give an explanation of the behavior of real fluids in the so-called inertial range. This is, as discussed in Section 7.3, an intermediate range of scales between those in which the energy (or the enstrophy in dimension two) is injected in the system and those in which the dissipation is dominant. In this range the viscosity coefficient and the forcing term should not influence the shape of the energy spectrum (or the structure of other relevant quantities) so that we can hope to explain this behavior by means of the Euler equation only. Moreover, the necessity of a statistical description suggests looking for invariant measures of the Euler flow or, equivalently, for stationary solutions of (4.17). In doing this we try to take advantage of the Hamiltonian structure of the Euler equation that allows us to follow the Gibbs prescription.

We will deal with the easiest two-dimensional case, so that, from now on, we will assume $d = 2$. Recall that a canonical Gibbs measwe depends only on a few parameters such as the temperature and other Lagrangian multipliers associated to other first integrals of motion different from the energy. In our case we have many first integrals

$$H = \frac{1}{2} \int_\Lambda dx \, \omega(x) \Delta^{-1} \omega(x) = \frac{1}{2} \int_\Lambda dx \, u^2(x) \qquad \text{(energy)}, \qquad (4.18)$$

$$\mathscr{E} = \frac{1}{2} \int_\Lambda dx \, \omega^2 \qquad\qquad\qquad \text{(enstrophy)}. \qquad (4.19)$$

More generally, the integral of any function $\phi \in C(\mathbb{R})$ of the vorticity

$$\int_\Lambda dx \, \phi(\omega)(x) \qquad\qquad\qquad (4.20)$$

is preserved by the Euler flow.

The Euler equation in terms of the Fourier transform of the vorticity reads as

$$\frac{d}{dt} \omega_t^\wedge(k) = B_k(\omega_t^\wedge), \qquad\qquad\qquad (4.21)$$

where

$$\omega_t^\wedge = \{\omega_t^\wedge(k) | k \in \mathbb{Z}^2\} \qquad\qquad\qquad (4.22)$$

and

$$B_k(\omega_t^\wedge) = i \sum_{h \in \mathbb{Z}^2} \omega_t^\wedge(k - h) \cdot h \omega_t^\wedge(h). \qquad\qquad (4.23)$$

The basic remark is the following. We know from Chapter 1 that in a flat torus Λ, the only admissible vorticity distributions are those for which

$$\int_\Lambda dx \, \omega(x) = \omega^\wedge(0) = 0 \qquad\qquad\qquad (4.24)$$

(recall that otherwise the Laplacean operator is not invertible) we then conclude that

$$\frac{\partial}{\partial \omega^\wedge(k)} B_k(\omega^\wedge) = 0. \qquad\qquad\qquad (4.25)$$

Thus the measure

$$\prod_{h \in \mathbb{Z}^2} d\omega^\wedge(k) \equiv \prod_{h \in \mathbb{Z}^2} d \, \text{Re} \, \omega^\wedge(k) \, d \, \text{Im} \, \omega^\wedge(k)$$

is (formally) invariant for the Euler flow since the infinite-dimensional vector field B is divergence-free.

As a consequence, we are led to introduce a probability measure proportional to

$$e^{-\beta H - \lambda E} \prod_{k \in \mathbb{Z}^2} d\omega^\wedge(k). \tag{4.26}$$

Such a measure may be rigorously defined. For a function $f \in C$ based on Λ_M we define

$$\mathbb{E}_{\beta,\lambda}(f) = Z^{-1}(\beta, \lambda, M) \int \prod_{k \in \Lambda_M} d\omega^\wedge(k) \exp\{-\beta k^{-2}\omega^\wedge(k)^2 - \lambda\omega^\wedge(k)^2\}f(\omega^\wedge), \tag{4.27}$$

where

$$Z(\beta, \lambda, M) = \int \prod_{h \in \Lambda_M} d\omega^\wedge(k) \exp\{-\beta k^{-2}\omega^\wedge(k)^2 - \lambda\omega^\wedge(k)^2\} \tag{4.28}$$

is an explicitly computable integral. From the values $\mathbb{E}_{\beta,\lambda}(f)$ we can easily construct a measure on a suitable space of the fields ω's. We do not make it in detail since such a construction is not relevant for what follows. We mention only that such a measure is Gaussian. It is an infinite-dimensional generalization of the ordinary Gaussian measures in $\mathbb{R}^n$. The Gaussian nature of the measure $\mu_{\beta,\lambda}$ (whose $\mathbb{E}_{\beta,\lambda}$ denotes the expectation) follows from the fact that the energy and the enstrophy are quadratic forms in $\omega^\wedge$ and the positivity of β and λ.

This measure is invariant with respect to the Euler flow. Indeed, the following identity holds for all $f \in C$:

$$\mathbb{E}_{\beta,\lambda}\left(\sum_{k \in \mathbb{Z}^d} \frac{\partial f}{\partial \omega^\wedge(k)} B_k(\omega^\wedge) \right) = 0. \tag{4.29}$$

The above identity can be easily derived by using the time invariance of H and $\mathscr{E}$ and (4.25). However, to give a rigorous meaning to (4.29), we have to prove that B_k, which is defined by means of a series, is at least $L_1(\mu_{\beta,\lambda})$. Actually, we can prove that $B_k \in L_2(\mu_{\beta,\lambda})$ so that (4.29) makes perfect sense.

After having constructed an invariant measure for the Euler flow, we pose the question of whether such a measure has some physical relevance in the theory of turbulence. We doubt this. The reasons are the following. First of all, an easy calculation shows that

$$\sum_{k \in \mathbb{Z}^d} \mathbb{E}_{\beta,\lambda}\left(\frac{|\omega^\wedge(k)|^2}{k^2} \right) = +\infty. \tag{4.30}$$

This means that the mean energy is infinite. An explicit calculation also shows that the energy spectrum behaves like $|k|^{-1}$, so that there is no agreement with the dimensional considerations developed in the previous section. Moreover, the statistically relevant vorticity distributions bring infinite energy as follows by a simple analysis on the support of the measure $\mu_{\beta,\lambda}$. From this we even have trouble defining the flow T on a full measure set of the fields ω's.

Finally, the measure we are dealing with is, in a sense, trivial. All the modes $\omega^{\wedge}(k)$ are not interacting. It would be really surprising that a complex phenomenon such as turbulence could be described by a noninteracting model in which all the observables are explicitly computable. There is the possibility of trying to construct a non-Gaussian measure of the type

$$\text{const}\left\{\exp - \int_{\Lambda} dx\ \phi(\omega)(x)\right\}\mu_{\beta,\lambda}(d\omega^{\wedge}). \tag{4.31}$$

The problem of giving a sense to the measure (4.31) is a very hard and well-known problem in the domain of functional integration. This is similar to that posed by the Euclidean quantum field theory. A heuristic analysis of this problem has been given in [BPP 87] wherein a way to construct a measure of the type (4.31) by means of standard approaches, leading to a Gaussian, and hence trivial, measure, is shown.

We conclude this section by giving a list of references concerning the analysis presented here [Hop 52], [Kra 75], [Gal 76], [BoF 78], [ADH 79], [BoF 80], [AlH 81], [CaD 85], [BPP 87], and [AlC 88]. We do not know of any other attempt to construct invariant measures for the Euler flow. In the next section, we discuss an approach based on the vortex system which we believe more promising.

7.5. Statistical Mechanics of Vortex Systems

In Chapter 4 we have illustrated many connections between the vortex flow and the Euler equation and, in particular, we have shown that vortex dynamics can be considered as a finite-dimensional special solution of the Euler equation. On the other hand, the vortex system is an Hamiltonian flow with a finite number of degrees of freedom, so that we can apply to it the ordinary methods of equilibrium statistical mechanics avoiding the difficulties and ambiguities connected with the problem of dealing with Gibbs measures for fields.

Following Onsager [Ons 49] we can define the microcanonical ensemble for a system of N identical point vortices of intensity $\alpha > 0$ in a bounded connected region Λ. The Hamiltonian of the system is

$$H(x_1, \ldots, x_N) = \frac{\alpha^2}{2} \sum_{i \neq j} V(x_i, x_j) + \frac{\alpha}{2} \sum_{i=1}^{N} \gamma(x_i), \tag{5.1}$$

where

$$V(x, y) = -\frac{1}{2\pi} \log|x - y| + \gamma(x, y) \tag{5.2}$$

is the fundamental solution of the Poisson equation, γ its regular part, and

$$\gamma(x) = \gamma(x, y).\tag{5.3}$$

The microcanonical measure is defined in Λ^N as

$$\mu^E(dx_1, \ldots, dx_N) = \Omega(E)^{-1}\delta(H - E)\, dx_1, \ldots, dx_N,\tag{5.4}$$

where:

$$\Omega(E) = \int_{\Lambda^N} \delta(H - E)\, dx_1, \ldots, dx_N.\tag{5.5}$$

This measure gives equal probability to all points on the surface $H(x_1, \ldots, x_N) = E$. Notice that in the microcanonical ensemble there are no parameters other than the energy (beyond N and α which for the present time are thought of as fixed). The entropy is defined by

$$S(E) = \log \Omega(E)\tag{5.6}$$

and the inverse temperature is given by

$$\beta = \frac{\partial S(E)}{\partial E}.\tag{5.7}$$

Before going further in the analysis of the microcanonical measure, we have to say that we cannot produce any convincing physical argument in favor of this measure for the study of two-dimensional turbulence. On the contrary, we already know that the system we are considering is not ergodic. As shown in Chapter 4, we can exhibit nontrivial invariant sets of positive measure in which the motion is quasi-periodic. Thus a typical trajectory of the vortex system does not span, densely, the surface $H = E$, and there is no reason for assuming a uniform distribution on the surface $H = E$. Nevertheless, in view of the limit $N \to \infty$ we are going to consider, we can hope that such invariant sets become negligible for sufficiently large values of the energy.

Defining now

$$\Theta(E) = \int_{\Lambda^N} \chi(H < E)\, dx_1, \ldots, dx_N,\tag{5.8}$$

where $\chi(H < E)$ denotes the characteristic function of the set in which $H < E$, we find

$$\Theta(E) = \int dE\,\chi(E' < E)\int_{\Lambda^N} \delta(H - E')\, dx_1, \ldots, dx_N$$

$$= \int_{-\infty}^{E} dE'\,\Omega(E').\tag{5.9}$$

Therefore

$$\Omega(E) = \Theta'(E), \qquad \beta = \frac{\Theta''}{\Theta'}.\tag{5.10}$$

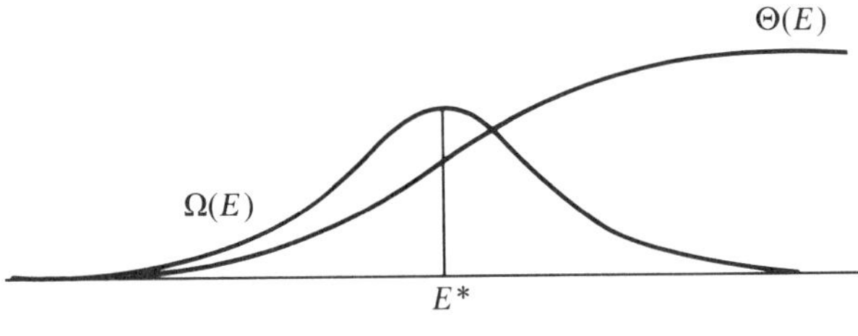

Figure 7.12

On the other hand, Θ is an increasing bounded (by Λ^N) function, for which we expect that there exists a point in which the concavity changes the sign. This means that negative temperature states can occur in the system. In other words, we expect the behavior shown in Fig. 7.12. Notice that E^* is the point in which Ω takes its maximum (Θ changes concavity) and the temperature changes sign.

As remarked by Onsager, who first introduced the statistical mechanics of point vortex systems in connection with the two-dimensional turbulence problem, there is no reason for considering positive temperature only. The physical occurrence of negative temperature can be explained in the following way. We can outline three different regimes. In the first the energy assumes negative large values and is dominating on the entropy. The vortices are likely near the boundary of Λ and are basically interacting with their mirror images. In this case the temperature is positive. Increasing E, the entropy becomes dominant. The vortices are, more or less, uniformly distributed. β is close to zero, which means that the temperature is positively very large. Increasing E, the vortices are forced to be close to each other. This is like a change in the sign of interaction and the temperature becomes negative. Numerical experiments seem to confirm this behavior. Clusters of vortices of the same sign are actually observed.

It is probably worthless, at this point, to underline that β is not the inverse temperature of the fluid, which is an inessential constant, but only an overall parameter describing general dynamical features of our system. So, considering β as negative does not contradict general principles. On the other hand, there are other examples of negative temperature states occurring in physics (see, for instance, [LaL 68]$_2$).

Notice finally that the occurrence of negative temperature is not, in our argument, peculiar to the logarithmic interaction. However, we remark that, for a usual gas of particles, this is not the expected behavior. In fact, in this case the Hamiltonian contains a kinetic part, the phase space is unbounded and $\Omega(E)$ is diverging when $E \to \infty$. Therefore no change in the concavity of Θ must appear. If we want to try to make the above arguments rigorous we encounter some difficulties, because the microcanonical ensemble is hard to deal with and so we introduce the more suitable canonical ensemble.

The canonical Gibbs measure is defined by

$$\mu^{\beta,N}(x_1, \ldots, x_N)\, dx_1, \ldots, dx_N = \frac{e^{-\beta H}\, dx_1, \ldots, dx_N}{Z(\beta, N)}, \tag{5.11}$$

where

$$Z(\beta, N) = \int_{\Lambda^N} e^{-\beta H}\, dx_1, \ldots, dx_N \tag{5.12}$$

is the partition function.

According to what has been seen before, we allow the maximum range of variability for β, the only restriction being the well-posedness of the object under consideration, which is the existence of the partition function $Z(\beta, N)$. Before analyzing the measure (5.11) from a mathematical point of view, we want to make some preliminary comments.

We passed, just as a matter of convenience, from the microcanonical to the canonical ensemble. Once again we are unable to justify fully this change of scenario. We only say that in the asymptotic regime in which we are interested, we expect that some equivalence of the two ensembles can be proved. Usually, the equivalence of the ensembles is valid in the thermodynamic limit only, i.e., when $N \to \infty$, meas $\Lambda \to \infty$ in such a way that $N/\text{meas }\Lambda \to \text{const.}$ At the moment, for a fixed N and Λ, the two measures, microcanonical and canonical, are really different. However, a limit in which a sort of equivalence of the two ensembles could be recovered must be considered for other reasons. In fact, if we try to compute the energy spectrum via the canonical Gibbs measure we find

$$\frac{\alpha^2 N}{|k|} + \text{other terms,}$$

where the other terms are not explicitly computable due to the nontriviality of the measure $\mu^{\beta,N}$. However, they have a better behavior than the first term. The physical meaning of $\alpha^2 N/|k|$ is clear. The hydrodynamical kinetic energy

$$\frac{1}{2} \int_\Lambda dx\, u^2(x) \tag{5.13}$$

is not the Hamiltonian of the vortex system which is, in general, not even positive. Actually, the velocity field u produced by a point vortex is not L_2 and the kinetic energy for point vortex systems is infinite. The Hamiltonian H differs from true kinetic energy (5.13) because of the self-interaction which is, obviously, logarithmically divergent. The term $\alpha^2 N/|k|$ is just the contribution to the energy spectrum due to the self-energy. Notice that this term disappears as

$$\alpha \to 0, \qquad N \to \infty, \qquad \alpha N \to \text{const.} \tag{5.14}$$

In this limit we expect the relevant point vortex configurations to become smooth, and the energy of the vortex system to converge to the hydrodynamical energy (5.13).

Notice that the limit (5.14) is of the same type as that discussed in Chapter 5, Section 3, the only difference being that, in that case, we were interested in the behavior of a single configuration in time, while here we are looking at the statistical properties of many of them.

Another argument for keeping the limit (5.14) is that there is no apparent reason for which α and N should parametrize our measure. Looking at the phenomenology, which is essentially computer simulations of the two-dimensional Euler flow, we observe the tendency to create coherent structures consisting of vortex blobs of various diameters and intensities. Such structures are stable on a rather large scale of time. Thus, the main point of two-dimensional turbulence is to understand how such self-organization takes place, namely, it would be of great interest to give a theory explaining the possible shapes of the blobs and their statistics. In this spirit, we want to exploit the limit (5.14) for the canonical measures $\mu^{\beta,N}$.

To begin a rigorous analysis we first need to investigate the existence of $\mu^{\beta,N}$. We observe that, due to the logarithmic divergence of V, the existence of the partition function is not assured for all values of β. Actually, an estimate of $Z(\beta, N)$ reduces to the evaluation of the following integral:

$$\int_{|x_l| < 1} dx_1, \ldots, dx_N \prod_{i=1}^{N} \prod_{j \neq i}^{N} |x_i - x_j|^{\beta\alpha^2/4\pi}. \tag{5.15}$$

A simple argument shows that the above integral is finite if and only if $\beta \in (-8\pi/\alpha^2 N, +\infty)$.

In view of the mean field limit (5.14) we have in mind, it is convenient to think of $\mu^{\beta,N}$ as defined on the space M_1 of the Borel probability measures on Λ equipped with the topology of the weak convergence of the measures. In other words, $\mu^{\beta,N}$ is concentrated on measures v_N of the type

$$v_N = N^{-1} \sum_{i=1}^{N} \delta_{x_i}(dx) \tag{5.16}$$

yielding the same statistical weight as to the configuration $\{x_1, \ldots, x_N\}$. More precisely, if $F: M_1 \to R$ is defined by

$$F(v) = \prod_{i=1}^{k} \int v\,(dx)\,\phi_i(x), \tag{5.17}$$

where ϕ_i are continuous and bounded functions from Λ to R, then

$$E_{\beta,N}(F) = N^{-k} \sum_{i_1=1}^{N} \cdots \sum_{i_k=1}^{N} \int_{\Lambda^N} \mu^{\beta,N}(x_1, \ldots, x_N)\phi_1(x_{i_1}), \ldots, \phi_k(x_{i_k})\,dx_1, \ldots, dx_N.$$

$$\tag{5.18}$$

Here $E_{\beta,N}$ denotes the expectation with respect to the measure $\mu^{\beta,N}$, thought of as a measure on M_1.

Now we want to study the asymptotic behavior of the measures $\mu^{\beta,N}$ in the limit $\alpha \to 0$, $N \to \infty$, $\alpha N \to 1$ (1 is just a matter of notational simplicity). Observe that, due to the relative compactness of Λ, we deduce that M_1 is relatively compact and hence $\mu^{\beta,N}$ has weak limits. Suppose then that μ is one of such weak limit

$$\lim_{N \to \infty} E_{\beta,N}(F) = \int F(v)\mu\,(dv), \tag{5.19}$$

where the limit has to be understood for some subsequence N_k.

We want to characterize the nature of μ. This measure is supported on a set of measures having the physical meaning of vorticity profiles. Suppose, for the moment, that μ concentrates on smooth profiles. Then μ as a weak limit of invariant measures for the vortex model is expected to be invariant for the Euler flow. However, because of the very special nature of the mean field limit we are doing, we expect μ to concentrate on a single profile. The reason is the following. Suppose, for the moment, to have a sequence of random variables $\{x_1, \ldots, x_N\}$ independent and identically distributed according to a distribution function ρ. Then, by the law of large numbers, the random variable

$$N^{-1} \sum_{i=1}^{N} \phi(x_i), \tag{5.20}$$

where $\phi \in C(\Lambda)$ converges to the number

$$\int \phi(x)\rho(x)\,dx \equiv \rho(\phi) \tag{5.21}$$

with large probability. More precisely

$$\Pr\left(\left| N^{-1} \sum_{i=1}^{N} \phi(x_i) - \rho(\phi) \right| > \varepsilon \right) \to 0 \qquad \text{as} \quad N \to \infty, \tag{5.22}$$

where the probability $\Pr$ is computed with respect to $\rho^{\otimes N}$.

In our vortex system the random variables $\{x_1, \ldots, x_N\}$ are not independent at all. However, the interaction between two tagged particles, say x_1 and x_2, is proportional to α^2, so that it is going to vanish in the mean field limit. Thus, in the limit, each random variable x_i is expected to become independent of the others. If so, for large N

$$\mu^{\beta,N}(x_1, \ldots, x_N) \cong \prod_{i=1}^{N} \rho(x_i)$$

$$\cong \prod_{i=1}^{N} \frac{\exp[-(\beta/2N^2)\sum_{j \neq i} V(x_i, x_j)]}{Z(\beta, N)} \tag{5.23}$$

(the term with γ is vanishing), so that

$$\rho(x) \cong \frac{\exp[-(\beta/N)\int \rho(y)V(x, y)\,dy]}{\int dx\,\exp[-(\beta/N)\int \rho(y)V(x, y)\,dy]}, \tag{5.24}$$

where the last step is justified by the fact that

$$\frac{1}{N} \prod_{j \neq i} V(x_i, x_j) \cong \int \rho(y) V(x, y) \, dy. \tag{5.25}$$

The factor β/N at the exponential says that, under this scaling, we are converging to the uniform distribution. To obtain something nontrivial we have to rescale also the inverse temperature β by $\beta \to \beta N$. After this additional scaling we expect the limiting measure μ, given by (5.19), to be a δ-measure concentrated on a single vorticity profile ω, which is the solution of the following equation:

$$\omega(x) = \frac{\exp[-\beta \int \omega(y) V(x, y) \, dy]}{\int dx \, \exp[-\beta \int \omega(y) V(x, y) \, dy]}. \tag{5.26}$$

In other words, what is expected to happen in the limit is the following. The vortices are distributed according to the Gibbs distribution. When N is large they fluctuate very little. With very large probability they arrange themselves to form a single, possibly smooth, profile ω, which solves (5.26).

Strictly speaking, this is true only if we can provide uniqueness of the solutions to (5.26), otherwise μ is expected to be a convex combination of δ-measures concentrated on the solutions of (5.26). We will discuss this point later. For the moment observe that, introducing the stream function ψ by $-\Delta \psi = \omega$, (5.26) becomes

$$-\Delta \psi = \frac{\exp(-\beta \psi)}{\int dx \, \exp(-\beta \psi)},$$
$$\psi = 0 \quad \text{on } \partial \Lambda. \tag{5.27}$$

This is a nonlinear elliptic problem which has a unique solution for $\beta > 0$. For negative β (actually, we are interested in the range of negative temperature $\beta \in (-8\pi, 0)$ only, as follows by statistical mechanics arguments), very little is known about the structure of this equation. We discuss this point later.

As regards the statistical mechanics problem, we mention that the above heuristic considerations can be made rigorous (see [CLM 92] and [Kie 93] which follow some ideas from [MeS 82] where the same problem was studied for bounded interaction). Actually we can prove the following theorem:

Theorem 5.1 ([CLM 92], [Kie 93]). *Consider the sequence $\mu^{\beta(N),N}$ of Gibbs measures, where $\beta(N) = \beta N$ (thought of as measures on M_1) in the limit $N \to \infty$. Then:*

(i) *If $\beta > 0$, $\mu^{\beta(N),N}$ converges weakly to a δ measure concentrated on the unique solution of (5.26). Moreover, such a solution minimizes the* *energy–entropy functional*

$$f(\omega) = \frac{1}{\beta} \int_{\Lambda} \omega \log \omega \, dx + \frac{1}{2} \int_{\Lambda^2} \omega(x)\omega(y) V(x, y) \, dx \, dy \qquad (5.28)$$

with the constraint that ω is a probability.

(ii) *If $\beta \in (-8\pi, 0)$, the weak cluster points of $\mu^{\beta(N),N}$ are convex combinations of solutions of (5.26) which maximize the energy–entropy functional (5.28) with the constraint that ω is a probability.*

We do not provide proof of the above theorem, which is rather technical, but limit ourselves to some additional comments.

Remark. We notice that Theorem 5.1 also provides an existence proof for the solutions of (5.26) or for the equivalent formulation (5.27) by means of statistical mechanics techniques in the temperature range $\beta \in (-8\pi, +\infty)$.

The solutions we have found are particular stationary solutions of the two-dimensional Euler equation which satisfy a variational principle. For negative β they obey an energy–entropy balance which is interesting in itself beyond the turbulence problem which initially motivated this analysis. Further discussion on this point, however, may bring us quite far from the main purpose of this chapter so that we only mention a few facts about such solutions. On the other hand, as we said before, not much is known about this argument which is still a current research topic.

The main problem concerning the solutions for negative β to the problem (2.27) is to understand what happens when $\beta \to -8\pi^+$ and if there is only one maximizing solution. We do not know the answer to this last problem. The behavior for $\beta \to -8\pi^+$ is also unclear. On one side we know, simply by exploiting the explicit radial solution on the circle, that there are situations in which the solutions concentrate, i.e., ω converges weakly to the δ function on the origin when $\beta \to -8\pi^+$. We could conjecture that such behavior is general. For instance, we might believe that, in a convex domain, ω does concentrate on the unique equilibrium point of a single vortex. It is proved that this is not true in general (see [CLM 92]), and so the behavior of the solutions for $\beta \to -8\pi^+$ is very sensitive on the geometry of the boundary of Λ. We do not know whether the presence–absence of a concentration of the solutions is a relevant problem for a better understanding of two-dimensional turbulence, so that we end our analysis here and address the reader to [CLM 92] for further details.

Another class of solutions which could play an important role in two-dimensional turbulence is that constructed with the same ideas in the whole plane with the angular momentum as an additional first integral. Namely, modifying the invariant measure (5.11) by adding the angular momentum

$$I = \sum_{i=1}^{N} x_i^2 \qquad (5.29)$$

which is invariant for the vortex motion in the plane, we arrive at

$$\mu^{\beta,N}(x_1, \ldots, x_N)\, dx_1 \ldots dx_N = \frac{e^{-\beta H - \lambda I}\, dx_1, \ldots, dx_N}{Z(\beta, N)} \qquad (5.30)$$

for $\lambda > 0$. The angular momentum plays the role of confining the particles around the origin. For positive λ this is a sort of natural boundary dictated by physical arguments. The rigorous analysis summarized by Theorem 5.1 can easily be rephrased in this context. The elliptic problem (5.27) becomes

$$-\Delta\psi = \frac{\exp(-\beta\psi - \lambda x^2)}{\int dx \, \exp(-\beta\psi - \lambda x^2)},$$

$$\psi \to 0 \qquad \text{as} \quad |x| \to \infty. \qquad (5.31)$$

The energy–entropy functional in this case is

$$f(\omega) = \frac{1}{\beta}\int_\Lambda \omega \log \omega \, dx + \frac{1}{2}\int_{\Lambda^2} \omega(x)\omega(y)V(x, y)\, dx\, dy + \frac{\lambda}{\beta}\int_\Lambda \omega x^2 \, dx. \qquad (5.32)$$

We can prove (see [CLM 92]):

Theorem 5.2. *The sequence of measures $\mu^{\beta(N),N}$ given by (5.30), with β replaced by $\beta(N) = \beta N$ and λ replaced by $\lambda(N) = \lambda N$, $\lambda > 0$, converges in the limit $N \to \infty$ to ω where $\omega = -\Delta\psi$ and ψ is the unique solution of (5.31). Moreover, such a solution is radially symmetric and minimizes, for $\beta > 0$, or maximizes, for $\beta \in (-8\pi, 0)$, the energy–entropy functional (5.32).*

The solution we have found is not explicitly computable but its qualitative behavior can be understood. In particular, we know that there is concentration for $\beta \to -8\pi^+$.

Let us now come back to the microcanonical description. For the canonical ensemble we were led to consider the variational problem associated with the functional (5.28) for a fixed β. The analogous problem in microcanonical language is the following. In the space of the absolutely continuous probability measures on Λ (once again a two-dimensional relatively compact, smooth domain) consider the two functionals

$$S(\rho) = -\int dx \, \rho \log \rho, \qquad E(\rho) = \tfrac{1}{2}\int dx\, dy\, \rho(x)V(x, y)\rho(y). \qquad (5.33)$$

We want to maximize $S(\rho)$ for a fixed value of the energy $E(\rho) = E$. If the above variational principle has a solution, this is expected to be a limiting state for a sequence of microcanonical measures for the vortex system in the limit (5.14). Standard heuristic arguments actually show that the micro-canonical measures do concentrate, in the limit (5.14), on those vortex configurations whose density ρ maximizes the entropy S (see the references quoted in the context of the sinh–Poisson equation which will be discussed later).

The microcanonical variational principle can be handled without difficulty. We can prove the following theorem.

Theorem 5.3. *Define*

$$S(E) = \sup_{\rho} S(\rho), \tag{5.34}$$

then:

(i) $S(E) < +\infty$ *and there exists ω such that $S(E) = S(\omega)$.*
(ii) $E \to S(E)$ *is a continuous function.*
(iii) *There exists $\beta = \beta(E)$ such that ω solves the mean field (5.26) for such a value of β.*
(iv) *The function $\beta(E)$ is bounded from below if the domain is starlike.*

We do not provide proof of the above theorem. Notice that even though the canonical and microcanonical solutions satisfy the same equation, this does not imply that they are the same. In fact, they satisfy two different variational principles and the coincidence of the two associated Euler–Lagrange equations simply means that a microcanonical solution is an extremal point for the free energy (5.28) and that a canonical solution, with a given energy E, is an extremal point for the entropy functional.

However, as for circular domains, if it is known to be a unique solution of (5.28) maximizing the free energy (actually for a disk there is only one solution of the equation which is also radially symmetric), then we can conclude (in the case of concentration at -8π!) the complete equivalence of the two sets of canonical and microcanonical solutions. In this case, the function $\beta \to -\beta f(\beta)$ is the Legendre transformation of $E \to S(E)$, as expected. In this case, we can also prove rigorously the existence of the limit (5.14) for the sequence of microcanonical measures for the vortex system. The general situation is more involved and constitutes the argument of current research.

Obviously, there is no reason for considering all vortices of the same sign, as we have done so far. Moreover, for special relevant domains such as the two-dimensional torus, we are obliged to consider neutral systems. We can carry out with minor modifications, at least at a heuristic level, our analysis for two species of vortices of opposite charges. If we denote by ω^+ and ω^- the densities of positive and negative vortices, respectively, we arrive at the two equations

$$\omega^+ = \frac{\exp(-\beta\psi)}{\int dx \, \exp(-\beta\psi)}, \qquad \omega^- = \frac{\exp(\beta\psi)}{\int dx \, \exp(\beta\psi)}, \tag{5.35}$$

where $\psi = \int dy \, V(\cdot, y)\omega(y)$ and $\omega = \omega^+ - \omega^-$.

Subtracting the two equations, we have

$$-\Delta\psi = \frac{\exp(-\beta\psi)}{Z^+} - \frac{\exp(\beta\psi)}{Z^-}, \tag{5.36}$$

where

$$Z^+ = \int dx \, \exp(-\beta\psi), \qquad Z^- = \int dx \, \exp(+\beta\psi). \tag{5.37}$$

Finally, putting

$$\psi = \phi + \log\frac{Z^+}{Z^-} \tag{5.38}$$

we obtain

$$-\Delta\phi = \alpha \sinh(-\beta\phi) \tag{5.39}$$

and α is a suitable constant.

Equation (5.39) is known as the sinh–Poisson equation and has been widely investigated in the literature from a heuristic point of view, in connection with the maximum entropy principle and two-dimensional turbulence ([JoM 73], [MoJ 74], [PoL 76]$_1$, [PoL 76]$_2$, [LuP 77]$_1$, [LuP 77]$_2$, [KrM 80]). We only remark that the sinh formulation given by (5.39), while useful for periodic boundary conditions, is not particularly appropriate, with respect to the formulation (5.35), in the case of a domain with boundary for which we have to satisfy $\psi = 0$ on $\partial\Lambda$.

Let us now come back to the turbulence problem as presented by the reality. By this we mean numerical simulations of fluid dynamical equations. At this point we have to be a bit more precise. We can simulate (see Chapter 5, Section 3) the Euler equation by means of the vortex dynamics. For suitable values of the energy and the number of vortices, solutions such as those described by the statistical theories we have discussed so far, have actually been observed. This is not very surprising since those solutions have been derived by a statistical ansatz on the vortex system.

Let us now see what happens in the simulation of the Navier–Stokes equation (which is not even Hamiltonian) by means of suitable spectral methods. Such numerical simulations (see, for instance, [BPS 87] and [BPS 88]) show that a two-dimensional turbulent flow has the tendency to create vortical blobs which are rather stable on a reasonable scale of time. It is tempting to interpret these blobs as a way by which the vorticity field self-organizes itself in a smooth way, that is, such blobs are the solutions of (5.31) (or similar mean field equations) with suitable values of β and λ. If this is true, the parameters β and λ have a local character only. Moreover, as shown by a remarkable numerical simulation of the Navier–Stokes on a torus, due to Montgomery et al. [MMS 91], [MMS 92], these blobs eventually merge to form a rather stable (on a suitable scale of time) configuration in very good agreement with a solution of the sinh–Poisson equation. Obviously, the solution is attracted by the trivial one, on a longer scale of time. However, even in the presence of a trivial attractor, a solution of the Navier–Stokes equation spends a large part of time close to a special stationary solution of the Euler equation. This is a very interesting and, in a sense, unexpected feature. Why so? There is no rigorous justification of this behavior, although a rough explanation can be given in the following terms. Notice first that the entropy increases (and the energy decreases) in the Navier–Stokes flow. Then it is conceivable to have suitable initial conditions for which the entropy rate is much larger than the energy rate, so that the Navier–Stokes flow arrives at the solution which maximizes the entropy at energy practically constant, in other words, close to a solution of the sinh–Poisson equation. This argument is, at the moment, only speculative. A more rigorous explanation and other experimental confirmations would be of great interest.

A final remark. We know, in general, that a functional relation between the stream function ψ and the vorticity ω imposes that ω be a stationary solution of the Euler equation. This is the case in (5.27). Other choices, based on different point of view, are of course possible. Here we mention another approach based on statistical mechanics ideas avoiding the explicit introduction of the point vortex system. For simplicity, we first consider the case in which ω can assume two values only, namely 0 and q. This corresponds to the attempt of constructing a statistical mechanics over a configurational set of vortex patches of intensity q. Ob   have the first integrals

$$(5.40)$$

$$\Omega = \int_{\Lambda} \tag{5.41}$$

We make now the hypothesis that a sort of ergodic propeny holds, so that the system is driven to an equilibrium state which maximizes the entropy

$$S(p) = -\int_{\Lambda} [p \log p + (1 - p) \log(1 - p) \, dx, \tag{5.42}$$

where $p = p(x)$ is the probability of having the vorticity value q in the point x. The maximum entropy must be chosen with the constraints that the total energy E and the total vorticity Ω are fixed.

The variational principle can be solved to yield (as well known in statistical mechanics)

$$p(x) = \frac{\exp[-\alpha - \beta q \psi(x)]}{1 + \exp[-\alpha - \beta q \psi(x)]}, \tag{5.43}$$

where α and β are two parameters which are the Lagrangian multipliers associated with the constraints Ω and E. Moreover, we assume that the probability $p(x)$ can be confused with the actual vorticity in the point x (and this is a sort of mean field hypothesis) for which

$$p(x) = -\Delta\psi(x). \tag{5.44}$$

Hence we obtain an equation in ψ. Equation (5.44) although reminiscent of (5.27) is very different. Here we have an a priori bound on the admissible values of ω. In terms of vortex systems, this would imply a sort of hard core condition avoiding concentrations.

The above theory can be generalized, allowing ω to have a more general set of values. Suppose that $\omega(x) \in [-q, q]$. Let $\rho(x, \xi)$ and $\pi(d\xi)$ be such that

$$\int_\Lambda \rho(x, \xi)\, dx = \text{vol } \Lambda, \tag{5.45}$$

$$\int_A \rho(x, \xi)\, d\pi(\xi) \text{ is the probability that } \omega \text{ assumes a value in } A. \tag{5.46}$$

Here $d\pi$ denotes a reference measure on the set of all admissible values for ω. The maximum entropy variational principle says that $\rho(x, \xi)$ satisfies

$$\rho(x, \xi) = \frac{\exp[-\alpha(\xi) - \beta\psi(x)]}{\int \pi(d\xi)\exp[-\alpha(\xi) - \beta\psi(x)]}. \tag{5.47}$$

Finally we obtain an equation by imposing

$$-\Delta\psi(x) = \int_{-q}^{q} \rho(x, \xi)\pi\,(d\xi). \tag{5.48}$$

For more details on this approach see [Rob 91].

We conclude this section with a general comment. Any approach to two-dimensional turbulence has necessarily an intrinsic limitation. Two-dimensional coherent structures can survive only on a suitable time scale, before three-dimensional asymmetries occur. Thus, even if we can say much more in two dimensions than in three, we conclude with the following question raised in the Onsager paper [Ons 49], "How soon the vortices will discover that there are three dimensions rather than two?"

7.6. Three-Dimensional Models for Turbulence

The larger complexity of the motion in three dimensions, compared with the two-dimensional case, increases the difficulty in constructing and analyzing reasonable statistical models for turbulent flows. A rather natural generalization of a gas of point vortices to the three-dimensional case would be a gas of vortex filaments. However, as we have seen in Chapter 4, Section 5, a vortex filament, i.e., a line in $\mathbb{R}^3$ in which the (vector) vorticity is concentrated as a δ function, is not a well-defined object, so that we are forced to consider filaments with a finite core. On the other hand, the study of a gas of vortex tubes from the point of view of statistical mechanics is, of course, something almost impossible to handle. Therefore it is very natural to introduce approximate models, which are, at least in principle, tractable from a mathematical point of view, and preserve, as much as possible, all the relevant physical features of real fluids. Following this philosophy A. Chorin proposed models of vortex filaments in a lattice which we are going to discuss.

Consider a vortex tube approximately supported in a cubic lattice. We recall that the general expression for the kinetic energy in terms of the vorticity is

$$E = \frac{1}{8\pi} \int dx \int dx' \frac{\omega(x) \cdot \omega(x')}{|x - x'|}. \tag{6.1}$$

Now suppose the tube to be a union of disjoint cylinders I_i, $i = 1, \ldots, N$, each of them having the symmetry axis in a lattice bond, then the kinetic energy takes the form

$$E = \frac{1}{8\pi} \sum_{i=1}^{N} \sum_{j=1}^{N} \int_{I_i} dx \int_{I_j} dx' \frac{\omega(x) \cdot \omega(x')}{|x - x'|} + \frac{1}{8\pi} \sum_{i=1}^{N} \int_{I_i} dx \int_{I_i} dx' \frac{\omega(x) \cdot \omega(x')}{|x - x'|}. \tag{6.2}$$

We assume $\omega(x)$ directed like the lattice bonds, and constant on the bound so that the interaction energy among the cylinders I_i can be further approximated by

$$\frac{1}{8\pi} \sum_{i=1}^{N} \sum_{j=1, j \neq i}^{N} \frac{t_i \cdot t_j}{|x_i - x_j|}, \tag{6.3}$$

where t_i are vectors along the bonds (of constant intensity) and $|x_i - x_j|$ denotes the distance of the two bonds in which t_i and t_j are localized, computed by means of the middle points. Under such restrictive hypotheses the self-energy of the vortex tube, i.e., the second member in the right-hand side of (6.2), is a constant not dependent on the geometry of the vortex tube, inessential from the point of view of statistical mechanics.

To summarize we have a vortex filament (henceforth called a vortex filament rather than a vortex tube because we are implicitly assuming that the cross section of the filament is very small and not playing, for the moment, any significant role) in a lattice. This is a connected set of N oriented links t_i, $i = 1, \ldots, N$. We require also that the vortex filament be self-avoiding which means that no vertex of the lattice is the end point of more than two links. Moreover, we also assume that the vortex filament is a closed path.

Observe that this last property, together with the connectivity and the fact that the t_i's are vectors of constant intensity, are a consequence of the conservation of the vorticity. Self-avoidance is a consequence of the fact that the overlapping of a vortex filament in a lattice would mean a bifurcation of the vortex tube with some singularity in the vorticity field. We will discuss this important point later on.

Figure 7.13 is a two-dimensional version of a vortex filament. To each vortex filament in the lattice we associate an energy given by

$$E = \frac{1}{8\pi} \sum_{i=1}^{N} \sum_{j=1, j \neq i}^{N} \frac{t_i \cdot t_j}{|x_i - x_j|}, \tag{6.4}$$

so that we can introduce a Gibbs measure at inverse temperature β on the space of the self-avoiding vortex filaments

$$\mu = \frac{e^{-\beta E}}{\text{Norm}}. \tag{6.5}$$

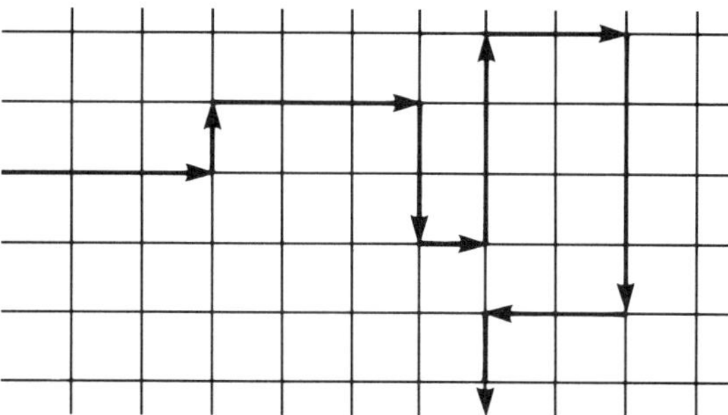

Figure 7.13

This is a well-defined statistical mechanical model which could be further complicated by considering many vortex filaments and the interaction among them. Observe however that, even considering a single vortex filament, the Gibbs measure (6.5) is very difficult to handle. To fix the ideas, consider the simplest case $\beta = 0$. Then the interaction energy among the links plays no role. The self-avoidance however, is a sort of nontrivial interaction making the model difficult to study. From the point of view of statistical mechanics this is equivalent to a one-dimensional spin system with very long-range interaction. In fact, all the admissible configurations of the vortex filaments are in one-to-one correspondence with a sequence of N vectors t_i with the constraint that, given t_i, $i = 1, \ldots, n$, t_{n+1} cannot be chosen among all six possible directions with equal probability, but only among those directions not creating overlapping. To do this, we need to know the whole "past" story t_i, $i = 1, \ldots, n$, so that we are dealing with a strongly non-Markovian process, very different from the usual random walk in which the random variable t_i is chosen independently from the past along the six possible directions.

This situation is well known in the theory of the so-called polymer statistics or Self-avoiding Random Walks (SRW). Here the same problem was approached heuristically and numerically in the absence of rigorous results which seem very difficult to obtain. For a SRW starting from the origin consider the quantity

$$r(N) = \langle d(N)^2 \rangle^{1/2}, \tag{6.6}$$

where $\langle \ \rangle$ means average and $d(N)$ is the distance from the origin of the SRW.

Thanks to phenomenological theories and numerical simulations the following behavior, for large N, is expected

$$r(N) \cong N^\mu, \qquad \mu \cong \tfrac{3}{5}, \tag{6.7}$$

μ is called the Flory exponent.

Now, given a polymer starting from the origin, suppose that it hits the sphere (in the lattice) of radius r. Then, in such a sphere there are, on average,

$N \cong r^{1/\mu}$ occupied sites. Between r and $r + dr$ there are something proportional to $r^{1/\mu-1}$ occupied sites which implies that the vorticity density is

$$\rho(r) = Cr^{1/\mu-3}. \tag{6.8}$$

The Fourier transform of this object behaves like

$$|k|^{-1/\mu}. \tag{6.9}$$

This is also the energy spectrum. Indeed, we have to multiplicate by $|k|^{-2}$ to obtain the Fourier transform of the velocity correlation, and by $|k|^2$ to finally get the energy spectrum. Thus we have found, although heuristically, the Kolmogorov spectrum starting from the polymer statistics, which has nothing to do, in principle, with the Navier–Stokes equation. This does not seem to us a trivial fact and this argument, even far from being conclusive, seems to indicate the interest of the connections between the turbulent motion of a vortex tube and the polymer statistics.

What we discussed so far and other considerations and numerical simulations on the matter (see [Cho 86], [Cho 88]$_1$, [Cho 88]$_2$, [Cho 90], [Cho 91]$_1$) constitute an interesting and promising part of the current research into the theory of fully developed turbulence. It is clear, however, that we are far from a logically satisfactory theory. From one side, we would like to give convincing arguments justifying the introduction of simplified models in place of the Euler or Navier–Stokes equations, from the other, a mathematically rigorous analysis of such models is needed.

Notice that such speculation did not make use of the finiteness of the cross section of a vortex tube (see [ChA 91]$_1$ and [Cho 91]$_2$ for corrections due to the finiteness of the cross section and further developments) and that they disregarded the fact that the vortex filament is in the reality a loop. Other considerarations, taking into account the fractal dimension of the physical region actually occupied by the vortex tube, have been taken into account (see the above references).

To compute averages with respect to the Gibbs measure (6.5) it is convenient to use Monte Carlo algorithms. Actually, the Gibbs measure (6.5) is the stationary state of a stochastic dynamics which can be numerically implemented taking into account self-avoidance. In this way positive and negative temperature states can be exploited.

An analysis of the Euler dynamics of a vortex tube (see [Cho 82]) is very delicate due to the divergence of the velocity field, and in all cases very costly. For the case of a vortex filament in a lattice it is more convenient to use a stochastic dynamics which has the property of preserving the energy and increasing the enstrophy just to take into account the really observed stretching feature ([Cho 86], [Cho 88]).

References

[ADH 79]   Albeverio, S., De Faria, M., Hoegh-Krohn, R. *J. Statist. Phys.* **20**, 585–595 (1979).

[AlC 88]   Albeverio, S., Cruzeiro, A. Global flow with invariant (Gibbs) measures for Euler and Navier–Stokes two-dimensional fluids. Preprint, Bochum (1988).

[AlH 81]   Albeverio, S., Hoegh-Krohn, R. *Phys. Rep.* **77**, 193–214 (1981).

[AnG 85]   Anderson, C.R., Greengard, C. On vortex methods. *SIAM J. Numer. Anal.* **22**, 413–440 (1985).

[ANO 85]   Aksman, M.J., Novikov, E.A., Orszag, S.A. Vorton method in three-dimensional hydrodynamics. *Phys. Rev. Lett.* **54**, 2410–2413 (1985).

[AmS 90]   Ambrosetti, A., Struwe, M. Existence of steady rings in an ideal fluid. *Arch. Rational Mech. Anal.* **108**, 97–108 (1989).

[ArA 68]   Arnold, V.I., Avez, A. *Ergodic Problems in Classical Mechanics.* Benjamin, New York (1968).

[Are 79]   Aref, H. Motion of three vortices. *Phys. Fluids* **22**, 393–400 (1979).

[Are 83]   Aref, H. Integrable, chaotic and turbulent vortex motion in two-dimensional flows. *Ann. Rev. Fluid Mech.* **15**, 345–389 (1983).

[Arn 65]   Arnold, V.I. Conditions for nonlinear stability of stationary plane curvilinear flows of an ideal fluid. *Dokl. Nat. Nauk* **162**, 773–777 (1965).

[Arn 66]$_1$ Arnold, V.I. Sur un principe variationel pour les écoulements stationnaires des liquides parfaits et ses applications aux problèmes de stabilité non linéaires. *J. Mécanique* **5**, 29–43 (1966).

[Arn 66]$_2$ Arnold, V.I. Sur la géométrie differentielle des groupes de Lie de dimension infinie et ses applications à l'hydrodynamique des fluids parfaits. *Ann. Inst. Fourier (Grenoble)* **16**, 319–361 (1966).

[Arn 69]   Arnold, V.I. On an apriori estimate in the theory of hydrodynamical stability. *Amer. Math. Soc. Transl.* **79**, 267–269 (1969).

[Arn 78]   Arnold, V.I. *Mathematical Methods of Classical Mechanics.* Graduate Texts in Mathematics, Vol. 60. Springer-Verlag, Heidelberg (1978).

[BaG 77]   Baouendi, M.S., Goulaouic, E.C. Remarks on the abstract form of the nonlinear Cauchy–Kowaleski theorem. *Comm. Partial Differential Equations* **2**, 1151–1162 (1977).

272

[Bar 72]   Bardos, C. Existence et unicité de la solution de l'équation d'Euler en dimension deux. *J. Math. Anal. Appl.* **40**, 769–790 (1972).

[BaR 93]   Batt, G., Rein, G. A rigorous stability result for the Vlasov–Poisson equation in three dimensions. *Ann. Math. Pure Appl.* (in press) (1993).

[BaS 90]   Baker, G.R., Shelley, M.J. On the connection between thin vortex layers and vortex sheets. *J. Fluid Mech.* **215**, 161–194 (1990).

[Bat 67]   Batchelor, K.G. *An Introduction to Fluid Dynamics.* Cambridge University Press, Cambridge, UK (1967).

[BeM 82]$_1$ Beale, J.T., Majda, A. Vortex methods, I: Convergence in three dimensions. *Math. Comput.* **39**, 1–27 (1982).

[BeM 82]$_2$ Beale, J.T., Majda, A. Vortex methods, II: High-order accuracy in two and three dimensions. *Math. Comput.* **39**, 29–52 (1982).

[BeP 91]   Benedetto, D., Pulvirenti, M. Fom vortex layers to vortex sheets. *SIAM J. Appl. Math.* **52**, 1041–1056 (1992).

[BKM 84]   Beale, J.T., Kato, T., Majda, A. Remarks on the breakdown of smooth solutions for the 3-D Euler equation. *Commun. Math. Phys.* **94**, 61–66 (1984).

[BEP 85]   Benfatto, G., Esposito, R., Pulvirenti, M. Planar Navier–Stokes flow for singular initial data. *Nonlinear Anal.: T.M.A.* **9**, 533–545 (1985).

[Bol 79]   Boldrighini, C. *Introduzione alla Fluidodinamica.* Quaderni del C.N.R., Roma (1979).

[BPP 87]   Benfatto, G., Picco, P, Pulvirenti, M. *J. Statist. Phys.* **46**, 729 (1987).

[BPS 87]   Benzi, R., Paternello, S., Santangelo, P. *Europhys. Lett.* **3**, 729 (1987).

[BPS 88]   Benzi, R., Paternello, S., Santangelo, P. *J. Phys. A* **21**, 1221–1237 (1988).

[BPV 82]   Benzi, R., Pierini, S., Vulpiani, A., Salusti, E. On nonlinear hydrodynamic stability of planetary vortices. *Geophys. Astrophys. Fluid Dynamics* **20**, 293–306 (1982).

[BeC 92]   Bertozzi, A.L., Constantin, P. Global regularity for vortex patches. Department of Mathematics, University of Chicago. Preprint (1992).

[BoF 79]   Boldrighini, C., Franceschini, V. A five-dimensional truncation of plane Navier–Stokes equations. *Commun. Math. Phys.* **64**, 159–170 (1979).

[BoF 78]   Boldrighini, C., Frigio, S. *Atti Sem. Mat. Fis. Univ. Modena* **27**, 106–125 (1978).

[BoF 80]   Boldrighnini, C., Frigio, S. *Commun. Math. Phys.* **20**, 55–76 (1980).

[BoB 74]   Bourguignon, J.P., Brezis, H. Remarks on the Euler equations. *J. Funct. Anal.* **15**, 341–363 (1974).

[BrH 77]   Braun, W., Hepp, K. *Commun. Math. Phys.* **56**, 1 (1977).

[CaD 85]   Caprino, S., De Gregorio, S. *Math. Methods Appl. Sci.* **7**, 55 (1985).

[Caf 90]   Caflisch, R.E. A simplified version of the abstract Cauchy–Kowaleski theorem with weak singularity. *Bull. Amer. Math. Soc.* **23**, 495–500 (1990).

[CaL 89]   Caflicsh, R.E., Lowengrub, J. Convergence of the vortex methods for vortex sheets. *SIAM J. Numer. Anal.* **26**, 1060–1080 (1989).

[CaM 86]   Caprino, S., Marchioro, C. On nonlinear stability of stationary plana euler flows in an unbounded strip. *Nonlinear Anal.: T.M.A.* **10**, 1263–1275 (1986).

[CaM 88]   Caprino, S., Marchioro, C. On nonlinear stability of stationary euler flows on a rotating sphere. *J. Math. Anal. Appl.* **129**, 24–36 (1988).

[CaO 86]   Caflisch, R.E., Orellana, O.F. Long-time existence for a slightly perturbed vortex sheet. *Commun. Pure Appl. Math.* **39**, 807–838 (1986).

[CaO 89]   Caflisch, R.E., Orellana, O.F., Singular solutions and ill-posedness for the evolution of vortex sheets. *SIAM J. Math. Anal.* **20**, 293–307 (1989).

[Cer 88]   Cercignani, C. *The Boltzmann Equation and its Applications.* Applied Mathematical Sciences, Vol. 67. Springer-Verlag, New York (1988).

[CFM 85]   Costantin, P., Foias, C., Manley, O., Temam, R. Determining modes and fractal dimension of turbulent flows. *J. Fluid Mech.* **150**, 427–440 (1985).

[CFT 85]   Costantin, P., Foias, C., Temam, R. Attractors representing turbulent flows. *Mem. Amer. Math. Soc.* **53**, 114 (1985).

[CFT 88]   Costantin, P., Foias, C., Temam, R. On the dimensions of the attractors in two-dimensional turbulence, *Phys. D* **30**, 284–296 (1988).

[Che 91]   Chemin, J.Y. Persistence de structure geometriques dans les fluids incompressibles bidimensionels, Preprint (1991).

[ChM 79]   Chorin, A.J., Marsden, J.E. *A Mathematical Introduction to Fluid Mechanics*. Springer-Verlag, Heidelberg (1979).

[Cho 73]   Chorin, A.J. Numerical study of slightly viscous flow. *J. Fluid Mech.* **57**, 785–796 (1973).

[Cho 82]   Chorin, A.J. The evolution of a turbulent vortex. *Commun. Math. Phys.* **83**, 517–535 (1982).

[Cho 86]   Chorin, A.J. Turbulence and vortex stretching on a lattice. *Commun. Pure Appl. Math.* **39** (Special Issue), S47–365 (1986).

[Cho 88]$_1$ Chorin, A.J. Scaling laws in the lattice vortex model of turbulence. *Commun. Math. Phys.* **114**, 167–176 (1988).

[Cho 88]$_2$ Chorin, A.J. Spectrum, dimension and polymer analogies in fluid turbulence. *Phys. Rev. Lett.* **60**, 1947–1949 (1988).

[Cho 90]   Chorin, A.J. Constrained random walks and vortex filaments in turbulence theory. *Commun. Math. Phys.* **132**, 519–536 (1990).

[Cho 91]$_1$ Chorin, A.J. Equilibrium statistics of a vortex filament with applications. *Commun. Math. Phys.* **141**, 619–631 (1991).

[Cho 91]$_2$ Chorin, A.J. Vortices, turbulence and statistical mechanics. In: *Vortex Flows and Vortex Methods* (Gustafson, K., Sethian, J., eds.). SIAM, Philiadelphi, Pa (1991).

[ChA 91]   Chorin, A.J., Akao, J. Vortex equilibria in turbulence theory and quantum analogues. *Phys. D* **52**, 403–414 (1991).

[Che 91]   Chemin, J.Y. Existence globale pour le problém de poches de tourbillon. *C.R. Acad. Sci. Paris* **312**, 803–806 (1991).

[CLM 92]   Caglioti, E, Lions, P.L., Marchioro, C., Pulvirenti, M. A special class of stationary flows for two-dimensional Euler equations: A statistical mechanics description. *Commun. Math. Phys.* **143**, 501–525 (1992).

[CMN 93]   Castilla, M.S., Moauro, V., Negrini, P., Oliva, W.M. The four positive vortices problem: Regions of chaotic behaviour and the nonintegrability. *Ann. Inst. H. Poincaré. Phys. Théor.* (in press) (1993).

[CoH 37]   Courant, R., Hilbert, D. *Methods of Mathematical Physics*, Vols. 1, 2. Interscience, New York (1937).

[CoS 88]   Cottet, G.H., Soler, J. Three-dimensional Navier–Stokes equation for singular filament initial data, *J. Differential Equations* **74**, 234–253 (1988).

[CoT 87]   Constantin, P., Titi, E.S. On the evolution of nearly circular vortex patches. *Commun. Math. Phys.* **119**, 177–198 (1988).

[Cot 83]   Cottet, G.H. Methodes particulaires pour l'equation d'Euler dans le plan. These 3émè Cycle, Université P. et M. Curie, Paris (1983).

[Cot 86]   Cottet, H.G. Equations de Navier–Stokes dans le plan avec tourbillon initial measure, *C. R. Acad. Sci. Paris Sér. I Math.* **303**, 105–108 (1986).

[DEG 82]   Dodd, R.K., Eilbeck, J.C., Gibbon, J.D., Morris, H.C. *Solitons and Nonlinear Wave Equations*. Academic Press, New York (1982).

[DeH 78]   Del Prete, V.M., Hald, O.H. Convergence of vortex methods for Euler's equations. *Math. Comput.* **32**, 791–809 (1978).

[DeP 91]   De Masi, A. Presutti, E. *Mathematical Methods for Hydrodynamic Limits*. Lecture Notes in Mathematics, Vol. 1501. Springer-Verlag, New York, (1991).

[DiM 87]₁   Di Perna, R., Majda, A. Concentrations in regularizations for 2-D incompressible flow. *Commun. Pure Appl. Math.* **40**, 301–345 (1987).

[DiM 87]₂   Di Perna, R., Majda, A. Oscillations and concentrations in weak solutions of the incompressible fluid equations. *Commun. Math. Phys.* **108**, 667–689 (1987).

[DiM 88]   Di Perna, R. Majda, A. Reduced Hausdorff dimension and concentration cancellation for 2-D incompressible flow. *J. Amer. Math. Soc.* **1**, 59–96 (1988).

[Dob 79]   Dobrushin, R.L. *Soviet J. Funct. Anal.* **13**, 115 (1979).

[Dri 88]₁   Dritschel, D.G. Nonlinear stability bounds for inviscid two-dimensional, parallel or circular flows with monotonic vorticity, and the analogous three-dimensional quasi-geostrophic flows. *J. Fluid Mech.* **191**, 575–581 (1988).

[Dri 88]₂   Dritschel, D.G. Strain induced vortex stripping. *Proceedings of the Workshop "Mathematical Aspects of Vortex Dynamics,"* Leesburg, Virginia (R. Caflisch, ed.). SIAM, Philadelphia, Pa, (1988).

[Dri 91]   Dritschel, D.G. Generalized helical Beltrami flows in hydrodynamics and magnetohydrodynamics. *J. Fluid Mech.* **222**, 525–541 (1991).

[DrR 81]   Drazin, P.G., Reid, W.H. *Hydrodynamic Stability.* Cambridge University Press, Cambridge, UK (1981).

[DuP 82]   Durr, D., Pulvirenti, M. On the vortex flow in bounded domains. *Commun. Math. Phys.* **85**, 265–273 (1982).

[DuR 86]   Duchon, J., Robert, R. Solution globales avec nappe tourbillionaire por les equations d'Euler dans le plan. *C. R. Acad. Sci. Paris* **302**, 163–186 (1986).

[DuR 88]   Duchon, J., Robert, R. Global vortex sheet solutions of Euler equations in the plane. *J. Differential Equations* **73**, 215–224 (1988).

[EbM 70]   Ebin, D., Marsden, J. Groups of diffeomorphisms on the motion of an incompressible fluid. *Ann. of Math.* **92**, 102–163 (1970).

[Ebi 88]   Ebin, D.G. Ill-posedness of the Rayleigh–Taylor and Helmholtz problems for incompressible fluids. *Commun. Partial Differential Equations* **13**, 1265–1295 (1988).

[EMP 88]   Esposito, R, Marra, R. Pulvirenti, M. Sciarretta, C. A stochastic Lagrangian formalism for the 3-D Navier–Stokes flow. *Commun. Partial Differential Equations* **13**, 12, 1601 (1988)

[EsP 89]   Esposito, R., Pulvirenti, M. Three-dimensional stochastic vortex flows. *Math. Methods Appl. Sci.* **11**, 431–445 (1989).

[Eul]   Euler, L. *Mémoires de l'Académie des Sciences.* Berlin (1755).

[FrB 74]   Fraenkel, L.E., Berger, M.S. A global theory of steady vortex rings in an ideal fluid. *Acta Math.* **132**, 13–51 (1974).

[FrT 85]   Franceschini, V., Tebaldi, C. Truncations to 12, 14 and 18 modes of the Navier–Stokes equations on two-dimensional torus. *Meccanica* **20**, 207–230 (1985).

[Gal 76]   Gallavotti, G. *Problemes Ergodiques de la Mechanique Classique.* Cours de 3 Cycles, EPF, Lausanne (1976).

[GHL 88]   Goodman, J. Hou, T.Y., Lowengrub, J. Convergence of the point vortex method for 2-D Euler equations. *Commun. Pure Appl. Math.* **43**, 415–430 (1990).

[GMO 86]   Giga, Y., Miyakawa, T., Osada, H. Two-dimensional Navier–Stokes flow with measure of initial vorticity. Preprint of the Institute of Mathematical Applications, no. 259. University of Minnesota, Minneapolis, MA (1986).

[Goo 87]   Goodman, J. Convergence of the random vortex method. *Commun. Pure Appl. Math.* **40**, 189–220 (1987).

[GrT 88]₁　Greengard, C., Thomann, E. On Di Perna–Majda concentration sets for two-dimensional incompressible flows. *Commun. Pure Appl. Math.* **41**, 295–304 (1988).

[GrT 88]₂　Greengard, C., Thomann, E. Singular vortex systems and weak solutions of the Euler equations. *Phys. Fluids* **31**, 2810–2813 (1988).

[GuH 83]　Guckenheimer, J, Holmes, P. *Nonlinear Oscillations, Dynamical Systems and Bifurcations of Vector Fields*, Springer-Verlag, New York (1983).

[Gus 79]　Gustaffson, B. On the motion of a vortex in two-dimensional flow of an ideal fluid in simply and multiply connected domains. Technical Report, Department of Mathematics, Royal Institute of Technology, Stockholm (1979).

[Hae 51]　Haegi, H.R. *Compositio Math.* **8**, 81–111 (1951).

[Hal 87]　Hald, O.H. Convergence of vortex methods for Euler's equations, III. *SIAM J. Numer. Anal.* **24**, 538–582 (1987).

[Hel]　　　Helmholtz, H. Uber Integrale der hydrodynamischen Gleichungen, welche den Wirbelbewegungen entsprechen, *Crelles J.* **55**, 25 (1858). Translated in: On the integral of the hydrodynamical equations which express vortex motion, *Phil. Mag.* **33**, 485–513 (1867).

[HiS 74]　Hirsch, M.W., Smale, S. *Differential Equations, Dynamical Systems and Linear Algebra*. Academic Press, New York (1974).

[HMR 85]　Holm, D.D., Marsden, J.E., Ratiu, T., Weinstein, A. Nonlinear stability of fluid and plasma equilibria. *Phys. Rep.* **123**, 1–116 (1985).

[HoL 90]　Hou, T.Y., Lowengrub, J. Convergence of the point vortex method for 3-D Euler equations. *Commun. Pure Appl. Math.* **43**, 965–981 (1990).

[Hop 52]　Hopf, E. *J. Rational Mech. Anal.* **1**, 87–123 (1952).

[HoL 88]　Hou, T.Y. Lowengrub, J. Convergence of the point vortex method for the 3-D Euler equations. Preprint (1988).

[Hua 63]　Huang, K. *Statistical Mechanics.* Wiley, New York (1963).

[HuM 76]　Hughes, T.J.R., Marsden, J.E. *A Short Course in Fluid Mechanics*. Publish or Perish, Berkeley (1976).

[JoM 73]　Joyce, G., Montgomery, D. Negative temperature states for the two-dimensional guiding-centre plasma. *J. Plasma Phys.* **10**, part 1, 107–121 (1973).

[KaN 79]　Kano, T, Nishida, T. Sur les ondes de surface de l'equ avec una justification mathématique des équations des ondes en eau peu profonde. *J. Math. Kyoto Univ.* **19**, 335–370 (1979).

[KaN 86]　Kano, T., Nishida, E. A mathematical justification for the Korteweg–De Vries equation and Boussineq equation of water surface waves. *Osaka J. Math.* **23**, 389–413 (1986).

[KaP 88]　Kato, T., Ponce, G. Commutator estimates and Navier–Stokes Equations. *Commun. Pure Appl. Math.* **41**, 893–907 (1988).

[Kat 67]　Kato, T. On classical solution of the two-dimensional non-stationary Euler equation. *Arch. Rational Mech. Anal.* **25**, 188–200 (1967).

[Kat 72]　Kato, T. Non-stationary flows of viscous and ideal fluids in $\mathbb{R}^3$. *J. Funct. Anal.* **9**, 296–309 (1972).

[Kat 75]　Kato, T. *Quasi-Linear Equation of Evolution with Application to Partial Differential Equations*. Lecture Notes in Mathematics, No. **448**, Springer-Verlag, New York (1975), pp. 25–70.

[Kat 92]　Kato, T. A remark on a theorem of C. Bardos on the 2D-Euler equation. Department of Mathematics, University of California. Preprint (1992).

[Kha 82]　Khanin, K.M. Quasi-periodic motions of vortex systems. *Phys.* **4D**, 261–269 (1982).

[Kel 10]　Kelvin, L. *Mathematical and Physical Papers*. Cambridge University Press, Cambridge, UK (1910).

[Kel 53]   Kellog, O.D. *Foundation of Potential Theory*. Dover, New York (1953).

[Kie 93]   Kiessling, M.K.H. Statistical mechanics of classical particles with logarithmic interactions. *Commun. Pure Appl. Math.* (in press) (1993).

[Kir]   Kirchhoff, G. *Vorlesungen Ueber Math. Phys.* Teuber, Leipzig (1876).

[Kol 41]   Kolmogorov, A.N. *Dokl. Akad. Nauk SSSR* **30**, 299 (1941).

[Kra 75]   Kraichnan, R.H. *J. Fluid Mech.* **67**, 155 (1975).

[Kra 86]   Krasny, R. On singularity formation in a vortex sheet and the point vortex approximation. *J. Fluid Mech.* **167**, 65–93 (1986).

[KrM 80]   Kraichnan, R.H., Montgomery, D. Two-dimensional turbulence. *Rep. Progr. Phys.* **43**, 547–619 (1980).

[Lad 69]   Ladyzhenskaya, O.A. *The Mathematical Theory of Viscous Incompressible Flows*, 2nd ed. Gordon and Breach, New York (1969).

[Lam 32]   Lamb, H. *Hydrodynamics*, 6th ed. Cambridge University Press, Cambridge, UK (1932).

[LaL 68]$_1$ Landau, L.D., Lifshitz, E.M. *Fluid Mechanics*. Pergamon Press, Oxford, UK (1968).

[LaL 68]$_2$ Landau, L.D., Lifshitz, E.M. *Statistical Mechanics*. Pergamon Press, Oxford, UK (1968).

[Leo 80]   Leonard, A. Vortex methods for flow simulation. *J. Comput. Phys.* **37**, 289–335 (1980).

[Lin 43]   Lin, C.C. *The Motion of Vortices in Two Dimensions*. University of Toronto Press, Toronto (1943).

[Lon 88]$_1$ Long, D.G. Convergence of the random vortex method in two dimensions. Preprint (1988).

[Lon 88]$_2$ Long, D.G. Convergence of the random vortex method in three dimensions. Preprint (1988).

[LuP 77]$_1$ Lundgren, T.S., Poyntin, Y.B. *Phys. Fluids* **20**, 356–363 (1977).

[LuP 77]$_2$ Lundgren, T.S., Poyntin, Y.B. *J. Statist. Phys.* **17**, 323–355 (1977).

[Maj 84]   Majda, A. *Compressible Fluid Flow and Systems of Conservation Laws in Several Space Variables*. Applied Mathematical Sciences, Vol. 53. Springer-Verlag, New York (1984).

[Maj 86]   Majda, A. Vorticity and the mathematical theory of incompressible fluid flow. *Commun. Pure Appl. Math.* **39**, S187–S220 (1986).

[MaM 87]   Marchioro, C., Moauro, V. A remark on the stability problem in fluid dynamics. *J. Math. Anal. Appl.* **128**, 413–418 (1987). Addendum, *J. Math. Anal. Appl.* **139**, 301 (1989).

[MaP 86]   Marchioro, C., Pagani, E. Evolution of two concentrated vortices in a two-dimensional bounded domain. *Math. Methods Appl. Sci.* **8**, 328–344 (1986).

[MaP 82]   Marchioro, C., Pulvirenti, M. Hydrodynamics in two dimensions and vortex theory. *Commun. Math. Phys.* **84**, 483 (1982).

[MaP 83]   Marchioro, C., Pulvirenti, M. Euler evolution for singular initial data and vortex theory. *Commun. Math. Phys.* **91**, 563–572 (1983).

[MaP 84]   Marchioro, C., Pulvirenti, M. *Vortex Methods in Two-Dimensional Fluid Dynamics*. Lecture Notes in Physics, No. **203**. Springer-Verlag, Berlin (1984).

[MaP 85]   Marchioro, C., Pulvirenti, M. Some consideration on the nonlinear stability of stationary planar Euler flows. *Commun. Math. Phys.* **100**, 343–354 (1985).

[MaP 91]   Marchioro, C., Pulvirenti, M. *On the Vortex-Wave System, Mechanics, Analysis and Geometry: 200 Years after Lagrange* (M. Francaviglia, ed.). Elsevier Science, Amsterdam (1991), pp. 79–95.

[MaP 93]   Marchioro, C., Pulvirenti, M. Vortices and localitation in Euler flows. *Commun. Math. Phys.* **154**, 49–61 (1993).

[Mar 86]   Marchioro, C. An example of absence of turbulence for any Reynolds number. *Commun. Math. Phys.* **105**, 99–106 (1986).

[Mar 86]   Marchioro, C. An example of absence of turbulence for any Reynolds number: II. *Commun. Math. Phys.* **108**, 647–651 (1987).

[Mar 88]   Marchioro, C. Euler evolution for singular initial data and vortex theory: A global solution. *Commun. Math. Phys.* **116**, 45–55 (1988).

[Mar 89]   Marchioro, C. Incompressible fluids with a potential field and vortex theory. *Rend. Mat.* **9**, 427–443 (1989).

[Mar 90]   Marchioro, C. On the vanishing viscosity limit for two-dimensional Navier–Stokes equations with singular initial data. *Math. Methods Appl. Sci.* **12**, 463–470 (1990).

[MBO 82]   Meiron, D.I., Baker, G.R., Orzag, S.A. Analytic structure of vortex sheet dynamics, Part 1. Kelvin–Helmholtz in stability. *J. Fluid Mech.* **114**, 283–298 (1982).

[McG 68]   McGrath, F.J. Non-stationary plane flow of viscous and ideal fluids. *Arch. Rational Mech. Anal.* **27**, 329 (1968).

[McK 69]   McKean, H.P. *Lectures in Differential Equations*, Vol. 2. (Aziz, ed.). Von Nostrand, New York (1969), p. 177.

[MeS 82]   Messer, J., Spohn, H. Statistical mechanics of the isothermal Lane–Emden equation. *J. Statist. Phys.* **29**, 561–578 (1982).

[Mey 81]   Meyer, R.E. *Introduction to Mathematical Fluid Dynamics*, Wiley, New York (1981).

[MiT 60]   Milne, L.M., Thomson, C.B.E. *Theoretical Hydrodynamics*. Macmillan, London (1960).

[MMS 91]   Montgomery, D., Matthaeus, W.T., Martinez, D., Oughton, S. *Phys. Rev. Lett.* **66**, 2731 (1991).

[MMS 92]   Montgomery, D., Matthaeus, W.T., Martinez, D., Oughton, S. Relaxation in two dimensions and the "Sinh–Poisson" equation. *Phys. Fluids A*, **4**, 3–6 (1992).

[MMZ 87]   Melander, M.V., McWilliams, J.C., Zabusky, N.J. Axisymmetrization and vorticity gradient intensification of an isolated two-dimensional vortex through filamentation. *J. Fluid Mech.* **146**, 2143 (1987).

[MoJ 74]   Mortgomery, D., Joyce, G. Statistical mechanics of "negative temperature" states. *Phys. Fluids* **19**, 1139–1145 (1974).

[Moo 79]   Moore, D.W. The spontaneous appearance of the singularity in the shape of an evolving vortex sheet. *Proc. Roy. Soc. London Ser. A* **365**, 105–119 (1979).

[Moo 84]   Moore, D.W. Numerical and analytical aspects of Helmholtz instability. In: *Theoretical and Applied Mechanics, Proceedings of the XVI International Congress of Theoretical Applied Mechanics* (F.I. Niordson, N. Olhoff, eds.), North-Holland, Amsterdam (1984), pp. 629–633.

[Nav 69]   Nalimov, V.I. A priori estimates of the solutions of elliptic equations with application to the Cauchy–Poisson problem. *Dokl. Akad. Nauk. URSS* **189**, 45–48 (1969).

[Neu 81]   Neunzert, H. An Introduction to the Non-Linear Boltzmann–Vlasov Equation. Lectures given at Summer School "Kinetic Theories and Boltzmann Equation." CIME, Montecatini, Italy (1981).

[Nis 77]   Nishida, T. A note on a theorem of Niremberg. *J. Differential Geom.* **12**, 629–633 (1977).

[Nir 72]   Niremberg, L. An abstract form for the nonlinear Cauchy–Kowaleski theorem. *J. Differential Geom.* **6**, 561–576 (1972).

[Nov 75]   Novikov, E.A. Dynamics and statistics of a system of vortices, *Zh. Èksper. Teoret. Fiz.* **68**, 1869 (1975). (English transl. *Soviet Phys. JEPT* **41**, 937–943 (1976).)

[Nov 83]   Novikov, E.A. Generalized dynamics of three-dimensional vortical singularities (vortons). *Zh. Eksper. Teoret. Fiz.* **84**, 975–981 (1983). (English transl. *Soviet Phys. JEPT* **57**, 566–569 (1983).)

[Ons 49]   Onsager, L. Statistical hydrodynamics. *Supplemento al Nuovo Cimento* **6**, 279–287 (1949).

[Osa 86]   Osada, H. Propagation of chaos for the two-dimensional Navier–Stokes equation. *Proc. Japan Acad. Ser. A Math. Sci.* **62**, 8–11 (1986).

[Ovs 71]   Ovsjannikov, L.V. Problème de Cauchy non-linéaire dans l'échelle des espace de Banach. *Dokl. Akad. Nauk URSS* **200**, 789–792 (1971).

[Ovs 74]   Ovsjannikov, L.V. To the shallow water theory foundation. *Arch. Mech. (Warsaw)* **26**, 407–422 (1974).

[Ovs 76]   Ovsjannikov, L.V. *Cauchy Problem in a Scale of Banach Spaces and its Application to the Shallow Water Justification.* Lecture Note in Mathematics, No. 503, Springer-Verlag, Berlin (1976), pp. 426–437.

[Ped 79]   Pedlosky, J. *Geophysical Fluid Dynamics.* Springer-Verlag, Heidelberg (1979).

[Poi]   Poincaré, H. *Theories des Tourbillons.* George Carré (1893).

[PoL 76]₁   Poyntin, Y.B., Lundgren, T.S. *Phys. Rev. A* **13**, 1274–1275 (1976).

[PoL 76]₂   Poyntin, Y.B., Lundgren, T.S. *Phys. Fluids* **19**, 1459–1470 (1976).

[Pul 89]   Pullin, D.I. On similarity flows containing two-branched vortex sheets. *Proceedings of the Workshop "Mathematical Aspects on Vortex Dynamics"* (R.E. Caflisch, ed.). SIAM, Philadelphia, Pa (1988).

[Rav 83]   Raviart, P.A. An Analysis of Particle Methods. CIME Course, Como, Italy (1983).

[Rob 91]   Robert, R. A maximum-entropy principle for two-dimensional perfect fluid dynamics. *J. Statist. Phys.* **65** 531–553 (1991).

[Rue 69]   Ruelle, D. *Statistical Mechanics, Rigorous Results.* Benjamin, New York (1969).

[Rue 84]   Ruelle, D. *Thermodynamic Formalism, Encyclopedia of Mathematics and its Applications*, vol. 5, Cambridge University Press, Cambridge, UK (1984).

[Rue 87]   Ruelle, D. *Chaotic Evolution and Strange Actractors.* Cambridge University Press, Cambridge, UK (1987).

[RuT 71]   Ruelle, D., Takens, F. On the nature of turbulence. *Commun. Math. Phys.* **21**, 12 (1971).

[SaM 86]   Saffman, P.G., Meiron, D.I. Difficulties with three-dimensional weak solutions for inviscid incompressible flow. *Phys. Fluids* **29**, 2373–2375 (1986).

[Ser 59]   Serrin, J. *Mathematical Principles of Classical Fluid Mechanics.* Handbook der Physik, No. 8, (1959), p. 125.

[ShB 88]   Shelley, M.J., Baker, G.R. On the connection between thin vortex layers and vortex sheets. *Proceedings of the Workshop "Mathematical Aspects on Vortex Dynamics"* (R.E. Caflisch, ed.) SIAM, Philadelphia, Pa (1988).

[ShB 90]   Shelley, M.J., Baker, G.R. On the connection between thin vortex layers and vortex sheets, Part II. Numerical study. *J. Fluid Mech.* **213**, 161–194 (1990).

[Shi 73]   Shimbrot, M. *Lectures in Fluid Mechanics.* Gordon and Breach, New York (1973).

[Shi 76]   Shimbrot, M. The initial value problem for surface waves under gravity, I. The simplest case. *Indiana Univ. Math. J.* **25**, 281–300 (1976).

[ShR 76]   Shimbrot, M, Reeder, J, The initial value problem for surface waves under gravity, II. The simplest three-dimensional case. *Indiana Univ. Math. J.* **25**, 1049–1071 (1976).

[SSB 81]   Sulem, C., Sulem, P.L., Bardos, C., Frisch, U. Finite time analyticity for two- and three-dimensional Kelvin–Helmoholtz instability. *Commun. Math. Phys.* **80**, 485–516 (1981).

[Swa 71] Swann, H. The convergence with vanishing viscosity of stationary Navier–Stokes flow to ideal flow in R_3. *Trans. Amer. Math. Soc.* **157**, 373–397 (1971).

[Tem 75] Temam, R. On the Euler equation of incompressible perfect fluids. *J. Funct. Anal.* **20**, 32–43 (1975).

[Tem 76] Temam, R. *Local Existence of C^∞ Solution of the Euler Equations of Incompressible Perfect Fluids*, Lecture Notes in Mathematics, vol. 565. Springer-Verlag, New York (1976).

[Tem 84] Temam, R. *Navier–Stokes Equations, Theory and Numerical Analysis*, 3th ed. North-Holland, Amsterdam (1984).

[Tem 86] Temam, R. Remarks on the Euler equations. In: *Nonlinear Functional Analysis and its Applications* (F. Browder, ed.). AMS Proceedings of Symposia in Pure Mathematics, vol. 45 (1986), pp. 429–430.

[Tem 88] Temam, R. *Infinite Dimensional Dynamical Systems in Mechanics and Physics*. Springer-Verlag, New York (1988).

[Tho 72] Thompson, C.J. *Mathematical Statistical Mechanics*. Macmillan, New York (1972).

[Tur 87] Turkington, B. On the evolution of a concentrated vortex in an ideal fluid. *Arch. Rational Mech. Anal.* **97**, 75–87 (1987).

[VMF 71] Von Mises, R., Friedrichs, K.O. *Fluid Dynamics*. Springer-Verlag, Heidelberg (1971).

[Wan 86] Wan, Y.H. The stability of rotating vortex patches. *Commun. Math. Phys.* **107**, 1–20 (1986).

[WaP 85] Wan, Y.H., Pulvirenti, M. Nonlinear stability of circular vortex patches. *Commun. Math. Phys.* **99**, 435–450 (1885).

[WiL 88] Winckclmans, G., Lconard, A. Wcak solutions of the three-dimensional vorticity equation with vortex singularities. *Phys. Fluids* **31**, 1838–1839 (1988).

[Wol 33] Wolibner, W. Un theorème sur l'existence du mouvement plan d'un fluide parfait, homogène, incompressible, pendent un temps infinitent long. *Mat. Z.* **37**, 698–726 (1933).

[Yud 63] Yudovic, V.I. Non-stationary flow on an ideal incompressible liquid. *USSR Comput. Math. Math. Phys.* **3**, 1407–1456 (1963).

Index

Applied Mathematical Sciences

(continued from page ii)